Lecture Notes in Physics

For information about Vols. 1–151, please contact your bookseller or Springer-Verlag.

Vol. 152: Physics of Narrow Gap Semiconductors. Proceedings, 1981. Edited by E. Gornik, H. Heinrich and L. Palmetshofer. XIII, 485 pages. 1982.

Vol. 153: Mathematical Problems in Theoretical Physics. Proceedings, 1981. Edited by R. Schrader, R. Seiler, and D.A. Uhlenbrock. XII, 429 pages. 1982.

Vol. 154: Macroscopic Properties of Disordered Media. Proceedings, 1981. Edited by R. Burridge, S. Childress, and G. Papanicolaou. VII, 307 pages. 1982.

Vol. 155: Quantum Optics. Proceedings, 1981. Edited by C.A. Engelbrecht. VIII, 329 pages. 1982.

Vol. 156: Resonances in Heavy Ion Reactions. Proceedings, 1981. Edited by K.A. Eberhard. XII, 448 pages. 1982.

Vol. 157: P. Niyogi, Integral Equation Method in Transonic Flow. XI, 189 pages. 1982.

Vol. 158: Dynamics of Nuclear Fission and Related Collective Phenomena. Proceedings, 1981. Edited by P. David, T. Mayer-Kuckuk, and A. van der Woude. X, 462 pages. 1982.

Vol. 159: E. Seiler, Gauge Theories as a Problem of Constructive Quantum Field Theory and Statistical Mechanics. V, 192 pages. 1982.

Vol. 160: Unified Theories of Elementary Particles. Critical Assessment and Prospects. Proceedings, 1981. Edited by P. Breitenlohner and H.P. Dürr. VI, 217 pages. 1982.

Vol. 161: Interacting Bosons in Nuclei. Proceedings, 1981. Edited by J.S. Dehesa, J.M.G. Gomez, and J. Ros. V, 209 pages. 1982.

Vol. 162: Relativistic Action at a Distance: Classical and Quantum Aspects. Proceedings, 1981. Edited by J. Llosa. X, 263 pages. 1982.

Vol. 163: J. S. Darrozes, C. Francois, Mécanique des Fluides Incompressibles. XIX, 459 pages. 1982.

Vol. 164: Stability of Thermodynamic Systems. Proceedings, 1981. Edited by J. Casas-Vázquez and G. Lebon. VII, 321 pages. 1982.

Vol. 165: N. Mukunda, H. van Dam, L.C. Biedenharn, Relativistic Models of Extended Hadrons Obeying a Mass-Spin Trajectory Constraint. Edited by A. Böhm and J.D. Dollard. VI, 163 pages. 1982.

Vol. 166: Computer Simulation of Solids. Edited by C.R.A. Catlow and W.C. Mackrodt. XII, 320 pages. 1982.

Vol. 167: G. Fieck, Symmetry of Polycentric Systems. VI, 137 pages, 1982.

Vol. 168: Heavy-Ion Collisions. Proceedings, 1982. Edited by G. Madurga and M. Lozano. VI, 429 pages. 1982.

Vol. 169: K. Sundermeyer, Constrained Dynamics. IV, 318 pages. 1982.

Vol. 170: Eighth International Conference on Numerical Methods in Fluid Dynamics. Proceedings, 1982. Edited by E. Krause. X, 569 pages. 1982.

Vol. 171: Time-Dependent Hartree-Fock and Beyond. Proceedings, 1982. Edited by K. Goeke and P.-G. Reinhard. VIII, 426 pages. 1982.

Vol. 172: Ionic Liquids, Molten Salts and Polyelectrolytes. Proceedings, 1982. Edited by K.-H. Bennemann, F. Brouers, and D. Quitmann. VII, 253 pages. 1982.

Vol. 173: Stochastic Processes in Quantum Theory and Statistical Physics. Proceedings, 1981. Edited by S. Albeverio, Ph. Combe, and M. Sirugue-Collin. VIII, 337 pages. 1982.

Vol. 174: A. Kadić, D.G.B. Edelen, A Gauge Theory of Dislocations and Disclinations. VII, 290 pages. 1983.

Vol. 175: Defect Complexes in Semiconductor Structures. Proceedings, 1982. Edited by J. Giber, F. Beleznay, J. C. Szép, and J. László. VI, 308 pages. 1983.

Vol. 176: Gauge Theory and Gravitation. Proceedings, 1982. Edited by K. Kikkawa, N. Nakanishi, and H. Nariai. X, 316 pages. 1983.

Vol. 177: Application of High Magnetic Fields in Semiconductor Physics. Proceedings, 1982. Edited by G. Landwehr. XII, 552 pages. 1983.

Vol. 178: Detectors in Heavy-Ion Reactions. Proceedings, 1982. Edited by W. von Oertzen. VIII, 258 pages. 1983.

Vol. 179: Dynamical Systems and Chaos. Proceedings, 1982. Edited by L. Garrido. XIV, 298 pages. 1983.

Vol. 180: Group Theoretical Methods in Physics. Proceedings, 1982. Edited by M. Serdaroğlu and E. İnönü. XI, 569 pages. 1983.

Vol. 181: Gauge Theories of the Eighties. Proceedings, 1982. Edited by R. Raitio and J. Lindfors. V, 644 pages. 1983.

Vol. 182: Laser Physics. Proceedings, 1983. Edited by J. D. Harvey and D. F. Walls. V, 263 pages. 1983.

Vol. 183: J. D. Gunton, M. Droz, Introduction to the Theory of Metastable and Unstable States. VI, 140 pages. 1983.

Vol. 184: Stochastic Processes – Formalism and Applications. Proceedings, 1982. Edited by G. S. Agarwal and S. Dattagupta. VI, 324 pages. 1983.

Vol. 185: H. N. Shirer, R. Wells, Mathematical Structure of the Singularities at the Transitions between Steady States in Hydrodynamic Systems. XI, 276 pages. 1983.

Vol. 186: Critical Phenomena. Proceedings, 1982. Edited by F.J.W. Hahne. VII, 353 pages. 1983.

Vol. 187: Density Functional Theory. Edited by J. Keller and J.L. Gázquez. V, 301 pages. 1983.

Vol. 188: A. P. Balachandran, G. Marmo, B.-S. Skagerstam, A. Stern, Gauge Symmetries and Fibre Bundles. IV, 140 pages. 1983.

Vol. 189: Nonlinear Phenomena. Proceedings, 1982. Edited by K. B. Wolf. XII, 453 pages. 1983.

Vol. 190: K. Kraus, States, Effects, and Operations. Edited by A. Böhm, J. W. Dollard and W. H. Wootters. IX, 151 pages. 1983.

Vol. 191: Photon Photon Collisions. Proceedings, 1983. Edited by Ch. Berger. V, 417 pages. 1983.

Vol. 192: Heidelberg Colloquium on Spin Glasses. Proceedings, 1983. Edited by J. L. van Hemmen and I. Morgenstern. VII, 356 pages. 1983.

Vol. 193: Cool Stars, Stellar Systems, and the Sun. Proceedings, 1983. Edited by S. L. Balliunas and L. Hartmann. VII, 364 pages. 1984.

Vol. 194: P. Pascual, R. Tarrach, QCD: Renormalization for the Practitioner. V, 277 pages. 1984.

Lecture Notes in Physics

221

Quark Matter '84

Proceedings of the Fourth International Conference
on Ultra-Relativistic Nucleus-Nucleus Collisions
Helsinki, Finland, June 17–21, 1984

Edited by K. Kajantie

Springer-Verlag
Berlin Heidelberg GmbH

Editor

Keijo Kajantie
Department of Theoretical Physics, University of Helsinki
Siltavuorenpenger 20 C, SF-00170 Helsinki, Finland

ISBN 978-3-540-15183-8 ISBN 978-3-540-39260-6 (eBook)
DOI 10.1007/978-3-540-39260-6

Originally published by Springer-Verlag Berlin Heidelberg New York Tokyo in 1985

2153/3140-543210

PREFACE

Quark Matter '84, the Fourth International Conference on Ultra-Relativistic Nucleus-Nucleus Collisions, was held in Helsinki, 17 - 20 June, 1984. This conference brought together about 100 experimentalists and theorists to discuss the expected properties of bulk quark matter, matter in which colour is unconfined, and its detection in heavy-ion collisions.

This volume contains the text of 20 of the 21 plenary talks given at the conference; the talk on quark matter searches using ^{16}O beams at the CERN SPS by R. Stock (Darmstadt) is not included. In addition, 44 parallel-session talks, listed separately at the end of this volume , were delivered.

The conference was organised by the Research Institute for Theoretical Physics, University of Helsinki, and also received financial support from the Academy of Finland, the Finnish Ministry of Education and Nordita. This financial support is gratefully acknowledged. The organisers would also like to thank the session chairmen P. Landshoff, J. Kapusta, S. Nagamiya, I. Otterlund, H. Satz and L. Schroeder for their contributions.

This meeting was the fourth in a series. The following meetings will be organised at LBL in Berkeley and GSI in Darmstadt. It is evident that we are experiencing the beginnings of a new and fruitful line of research.

Helsinki, November 1984

Keijo Kajantie

CONTENTS

1. Theory of Quark Matter

2. Theory of Quark and Hadron Matter, Cosmology

3. Experimental Data

4. New Experiments

5. New Experimental Facilities

6. Closing Talk

Probes of the Quark-Gluon Plasma as It Might Be Produced in Ultra-Relativistic Nuclear Collisions

Larry McLerran
Theoretical Physics Department
Fermilab P. O. Box 500
Batavia Ill. 60510

Abstract: The energy densities which might be achieved in ultra-relativistic nuclear collisions are discussed. Using these estimates, promising probes of a quark-gluon plasma as it might be produced in such collisions are reviewed. I discuss in detail the emission of photons and di-leptons. The consequences of hydrodynamic expansion and a first order phase transition are explored for the transverse momentum spectrum of hadrons. Fluctuations in the rapidity distribution of hadrons are also discussed as a possible signal for a first order phase transition. The possibilty that copious production of strange particles may signal the production of a quark-gluon plasma is critically assessed.

I shall discuss the possible experimental probes of the quark-gluon plasma as it might be produced in ultra-relativistic nuclear collisions. I shall concentrate on the central region of collisions of large nuclei, $A \gtrsim 200$, for head-on collisions at extremely high energies, $E_{CM}/A > 50$ Gev/Nucleon. A picture of such a collision is shown in Fig. 1.[1-2] At some time τ_0 after the two nuclei pass through one another, matter begins to form between them. In the inside-outside cascade picture of the collision, this forming matter is assumed to be non-interacting until after the time τ_0. The rapidity of the particles which constitute newly forming matter is therefore

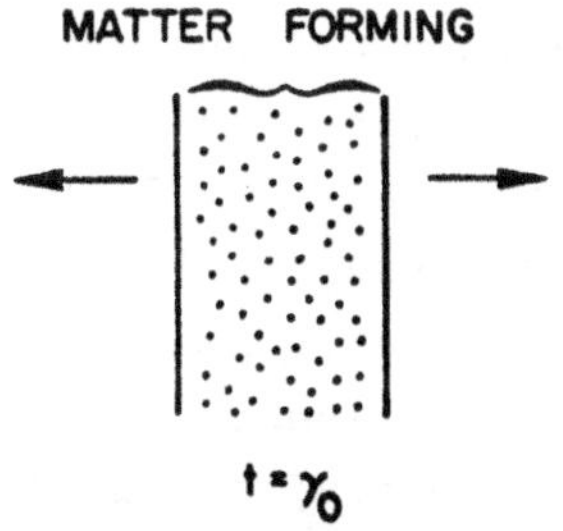

FIGURE 1 A nucleus-nucleus collision

$$y = \frac{1}{2} \ln\left\{\frac{1+v}{1-v}\right\} = \frac{1}{2} \ln\left\{\frac{t+x}{t-x}\right\} \qquad (1)$$

where v is a particle velocity, t is the time measured from the initial time of the collision, and x is the longitudinal coordinate measured from the position of the collision. This correlation between momentum and space-time persists after the time τ_o as a consequence of the hydrodynamic equations, and may be taken to be valid for all times.

The energy density of matter at the formation time τ_o is[1]

$$\rho = \frac{1}{\pi A^{2/3}} \frac{dN}{dx} \langle m_t \rangle = \frac{1}{\pi A^{2/3}} \frac{dN}{dy} \frac{\langle m_t \rangle}{\tau_o} \tag{2}$$

This result has been used to estimate the energy densities achieved in the ultra-relativistic collisions observed by the JACEE cosmic ray experiment.[3] If $\langle m_t \rangle \sim .4$ Gev and $\tau_o \sim 1$ Fm/c, the dN/dy distributions observed for intermediate A nuclei extrapolated to heavy nuclei such as uranium predict energy densities of 5-10 Gev/Fm^3. Such energy densities may be sufficient to produce a quark-gluon plasma.[4]

Recent results for hadron-nucleus collisions indicate that the formation time $\tau_o \sim 1$ Fm/c may be a little large.[5] The dependence of the energy density of Eqn. 2 upon τ_o is not trivial. By the uncertainty principle,

$$\langle m_t \rangle \gtrsim 1/\tau_o \tag{3}$$

so that

$$\rho \gtrsim \frac{1}{\pi A^{2/3}} \frac{dN}{dy} \frac{1}{\tau_o^2} \tag{4}$$

Since the energy density of a quark-gluon plasma scales as T^4 where T is the temperature, and the maximum achieved temperature is $T \sim \tau_o^{-1/2}$.

In a nice analysis presented by D. Kisielewska, the possible values of τ_o are extracted from hadron-nucleus and lepton-nucleus experimental data.[6] The range of values consistent with this data are determined to be $1/5 < \tau_o < 1$ Fm/c. The preferred values are 1/2 - 1/3 Fm/c. (It should be noted that in string models of nucleus-nucleus collisions, the formation time depends upon A, and may be considerably smaller for large A nuclei than is the case for hadron-nucleus collisions.)[7] If we consider a range $1/20 < \tau_o < 1$ Fm/c, the corresponding energy densities and temperatures are $\rho \sim 5 - 5000$ Gev/Fm^3 and $T \sim .2 - 1$ Gev.

Another method of estimating the energy densities achieved in ultra-relativistic nuclear collisions has been advocated by Matsui and Gyulassy.[8] They make use of the observation that the rapidity density $\frac{dN}{dy}$ is conserved in isentropic expansion. The hydrodynamic equations may be integrated backwards from the final time at which the system breaks up to the initial time τ . They derive[8]

$$\rho \sim 1.6 \left\{\frac{1}{\tau_0 \pi R^2} \frac{dN}{dy}\right\}^{4/3} \qquad (5)$$

The difference between this estimate and that of Eqn. 4 is that the $\frac{dN}{dy}$ distribution quoted here is that observed in the final state, whereas in Eqn. 4 it is only the initial rapidity density which appears, and this might be changed if the system produces entropy as it expands. Usually it is assumed that the system thermalizes at formation, that is, the formation time is of the order of the collision time and that the initial distributions are close to thermal distributions. In this circumstance, it should be a fair approximation to treat the expansion as isentropic, and the initial rapidity density may be identified with the final one. In general, this cannot be true since Eqns. 4-5 do not agree. For practical purposes, the agreement is quite good however. If the formation time varies between .1 and 1 Fm, the estimated energy density varies by two orders of magnitude, where the difference between the two different estimates varies only by a factor 3-5. As order of magnitude estimates, either relation is acceptable.

If one requires that the system instantly thermalize, then both Eqns. 4 and 5 must be valid. This can only be true if the formation time is A dependent. An estimate of this dependence is

$$\tau \sim .4\ A^{-1/6} - .6\ A^{-1/3} \qquad (6)$$

For heavy nuclei, formation times of .1-.2 Fm would be consistent with either relationship.

If the formation time is small and if the formation time is of the order of the collision time, then the condition that the system be thermalized and expand according to the equations of perfect fluid hydrodynamics, that is, non-viscous hydrodynamics, seems on much firmer ground than is the case for larger formation times. Notice that according to the hydrodynamic equations, the initial time and temperature and the final time and temperature are related as

$$\tau_f = \tau_i \left\{\frac{T_i}{T_f}\right\}^3 \qquad (7)$$

If the initial temperature is 250 Mev, the time the system takes to cool to a temperature of 150 Mev is only a factor of five times the initial time. If the initial temperature is a factor of two larger, this ratio increases by nearly an order of magnitude, and hydrodynamic methods are probably on a somewhat better foundation.

Since the width of the fragmentation region is given by

$$y_{frag} \sim \ln R_{nuc}/\tau_0 \qquad (8)$$

experimental measurements of the width of this region may aid in a resolution of τ_0. Such a measurement might be to determine the rapidity distribution of baryons minus anti-baryons, Fig. 2, or $\pi^+-\pi^-$ mesons, Fig. 3. The values of E_{CM}/A required to produce a baryon free central region depend upon τ_0, and for the values of τ_0 above are 15-300 GeV/Nucleon.

The value of τ_0 needed in the above analysis might be determined by using the Hanburry-Brown-Twiss effect.[(9-10)] The

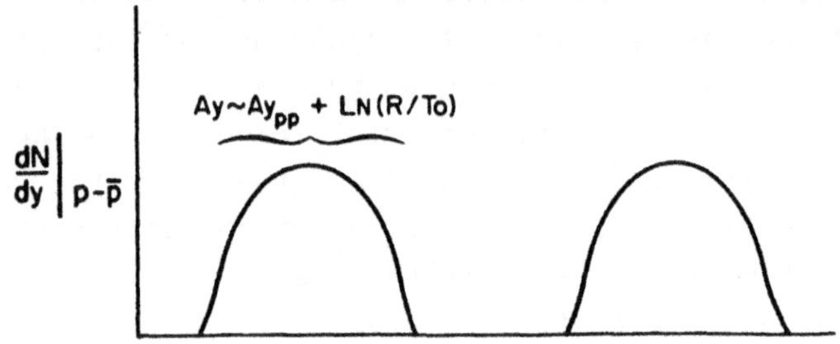

FIGURE 2 $^{dN}/dy$ for proton minus anti-proton

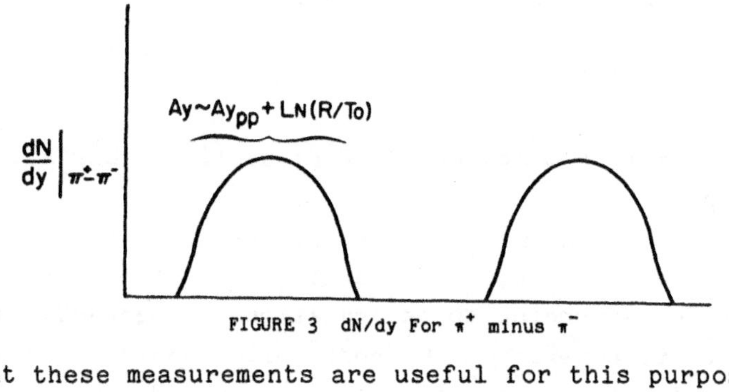

FIGURE 3 dN/dy For π^+ minus π^-

case that these measurements are useful for this purpose has not been made since hadronic final state interactions probably obscure a study of the formation process. Pion interferometry is probably most useful for studying the space-time dynamics of the hadronization process, or the break-up of the system as it freezes out of thermal equilibrium.[(9-10)] Again, there has been little theoretical analysis of this problem.

The formation time τ_0 might be measured and the validity of the formation time idea might be tested in hadron-nucleus collisions. As mentioned above, data analyzed by Busza and Goldhaber suggest that at available fixed target energies, projectile hadrons are stopped much more efficiently than might be expected from an inside-outside cascade picture with a formation time $\tau_0 \sim 1$ Fm.[(5)] The Kisielewska analysis is consistent with an inside-outside cascade picture if formation times somewhat smaller than 1 Fm are assumed.[(6)] Also, a new method of analysis has been developed by Bialas to measure the formation time for hadrons, a time which should be longer than the matter formation time for quarks and gluons.[11] When baryon production is analyzed, a baryon formation time of 1.5 ± .5 Fm results.

There is at present no consensus on methods for analyzing data from hadron-nucleus collisons and extracting a matter formation time. A very promising idea has been put forward by Hwa who suggests that the deviations of hadron-nucleus scattering from hadron-hadron in the hadron fragmentation region may cleanly isolate the effects of matter formation.[12] The argument is that within the context of an inside-outside cascade model, the probability that an inelastically produced particle forms inside a target nucleus is small if the particle is very energetic. The probability of rescattering is $P \sim e^{-x/\lambda}$ where λ is the mean free path for rescatternig. This mean free path is roughly the distance it takes a particle to form, since upon formation in the target, it has a large probability to rescatter, so that

$$\lambda \sim \frac{E\tau_0}{\langle m_t \rangle} \tag{9}$$

Since the particle has only a small chance to rescatter, its modification to hadron-hadron distributions, δF, should be small and computable in a single rescattering approximation,

$$\delta F \sim \frac{A^{1/3}}{\lambda} \tag{9}$$

A detailed, but still controversial analysis by Zahir and Hwa gives a λ which is consistent with the time dilation expected in the inside-outside cascade models.[12]

On the other hand, Glauber theory type models such as those employed by Kapusta and Csernai, and by Wong also seem to fit some of the hadron-nucleus data, and such models would be inconsistent with inside-outside cascade models unless the formation time was very small.[13-14]

A proper resolution of these theoretical models may require more experimental data. Experiments with $E_{lab} \sim 10 - 1000$ Gev, for a wide range of A, including hadron-proton and hadron- deuterium, would be useful. A wide range of x_f is probably best for the type of analysis suggested by Hwa,[13] but if good cascade models of hadronic interactions are developed along the lines of Kapusta and Csernai and of Wong, then coverage of the central and target fragmentation regions are also necessary.[13-14] The data on targets of various A including protons and deuterium should come from the same experiment to sort out systematic experimental biasis.

The study of the time development of the hadronic matter

distribution of matter as it is produced in ultra-relativistic heavy ion collisions has been initiated by the hydrodynamic computations of Bjorken,[1] by Kajantie, Raitio, and Ruuskanen.[15] In these computations, the longitudinal expansion of the matter is studied. In the computations of Kajantie et al, the fragmentation region as well as the central region is simulated. In later computations by Baym et al and Bialas et al, the transverse expansion is also computed.[16-17] The hydrodynamic treatment concludes that for formation time $\tau_0 \sim 1$ Fm, energy densities of 2-10 Gev/Fm^3 may be obtained in the central region, and 0-2 Gev/Fm^3 at various rapidities of the fragmentation region, with smallest values at the largest values of Feynman x. The compression of the baryon number density in the fragmentation region is 0-2 times that of ordinary nuclear matter. The transverse expansion calculations have been done only for the central region. Such expansion consists of transverse rarefraction and takes place over time scales large compared to longitudinal expansion for large $A \gtrsim 200$ nuclei.

In order for the hydrodynamic treatment to be valid, the mean free paths for quarks and gluons must be small compared to the spatial dimensions of the matter produced in the collision. Detailed computations seem to verify such a treatment for large A nuclei.[18-19] A much more stringent test of the validity of perfect fluid hydrodynamics is that the collision time be small compared to the longitudinal expansion time. Estimates of the collision time for appropriate energy densities are $\tau_c \sim .1-1$ Fm. Since a hydrodynamic treatment is valid if $\tau \gtrsim \tau_c$, the intrinsic error in these computations does not allow a good resolution of the time after which a hydrodynamic computation is reasonable. At the very least, viscous corrections to the perfect fluid hydrodynamic equations are probably important at early times in the hydrodynamic expansion, and there is certainly substantial entropy production for $\tau \sim \tau_c$.

Some measure of the degree of thermalization and the validity of a hydrodynamic treatment have been suggested by Shuryak.[20] In the transverse expansion of the matter produced in nuclear collisions, a transverse flow velocity develops. Particles of all masses flow with the velocity of the fluid. The more massive particles therefore have their transverse momentum enhanced more relative to lighter particles. A detailed treatment of this problem using the methods developed in Refs. 16-17 would be useful.

If there is a first order phase transition in hadronic matter, the transverse momentum distribution of hadrons may be drastically altered. Following Shuryak, the transverse momentum distributions receive a contribution due to transverse hydrodynamic expansion, and a thermal contribution due to the breakup of the system at some temperature

T,[20-21]

$$\langle p_t \rangle = \langle p_t \rangle_{hydro} + \langle p_t \rangle_{thermal} \qquad (10)$$

The hydrodynamic contribution arises from work which is done on constituents of the matter as the matter is driven into the vacuum. This work is produced from the pressure difference of the matter and the vacuum. Consider the work done as a function of the energy density achieved by the matter in a nuclear collision. As the energy density increases, the pressure will increase except when the energy density is in the region of a mixed phase of hadronic matter and quark-gluon plasma. For such energy densities, the pressure remains constant, as is shown in Fig. 4a. As the pressure increases, the transverse momentum due to hydrodynamic expansion increases. In the region of the phase transition, the pressure remains constant, as does the transverse momentum of hadrons.

This general feature of the hydrodynamic expansion coupled with a

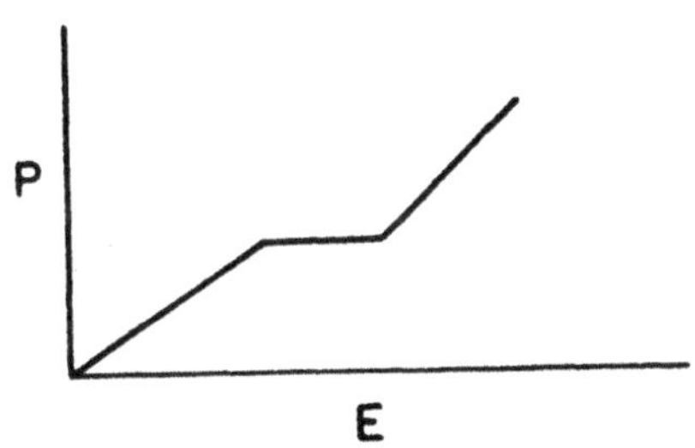

FIGURE 4a P vs E

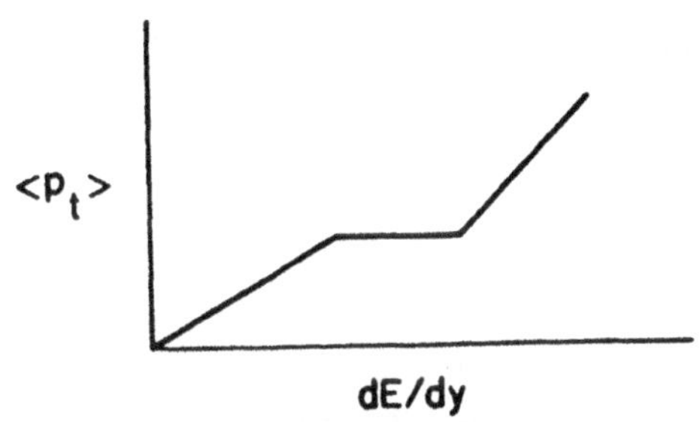

FIGURE 4b $\langle p_t \rangle$ vs dE/dy

phase transition may be explored by plotting measured transverse achieved energy densities, as is shown in Fig. 4b.[20-21]

Such a plot has been made by the JACEE cosmic ray collaboration, and is shown in Fig. 5.[22] Although the inferred energy density is somewhat model dependent, the sharp break in the transverse momentum distribution is quite suggestive.

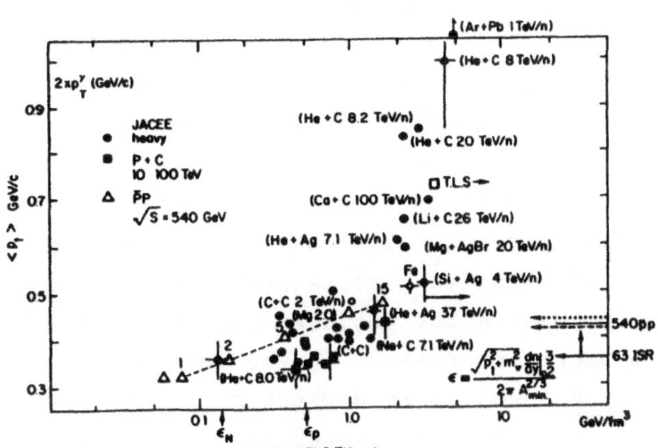

FIGURE 5

$\langle p_t \rangle$ vs dE/dy as measured by JACEE experiment

Studies by Heinz and by Bialas and Czyz suggest that color plasma oscilations may play an important role in non- equilibrium processes early in the expansion of matter produced in ultra- relativistic nuclear collisions.(23-24) A typical color oscillation is shown in Fig. 6. In such an oscillation, the local color charge density is analogous to the electromagnetic charge density in the electromagnetic plasma oscillation. For the oscillation shown in Fig. 6, both a color and electromagnetic oscillation is set up, and soft electromagnetic radiation may be emitted from the oscillating charge density. Bialas and Czyz estimate that for a not unreasonable spectrum of color plasma oscilations which may be characteristic of nuclear collisions, a substantial fraction of the hadronic energy may be emitted in low p_t photon radiation.(24)

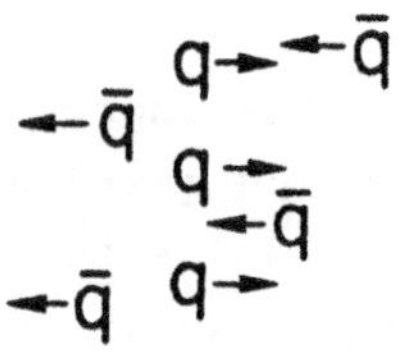

FIGURE 6 A color plasma oscillation

Electromagnetic probes of the nuclear collisions may be characterized by the value of the transverse mass, $M_t = \{p_t^2 + M^2\}^{1/2}$. For di-lepton and photon transverse masses $M_t \lesssim R_{nuc}^{-1} \sim 30$ Mev for Uranium, the photons and di-leptons are coherently produced.(25-26) These particles may be copiously produced in the nuclear fragmentation regions, or in the central region by fluctuations in the charge distributions of mesons. Detailed measurements of these distributions in correlation with measurements of the charge distributions, $\frac{dN}{dy}$, may probe the electromagnetic plasma oscillation, since as a consequence of this oscillation, radiation with frequency less than the electromagnetic plasma frequencv is stongly absorbed. The detailed computation of this absorption is complicated by the finite size of the matter produced in nuclear collisions, and since the matter density is a decreasing function of time, the plasma frequency is time dependent.

In the central region, these coherently produced low transverse mass photons are generated by charge fluctuations. In order to have a large number of photons, a large charge fluctuation must be generated. At sufficiently small p_t, however, the coherently produced photons will dominate over those arising from hadronic decays. In the fragmentation region, the net charge carried by the projectile nucleus generates a large number of low p_t photons.(25-26) The total number is $N_{tot} \sim Z^2\alpha \ln\frac{\omega_{max}}{\omega_{min}}$, where Z is the total number of struck nucleons, and ω_{max} and ω_{min} are maximum and minimum observed frequencies. These limiting frequencies are determined by detectors and backgrounds. Typically the maximum frequency is limited by π^0 decay photons and is $\omega \sim 15$ Mev in the rest frame of the struck nucleus. The low frequency cutoff is more difficult to estimate. The sensitivity of this result to Z suggests that measurements of these low p_t photons may provide a good impact parameter meter. The number and distribution of these emitted photons may also be used to infer the rapidity distributions of charged particles. The total number of photons emitted in a reasonable frequency range for reasonable esimates of the rapidity distributions of charged particles in head-on collisions of large A nuclei is $N_{tot} \sim 50 - 500$. Photons emitted from a beam projectile nucleus, or from colliding nuclei are in a small angular region $\Delta\theta \sim 1/\gamma$ where γ is the Lorentz γ factor of the nucleus. The energy of these photons is $E \sim \gamma\omega$. For a 100 Gev beam, a reasonable range of these parameters is $\Delta\theta < .6^o$ and $E <$ 1.5 Gev.

For larger values of the transverse mass, photons and di- leptons may be approximated as elementary probes with mean free paths large compared to the size of the matter produced in a nuclear collision. Such probes have advantages over hadrons, since hadrons strongly interact and their distribution is characteristic of matter either at the surface or at late times when the matter is at such a low density that hadrons cease interacting. Photons and di-leptons probe the matter at early times when it is hot and dense.[27]

A systematic study of photon and di-lepton production may start with a study of the thermal expectation value of the electromagnetic current-current correlation function,[27-29]

$$W^{\mu\nu}(q) = \int d^4x\, e^{iqx} \langle J^\mu(x)J^\nu(0)\rangle \tag{11}$$

The rate for thermal emission is related to this structure function as

$$\text{Rate/Volume} \sim e^2\, L^{\mu\nu}(q)\, W_{\mu\nu}(q) \tag{12}$$

where $L^{\mu\nu}$ is a computable lepton polarization tensor.

The properties of $W^{\mu\nu}$ are formally very similar to that of $W^{\mu\nu}$ for deep inelastic scattering of leptons from hadrons. The only difference is that there is a thermal expectation value here and not the matrix element between proton states, and that the photon momentum is either timelike or lightlike, not spacelike. The fluid four velocity u^μ is a timelike vector analogous to the proton four momentum of deep inelastic scattering. Because of these formal similarities, $W^{\mu\nu}$ may be written in terms of invariant structure functions A and B as

$$W^{\mu\nu} = \{q^2g^{\mu\nu} - q^\mu q^\nu\}\, A(q^2,u\cdot q,T,\Lambda) + \{g^{\mu\nu}(u\cdot q)^2 -$$

$$(u^\mu q^\nu + u^\nu q^\mu)\ u\cdot q + u^\mu u^\nu q^2\}\ B(q^2, u\cdot q, T, \Lambda) \tag{13}$$

The QCD scale parameter Λ is written explicitly in this equation.

The properties of this structure for large q may be studied and scaling behaviour analogous to that for deep inelastic scattering functions may be extracted. At very high temperatures, $T/\Lambda >> 1$, these structure functions may be computed in perturbation theory. In the large q limit, a detailed analysis of the asymptotic behaviour of A and B gives

$$A \to e^{\beta u\cdot q}\ \bar{A}\ ((u\cdot q)^2/q^2, \beta\Lambda, q^2/\Lambda^2) \tag{14}$$

$$B \to e^{\beta u\cdot q}\ \bar{B}\ ((u\cdot q)^2/q^2, \beta\Lambda, q^2/\Lambda^2) \tag{15}$$

Here $\beta = 1/T$. In the limit of large temperatures, the structure function B vanishes. The asymptotic scaling properties of A and B shown in Eqns. 14-15 is called thermal scaling. The Boltzmann weight factor $e^{\beta u\cdot q}$ and the dependence upon $(u\cdot q)^2/q^2$ are different than that of $W_{1,2}$, for deep inelastic scattering, which has no weight factor, and is a function of the Bjorken x variable, $x = q^2/(p\cdot q)$.

The structure function $W^{\mu\nu}$ must be folded into the hydrodynamic equations before experimental distributions of photons and di-leptons may be computed. I shall only state the results of such an analysis here. For photon and di-lepton transverse masses large compared to the temperature, the photon and di-lepton rapidities and transverse masses are closely correlated to the rapidity of the plasma from which they were emitted and the temperature of the plasma at the emission time. Direct computation gives

$$y_{photon} \sim y_{plasma} \tag{16}$$

$$M_t \sim \left(\frac{2}{v_s^2} + \frac{1}{2}\right) T \tag{17}$$

where v_s is the sound velocity of the matter. For sound velocities characteristic of an ideal quark-gluon plasma,

$$M_t \sim 6.5\ T \tag{18}$$

and the assumption that the transverse masses are large compared to the temperature seems a posteriori justified. For temperatures of $T \sim 100$ - 500 Mev, transverse masses of $M_t \sim .6 - 3$ Gev are emitted with greatest strength.

The absolute rate for photon and di-lepton emission is extremely sensitive to the maximum temperature achieved in the collision and to the sound velocity of the hadronic matter. Roughly three orders of magnitude result from each of these uncertainties. Probably, the rate for di-lepton production is $1 - 10^4$ orders of magnitude times the Drell-Yan rate extrapolated to this mass region. The uncertainty in the total rate is reflected in the A dependences of the total rates. A measurement of the A dependence of the total emission rate provides a check on dynamical assumptions used in computing the thermal emission spectrum.

The shape of the emission spectrum of di-leptons is quite different from that of Drell-Yan. The distribution is a pure power of M_t if the plasma acquires a large enough temperature,

$$\frac{dN}{dM^2 dy d^2 p_t} \sim \left\{ \frac{1}{M_t} \right\}^{\frac{2}{v_s^2}} \qquad (7)$$

As a consequence, the transverse momentum and the mass of the di-lepton pairs are strongly correlated. The structure function for di-lepton emission from a high temperature plasma, corrected for expansion involves only one structure function,

$$\Omega^{\mu\nu} \sim \{q^2 g^{\mu\nu} - q^\mu q^\nu\} \; \Omega_1 \qquad (7)$$

As a consequence of the correlation between temperature and transverse mass, any discontinuity as a function of temperature also appears as a function of transverse mass. If there is a first order phase transition, the electromagnetic current- current correlation function should be discontinous across the phase transition, and this may appear as a discontinuity in the transverse mass spectrum as shown in Fig. 7. If the system exists in a mixed phase for a long time, the thermal emission spectrum will attain a contribution of $e^{-M_t/T_{p.t.}}$, since the mixture of plasma and hadronic matter will emit at the phase transition temperature $T_{p.t.}$.

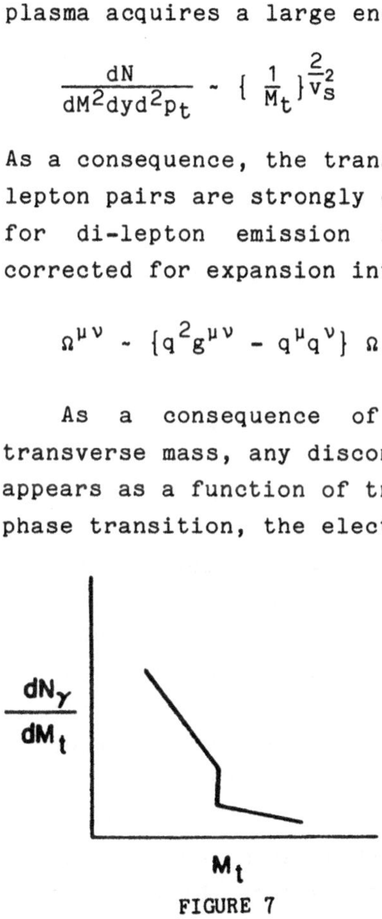

FIGURE 7

A possible transverse photon mass distribution

The physics which may be studied by photons and di-leptons is characterized by transverse mass values. The range of $M_t \sim .6 - 3$ Gev may be dominated by thermal emission. For very low M_t values, coherent emission processes dominate. For masses 30 Mev < M < 200 Mev, the effects of coherence begin to subside, and incoherent production processes begin to dominate. In the region $200 \lesssim M_t \lesssim 600$ Mev, incoherent processes should dominate. In these low mass regions it might be possible to probe the effects of chiral symmetry restoration.[30] The rate of production of such low transverse mass particles might be very sensitive to the constituent quark masses, since the di-lepton production process shown in Fig. 8 vanishes below the threshold $q^2 < 4m_q^2$. The outstanding problem in this low mass region is resolving background processes arising from hadronic decays. A thorough theoretical analysis of backgrounds and a comparison with emissions from a quark-gluon plasma has not yet been carried out.

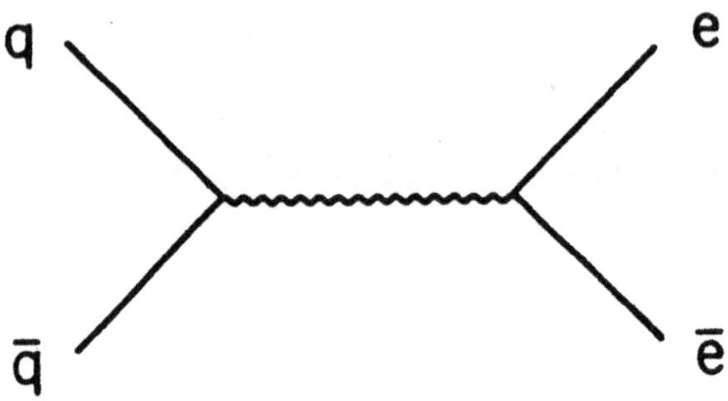

FIGURE 8 Quark annihilation into lepton pairs

At large trnasverse masses $1\text{-}5 \lesssim M_t \lesssim 10\text{-}20$ Gev, there should be corrections to the Drell-Yan emission rates arising from the pre-equilibrium distributions of quarks and gluons. At present a theory of these distribution is lacking, but the development of such a theory is necessary to put the production of a quark-gluon plasma in ultra-relativistic nuclear collisions on a stable foundation.

At transverse masses $M_t \gtrsim 10\text{-}20$ Gev, the Drell-Yan process should dominate.

The production of strange particles has long been suggested as a signal for the production of a plasma.[31] The ratio of strange to non-strange anti-baryons might retain some trace of an abundance of strange quarks and anti-quarks produced in a plasma. This conclusion is on somewhat shaky ground since in the hydrodynamic expansion of the plasma, the strange quarks and anti-quarks may become diluted. Also, a recent computation of Redlich suggests that the abundance of strangeness in a hadronic gas may not be so far different from that of a quark- gluon plasma.[32] A proper theoretical assesmant of strangeness production probably needs non-perturbative input from lattice Monte-Carlo computations, and a thorough analysis of the effects of hydrodynamuc expansion.

Charm particle production may also be important if sufficiently high plasma temperatures are achieved, $T \gtrsim 500$ Mev.[33] Corrections due to hydrodynamic expansion are probably less important for charmed particles than for strange particles since the charmed quark hadronic cross section is small $\sigma < 1$ mb.

An extremely spectulative experimental probe of quark-gluon plasma production may be in multi-particle correlations, and in large scale rapidity fluctuations. Such correlations and fluctuations may arise as the matter participating in a nuclear collision tries to negotiate a first order transition.[34-36] A variety of scenarios are possible all of which involve the production of large scale density flucuations over rapidity intervals $\Delta y \gtrsim 1$. In the collisions of heavy nuclei, such a rapidity interval may include several hundred to several thousand particles, and large scale fluctuations should be separable from statistical fluctuations. These density fluctuations may be generated by superheating, supercooling or the spinodal decomposition of the plasma. They might occur in baryon, anti-baryon or meson distributions.[37] There might also be p_t enhancements if the density fluctuations are accompanied by burning or explosive phenomenon. Backgrounds such as jet production may be ruled out by the azimuthal angle distributions.

The restoration of chiral symmetry may have striking consequences for the widths and masses of resonances produced in a quark-gluon plasma.[38] Since the chiral symmetry transition is expected to be abrupt, the masses and resonances may abruptly change their character as the energy density achieved in a collision increases. Correlations of the type described for $\langle p_t \rangle$ vs $\frac{dN}{dy}$ may be useful here. Also, anomalously large baryon or anti-baryon production might accompany chiral symmetry restoration. As a chiral symmetric world cools through a first order phase transition, light mass baryons in the chirally symmetric phase might become clustered into plasma droplets, as shown in Fig. 9. The large scale density fluctuations which characterize first order phase transitions might appear as large scale fluctuations in the rapidity distribution of baryons and anti-baryons.

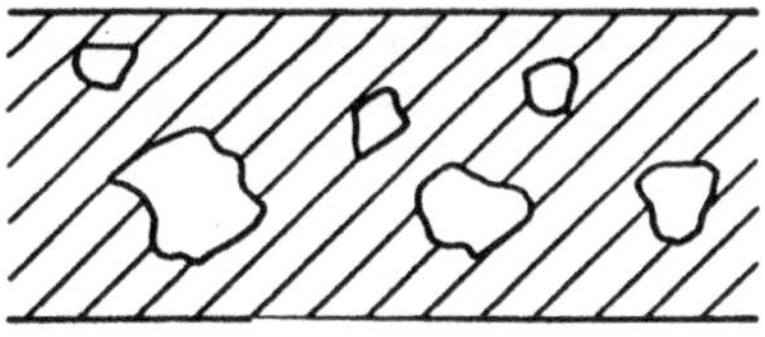

FIGURE 9

Large scale density fluctuations in the plasma

The distribution of jets produced in ultra-relativistic nuclear collisions may provide probes of the space-time evolution of the plasma, and the matter distribution produced in the collision. For example, the occurance of single jets, where one jet has been absorbed as it passes through the plasma, shown in Fig. 10, provides a measure of the mean free path of quarks and gluons in hadronic matter.[39]

Various speculations concerning the existence of exotic stable or metastable forms of matter have been suggested. Examples are Lee-Wick matter,[40] stable or metastable droplets of chirally symmetric strange

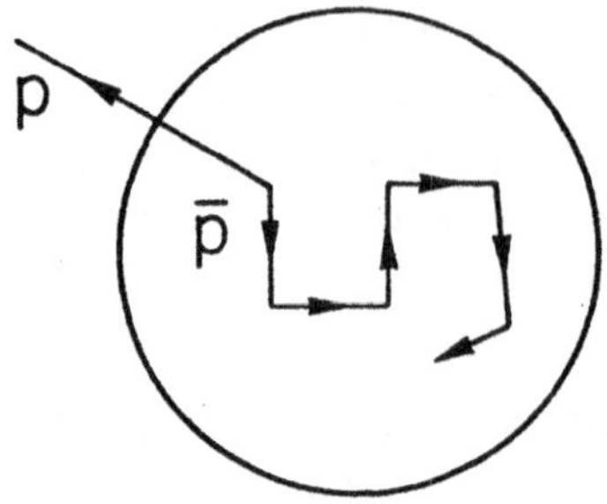

FIGURE 10 Single jet production

matter[41-43] or charmed matter,[44] and even metastable Higgs meson matter.[45] Also, arguments have been proposed that free quarks might be easier to produce in nuclear collisions than in e^+e^- or pp collisions.[46] The precise nature and the probability that such matter exists are difficult to determine, but the revolutionary character of its discovery justifies a generic search. Hints that such new forms of matter exist in cosmic rays have long been suggested by mountain-top emulsion chamber experiments.[47] Such new forms of matter fortunately have generic characteristics which distinguish it from ordinary nuclear matter. For free quarks, the charge is a signature. For particles whose production is associated with a conserved quantum number, the exotic particles should be produced in pairs. In general, the charge to mass ratios of exotic particles should bear no simple relation to that of ordinary nuclei. The penetrating power and cross sections are in general different for ordinary particles with the same Q or Q/A. If an exotic particle decays, the multiplicity might be anomalously large, and the p_t distribution might not be typical of either a nuclear break-up or a hadronic interaction. Since the exotic particle may carry charge, strangeness, or baryon number, the flavor composition and charge to neutral compostion of the final state may be anomolous. Secondaries of the decay may also themselves have anomalous interactions or decays.

References:

(1) J. D. Bjorken, Phys. Rev. D27, 140, (1983)

(2) K. Kajantie and L. McLerran, Nuc. Phys. B214, 261, (1983); Phys. Lett. 119B, 203, (1982)

(3) T. Burnett et al, Phys. Rev. Lett. 50, 2062, (1983)

(4) See Proceedings of Quark Matter 84, Helsinki, Finland, (1984)

(5) W. Busza and A. Goldhaber, Phys. Lett. 139B, 235, (1984)

(6) D. Kisielewska, Crakow Univ. Preprint (1984), to be published in Act. Phys. Polon.

(7) H. Ehtamo, J. Lindfors, & L. McLerran, Z. Phys. C18, 341, (1983)

(8) M. Gyulassy and T. Matsui, Phys. Rev. D29,419, (1984)

(9) A. Goldhaber, Nature, 275, 114, (1978)

(10) M. Gyulassy, S. K. Kauffmann, and L. W. Wilson, Phys. Rev. C20, 2267, (1979)

(11) A. Bialas, Max Planck Institute Preprint, MPI-PEA-PTh-33/84, (1984)

(12) R. Hwa and M. S. Zahir, Univ. of Oregon Preprint, OITS-256, April (1984)

(13) R. Hwa, Phys. Rev. Lett. 52, 492, (1984)

(14) J. Kapusta, Phys. Rev. C27, 2037, (1983); L. Csernai and J. Kapusta, Univ. of Minnesota Preprint, UMNTNP-70/1984, (1984); C. Y. Wong, Phys. Rev. Lett. 52, 1393 (1984)

(15) K. Kajantie, R. Raitio, and P. V. Ruuskanen, Nucl. Phys. B222, 152 (1983)

(16) A. Bialas and W. Czyz, Acta. Phys. Polon. B15, 229, (1984)

(17) G. Baym, B. L. Friman, J. P. Blaizot, M. Soyeur, and W. Czyz, Nucl. Phys. A407, 541, (1983)

(18) A. Hosoya and K. Kajantie, Helsinki University Preprint, HU-TFT-83-62, (1983)

(19) P. Danielowicz and M. Gyulassy, LBL Preprint, LBL-17278, (1984)

(20) E. V. Shuryak, Phys. Rep. 61, 71, (1980); E. V. Shuryak and O. V. Zhirov, Phys. Lett. 89B, (1979).

(21) L. van Hove, Phys. Lett. 118B, 138, (1982)

(22) O. Miyamura, Proceedings of Quark Matter 84, Helsinki, Finland (1984)

(23) U. Heinz, Phys. Rev. Lett. 51, 351, (1983)

(24) A. Bialas and W. Czyz, Univ. of Crakow Preprint

(25) J. D. Bjorken and L. McLerran, Fermilab-Pub-84/35-T

(26) J. Kapusta, Phys. Rev. C15, 1580, (1977)

(27) E. L. Feinberg, Nuovo Cim. 34A, 391, (1976)

(28) E. V. Shuryak, Phys. Lett. 78B, 150, (1978); Sov. J. Phys. 28, 408, (1978)

(29) L. McLerran and T. Toimela Fermilab-Pub-84/84-T

(30) J. Kapusta, Phys. Lett. 136B, 201, (1984)

(31) J. Rafelski & B. Muller, Phys. Rev. Lett. 48, 1066, (1982); T. S. Biro, B. Lukacs, J. Zimanyi & H. Barz, Nuc. Phys. A386, 617, (1982)

(32) K. Redlich, Technische Hochschule Darmstadt Preprint, IKDA 84/12

(33) J. Cleymans, Phys. Lett. 127B, 375, (1978); J. Cleymans and R. Philippe, Z. Phys. C22, 271, (1984); J. Cleymans and G. Vanderzande, Bielefeld Preprint, BI-TP-84/04, (1984)

(34) L. van Hove CERN-TH-3924, June 1984; Z. Phys. C21, 193,

(1983)

(35) M. Gyulassy, K. Kurki-Suonio, K. Kajantie, and L. McLerran, Nuc. Phys. B237, 477, (1984)

(36) J. Kapusta, Univ. of Minnesota Preprint, Print-84-0134 Jan. (1984)

(37) T. DeGrand Univ. of Colorado Preprint HU-TFT-84-20, Aug. (1984)

(38) R. Pisarski, Phys. Lett. 110B, 155, (1982); R. Pisarski and F. Wilczek, Phys. Rev. D29, 338, (1984)

(39) J. D. Bjorken, Fermilab Preprint, Fermilab-Pub-82/59-THY, (1982)

(40) T. D. Lee and G. Wick, Phys. Rev. D9, 2291, (1974)

(41) J. D. Bjorken and L. D. McLerran, Phys. Rev. D20, 2353, (1979)

(42) A. Kerman and S. A. Chin, Phys. Rev. Lett. 43, 1292, (1979)

(43) E. Witten, Phys. Rev. D30, 272, (1984)

(44) J. Klosinski, Rivers State University Preprint, (Port Harcourt, Nigeria) RSUST-Physics-83/1, (1983)

(45) R. Mackenzie, F. Wilczek, and A. Zee, ITP at Santa Barbara Preprint, NSF-ITP-84-99 (1984)

(46) G. L. Shaw and R. Slansky, Phys. Rev. Lett. 50, 1967, (1983)

(47) C. M. G. Lattes, Y. Fujimoto, and S. Hasegawa, Phys. Rep. 65, 151, (1980)

LATTICE MONTE CARLO CALCULATIONS OF FINITE TEMPERATURE QCD

Thomas DeGrand

University of Helsinki
Research Institute for Theoretical Physics
Siltavuorenpenger 20 C ,SF-00170 Helsinki 17, Finland
and
Physics Department
University of Colorado
Boulder, Colo. 80309, USA*

I. Introduction

Although the words "Monte Carlo" appear in the title of my talk, my goal will actually be somewhat broader. I want to discuss fairly generally the current status of the lattice description of the deconfinement transition and the properties of hadronic matter at high (and low) temperature T. An ultimate goal of these investigations is to learn whether or not QCD actually predicts the naive phase diagram shown in Fig.1 :

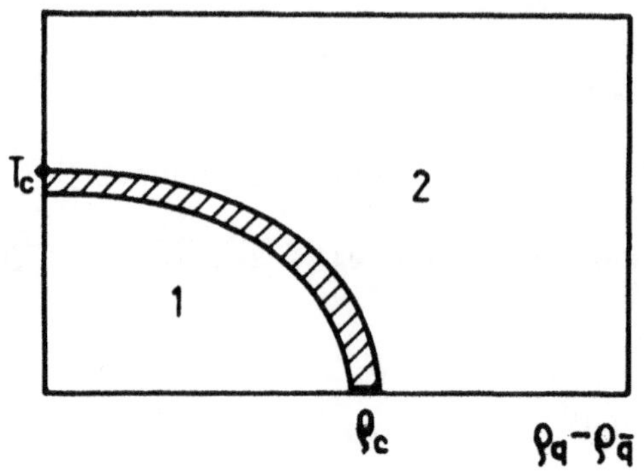

Fig.1 : "Naive" phase diagram of the strong interactions. Region 1 is a confined phase where chiral symmetry is broken. Region 2 is a quark-gluon plasma where chiral symmetry is restored. The shaded region represents one (or more) phase transitions. T_c is a few hundred MeV, ρ_c a few times $3/fm^3$.

confined hadronic matter at small temperature and quark density,bounded by some sort of phase transition into a quark-gluon plasma at high temperature and density. A more realistic goal, which is at present partially within our grasp, is to compute the static properties of QCD matter at T>0 from first principles. These include the order of phase transitions, critical temperatures T_c, critical exponents or latent heat, but not dynamical critical properties, such as the behavior of Green's functions near T_c. I know of no first- principles discussions of non-equilibrium properties of QCD, which would be required for a description of the experiments. In fact, experimentalists should think of the world studied by lattice or Monte Carlo methods as a little crystal in an oven whose temperature is kept constant in time.

I will break the talk down into three parts.I will begin by giving a short description of how we set up the finite-temperature field theory on a lattice to display the important parts of the calculation without going too much into details. Then I will discuss recent progress in our understanding of the glue world - pure gauge theories - and end by discussing the physically relevant case of fermions and gauge fields.

II. How?

To show how we perform lattice calculations at finite temperature let us study a sample problem, the simple harmonic oscillator in a heat bath. The partition function is

$$Z = \mathrm{Tr}\, e^{-\beta H} \tag{2.1}$$

$$= \sum_i \langle i|e^{-\beta H}\,|i\rangle$$

with $\beta = 1/kT$ - (not to be confused with the coupling $\beta = 6/g^2$!)

$$H = \frac{p^2}{2m} + \frac{1}{2} m\omega^2 x^2 = K + V$$

Expectation values of operators are computed by

$$\langle O\rangle = Z^{-1} \sum_i \langle i|\, O e^{-\beta H}\,|i\rangle \tag{2.4}$$

We "solve" this problem by introducing an imaginary time variable $\Delta\tau = \beta/N_\tau$ and write

$$Z = \sum_i \langle i|\, e^{-N_\tau \Delta\tau H}\,|i\rangle \tag{2.5}$$

which we latticize by putting in parentheses

$$Z = \sum_i \langle i|\, (e^{-\Delta\tau H})^{N_\tau}|i\rangle \tag{2.6}$$

and expanding in complete sets of states

$$Z = \sum_{x(\tau_1)} \sum_{x(\tau_2)} \ldots \sum_{x(\tau_N)} \langle x(\tau_1)| e^{-\Delta\tau H} | x(\tau_2)\rangle\langle x(\tau_2)| e^{-\Delta\tau H} |x(\tau_3)\rangle$$

$$\ldots \langle x(\tau_N)| e^{-\Delta\tau H} |x(\tau_1)\rangle \qquad (2.7)$$

where I have chosen a set of states diagonal in position. The interpretation of the integrand (2.7) is as follows: the oscillator begins in some location $x(\tau_1)$ at imaginary time τ_1 and the Hamiltonian carries it to site $x(\tau_2)$ at τ_2, then to $x(\tau_3)$ at τ_3, and so on, to return at last to the original location after leaving $x(\tau_N)$. One computes Z by summing over all locations $x(\tau_i)$. H is a quantum mechanical operator and the direct evaluation of $e^{-\Delta\tau H}$ can be tricky. However for small $\Delta\tau$ we can write

$$\langle x'| e^{-\Delta\tau H} |x\rangle = \sum_y \langle x'| e^{-\Delta\tau K} |y\rangle\langle y| e^{-\Delta\tau V} |x\rangle$$

$$= \sum \langle x'| e^{-\Delta\tau K} |x\rangle \; e^{-m\omega^2\Delta\tau x^2/2} \qquad (2.8)$$

and then assume that the particle propagates as a free particle between x and x'. Then

$$\langle x'| e^{-\Delta\tau H} |x\rangle \approx \exp[-\Delta\tau(\tfrac{1}{2}\, m(x-x')^2/\Delta\tau^2 + \tfrac{1}{2}\, m\omega^2 x^2)] \qquad (2.9)$$

The object in parentheses looks like the classical Lagrangian except for some minus signs (it is called the Euclidien Lagrangian L_E) Since it is a classical object Z is just a multi-dimensional integral

$$Z \approx \int dx_1 \ldots dx_{N_\tau} \quad \exp[-\Delta\tau \sum_i (\tfrac{1}{2}\, m(x_i - x_{i+1})^2/\Delta\tau^2 + \tfrac{1}{2}\, m\omega^2 x^2)] \qquad (2.10)$$

which can be computed numerically. In actual fact, one doesn't integrate directly - one generates sets of $(x_1 \ldots x_{N_\tau})$ with a probability $\exp(-\Delta\tau L_E)$ and measures operators using these sets of numbers.It's not very different from an event generator program. A field theory is just a quantum mechanical system with a large number of variables per time slice. To handle them, one usually has to introduce additional space lattices.The number of degrees gets unmanageable quickly - a typical lattice program nowadays is around a 10^6 -dimensional integral.

Finally we want to get rid of the lattice. After all, it was only a crutch to help in the calculation. For the harmonic oscillator we can do that by taking $\Delta\tau$ to zero while holding the temperature, or $N_\tau \Delta\tau (= 1/kT)$, fixed. For a field theory, like QCD, it is trickier. The spatial lattice spacing a sets the scale for physical excitations through the coupling constant g(a). For example

$$ma = f(g(a))$$

for the mass of some physical object (like a glueball). The continuum limit is achieved by taking a (measured in cm) to zero holding m (measured in MeV) fixed - which typically involves knowing f(g(a)) for many values of g. In QCD one needs to compute at small g(a) because from asymptotic freedom g is small at short distances. Unfortunately, for small g the correlation length becomes very large (in lattice units) and the finite size of the lattice becomes important. All this makes for a difficult numerical problem. That is why one sees 30% errors on lattice calculations of quantities such as T_c. The calculation at fixed lattice spacing of T_c (in units of 1/a) is not noisy, but the extrapolation is.

So before turning to examples I summarize the ingredients of a $T \neq 0$ lattice Monte Carlo calculation:

1) Introduce a time lattice of finite length and spacing $N_\tau \Delta\tau = \beta$. When you need a lattice in the space degrees of freedom which should be of infinite extent but is finite due to computer requirements, keep $N_s \gg N_\tau$ (or $N_s a \gg N_\tau \Delta\tau$), if possible.

2) Variables are periodic in imaginary time: $\phi(x, \tau) = \phi(x, \tau + \beta)$. This is a simple consequence of the fact $Z = \mathrm{Tr}\exp(-\beta H)$.

3) Convert H to the classical L_E and perform the integrals numerically.

4) Try to take the limits a, $\Delta\tau \to 0$ holding m, T fixed.

Now I turn to a discussion of recent results.

III. Results for Pure Gauge Theory

The signal for confinement in a pure gauge theory is given by the free energy of a heavy (static) quark-antiquark pair at spatial locations x and y[1-4]

$$e^{-\beta F(x,y)} = \langle L(x)L^+(y)\rangle$$

where $L(x) = T\exp(ig\int_0^\beta A_0(x,\tau)d\tau)$ measures the twinkling in color space of one of the quarks (which obeys a Schrödinger equation with $H = gA_0(x)$). It is called a "Wilson line". F (which is also the potential V(x-y)) goes to infinity for large x-y if quarks are confined, and approaches a constant if they are not. Equivalently one measures the expectation value of a single Wilson line $\langle L\rangle$; $\langle L\rangle = 0$ is the confinement signal.

In a gauge theory at $T \neq 0$ it is $A_\mu(x,\tau)$ which is periodic in τ.

However, it happens that gauge transformations are periodic in τ only up to an overall global phase factor ($G = \exp(2\pi i \ell/N), \ell = 1...N$ for SU(N)). The action is invariant under these transformations, but the Wilson line is not ($L' = GL$). Since the Wilson line carries all the interesting deconfinement physics we imagine integrating over all degrees of freedom to give an effective theory of interacting Wilson lines. The action will have a Z(N) symmetry associated with gauge transformations G. The combination of one Wilson line per point in d-dimensional space with a Z(N) invariant interaction means that this effective theory looks like a spin theory. The equilibrium state of the model can either respect the symmetry in which case the expectation value of the spin (=Wilson line) is zero (= confinement) - or it may break the symmetry. In that case the spin system exhibits spontaneous magnetization. This case corresponds to $\langle L \rangle \neq 0$, or deconfinement. The confinement-deconfinement transition is thus related to an order-disorder transition in an analog spin theory.

The results of these arguments are the predictions that the deconfinement transitions in lattice SU(2) and SU(3) pure gauge theory are respectively second[1-4] and first order[5]. These predictions are confirmed by Monte Carlo[6-11].

These predictions were reliable because the uniqueness of N = 2 or 3 models. For larger N possible models are not unique and one did not know whether the deconfinement transition was first or second order (exponents were predicted if the transition was second order[5]). A possible solution to the question was presented last fall by Green and Karsch.[12] They studied SU(N) gauge theories in the strong coupling limit in mean-field approximation and showed that the transition was first order for all N≥3. Mean field theory is not regarded as particularly reliable and so the calculation was not taken too seriously (by me at least!) until this spring, when Monte Carlo simulations for SU(4) gauge theory by Gross and Wheater[13], Batrouni and Svetitsky[14], and Goksch and Okawa[15] all showed a first-order deconfinement transition. The calculation must be done with some care since the T = 0 theory with the Wilson action has a (non-deconfining) first order phase transition which should not be confused with the deconfinement one. Batrouni and Svetitsky[1] avoid this difficulty by taking a more complicated (fundamental plus adjoint) form of the action which does not have this bulk transition. The SU(4) result showns that the simple mean-field result for the order of the transtition is probably correct.

There are mathematical proofs that the high temperature phase of an SU(2) gauge theory is in a plasma phase. If the proof of confinement by Tomboulis[17] holds true, then the pure gauge theory <u>must</u> possess some kind of confinement-deconfinement phase transition. Now back to SU(3).

What is T_c for pure glue SU(3) matter in the continuum? Unfortunately there is at present no reliable answer to this question : even the very large lattices studied so far do not show proper limit behaviour.

Recall (Section II) that a physical temperature is given in terms of the lattice spacing a by $T_c = 1/N_t a$. From dimensional transmutation a (in fermis) is a function of the coupling $\beta = 6/g^2$ for SU(3).

This function is only known (so far) through two loops in weak coupling

$$a = \frac{1}{\Lambda_L} f(\beta) \tag{3.1}$$

$$f(\beta) = (\frac{8\pi^2}{33} \beta)^{51/121} \exp(-\frac{4\pi^2}{33} \beta) \tag{3.2}$$

Λ_L is a constant cutoff parameter which may be measured by deep-inelastic scattering ($\Lambda_L = \Lambda_{MOM}/83.5$) or compared to the string tension (from-onium or from the Regge slope) by $\Lambda_L = C\sqrt{K}$ where C is another constant which must be extracted from Monte Carlo data. A reliable continuum measurement for T_c exists once the scaling behaviour (3.1) is satisfied, for only then is the ratio T_c/Λ_L (or $\sqrt{K}/\Lambda_L$) a renormalization group invariant.

To see if the data scales, let us define

$$T_c/\Lambda_L = 1/N_t f(\beta_c)$$

and plot it vs. β_c. As we see (Fig (2)) the ratio T_c/Λ_L is <u>not</u> constant except possibly at very large β_c, where the authors of Ref (11) fit the last two data points to $T_c = 53\ \Lambda_L$.

The parameter G has a similar history of not scaling below $6/g^2 \sim 6$[18]. Above that value C ~ 0.01. Taking $\sqrt{K} \sim 400$ MeV gives $T_c \sim 210$ MeV, a result I regard as being no better than two stars in the Particle Data Group sense.

There is an obvious solution to this problem : more money. The calculations are entirely straightforward but will take a lot of time since big lattices are required. The two last data points in Fig. (2)

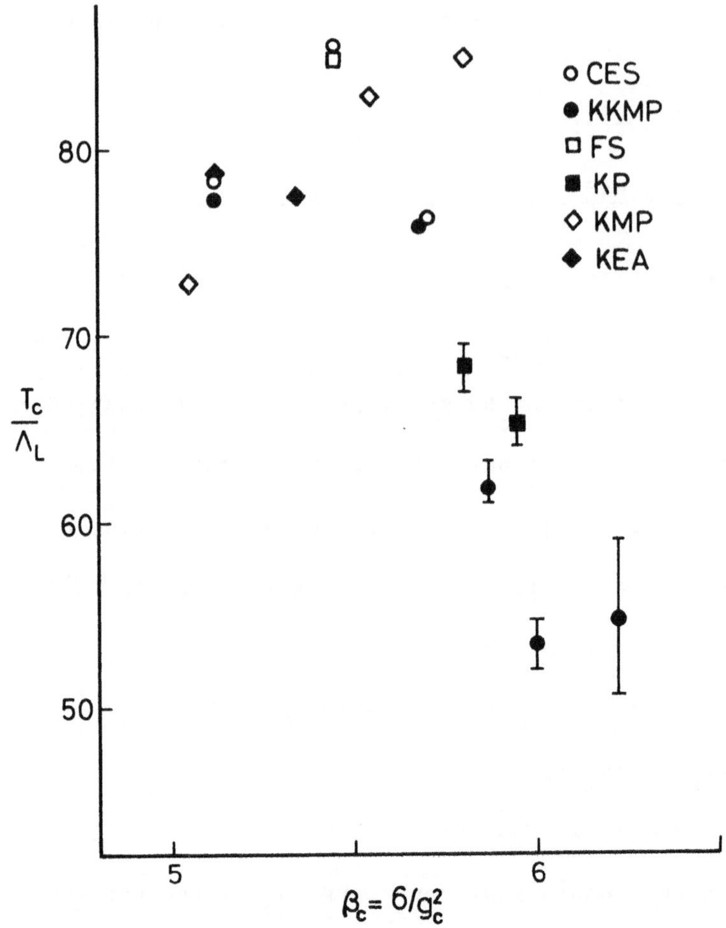

Fig.2 : Predictions of the ratio T_c/Λ_L .Data are CES-Ref (8), KKMP-(11), FS-(9), KP-(10), KMP-(6), and KEA-(7).

come from $7^3 \times 8$, $11^3 \times 8$, and $9^3 \times 10$ lattices. Does anyone have a free standby license on a Cyber 205?

The authors of Ref. (11) have taken data at many lattice volumes for each N_t and extrapolated $6/g^2$ to the infinite volume limit. $6/g^2$ scales very nicely like 1/volume; changes from $N_x = 7$ to infinity are typically 0.5%.

Rather than giving the latent heat directly, most authors prefer to quote the ratio $\Delta E/T_c^4$, which is expected to show better scaling properties than ΔE directly ($\Delta E \propto 1/a^4$). The data of Ref.(7, 8, 9, 10) give $\Delta E/T_c^4 \sim 7.5$, 6.5, 9.4, and 3.32 respectively. At this point interpretations differ. Ref(7-8) attempt to correct for finite lattice size effects by dividing their result by the ratio R= (internal energy for free massless lattice bosons)/(internal energy for free massless continuum bosons), reducing their number by a factor of about 1.8 at smaller N_τ to 1.23 for the 6×16^3 lattice of ref(10). Fucito and Svetitsky[(9)] augment their computation of the latent heat by the constraint that ΔP, the pressure difference, vanish at T_c. They argue that their result (which translates to $\Delta E \sim 1.89 \pm .1$ GeV/fm^3) is more reliable than that of Refs.(7-8) and point out that at T_c the quark-gluon plasma is not a gas of massless noninteracting particle as the correction argument assumes. My opinion is that the predictions are not in agreement and more data, taken in the scaling regime (which has not been found yet), is needed before ΔE is reliably known. (Are you reading this, Julius?)

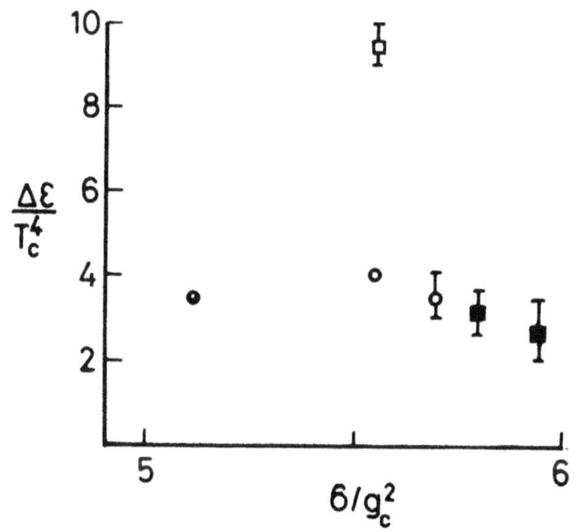

Fig.3: "Scaled" latent heat $\Delta\varepsilon/T_c^4$, with the same labels as in Fig.2.

IV. Gauge-Matter Systems

In contrast to pure gauge systems, where the outlines are clear, the situation regarding gauge-fermion systems is very confused. The regions of parameter space where we have reliable predictions are far away from the real world of weak coupling, light quarks, and small representations of the gauge group. It has taken a long time to develop Monte Carlo programs which approach the same speed for fermions as for bosons.

The root of the problem are the minus signs of Fermi statistics. Computers have a hard time keeping track of antisymmetry for a direct evaluation of the fermion path integral (it is not even a number!) The standard methods for handling fermions involve taking the fermion path integral

$$Z = \int[d\psi d\bar{\psi} dU] \exp (\bar{\psi}M\psi + S(U)), \qquad (4.1)$$

and formally integrating over the fermions to give

$$Z = \int[dU] \det M \exp S(U) \qquad (4.)$$

and the somehow evaluating detM - either directly[19] - time consumming! or approximately as a power series in $1/M_q$ (hopping parameter expansion[20], whose $M_q = \infty$ limit is called the quenched approximation[21]) or methods which assume M is a slowly varying function of U (pseudofermions). Rather different from this approach is the microcanonical ensemble of Polonyi and Wyld[23]. Anyway the resulting effective action is not local in U, and in the $m_q \to 0$ limit M has zero frequency modes which make its (often required) inversion delicate.

As a further complication, one of the important questions we want to answer is: Does $S_{eff}(U)$ have a phase transition or not, even though $S(U)$ does. Then the nature of approximations may be crucial.

One place where analytical calculations can be performed is for large but finite quark mass. In that limit one can argue that the Z(N) symmetry of the Wilson line effective action is always broken since the free energy of a static $Q\bar{Q}$ pair is never infinite - it is always possible to pop a dynamic $q\bar{q}$ pair out of the vacuum to screen the static pair. Losing the Z(N) symmetry means that the deconfinement transition need not persist for nonzero values of $1/m_q$. Banks and Ukawa[(24)] and DeTar and I[(25)] showed that at high temperature and large m_q the dynamic quarks act like an external magnetic field in the Z(N) theory, so that if the deconfinement transition is second order for the pure gauge theory (as is the case for SU(2)), it is smoothed out by any nonzero $1/m_q$. If the transition is first order for the pure gauge theory, then there will be a smallest M^* such that for $m_q > M^*$, the transition remains first order, for $m_q = M^*$ the transition is second order, and for $m_q < M^*$ the transition is smooth. Unfortunately M^* is not a universal quantity and its actual value can only be determined by Monte Carlo - it may be that the approximations which lead to a prediction of M^* break down for some $m_q > M^*$.

These arguments are on firmer footing for gauge-boson systems where $1/m_q$ is equivalent to the boson-gauge field coupling β_h. One can make Fradkin-Shenker arguments[(26)] to show that some β^*_m exists and that for $\beta > \beta^*_m$ only a smooth transition occurs.

Chiral symmetries complicate the light-fermion sector. Tomboulis and Yaffe have recently proven that chiral symmetry is restored at high temperature[27], so that if $m_q = 0$ there must be a chiral symmetry phase transition at some T_χ. If that transition is first order, it will persist for $m_q \neq 0$ (the case in the real world). If it is second order, $m_q \neq 0$ will smooth it out. If the chiral and deconfinement transitions are both first order, they may persist to such high (or low) values of m_q that they couple as shown in Fig.(4). One may encounter no transition at all, (Fig.(4a) two separate transitions (Fig(4b)) or one "combined" transition (4c). It may be hard to distinguish (b) and (c) in an experiment, Monte Carlo or otherwise.

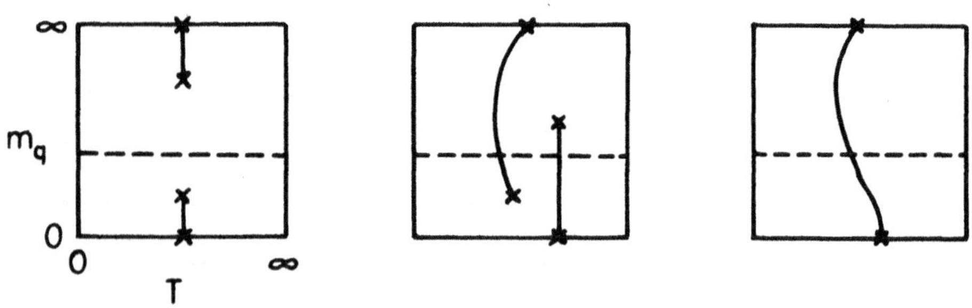

(a) (b) (c)

There are two generic kinds of models of the chiral transition, with all possible predictions. If the length scale of chiral symmetry breaking is much shorter than the confinement scale (as it would be if the quarks were to lie in very high representations of the color gauge group[28]) then a model of pions as fundamental particles at finite temperature might be appropriate. Very different versions of these models have been described in Ref(29-31). They share the property of having first-order transitions for number of <u>flavours</u> 3 or greater. The model of ref(30) is intriguing. Pisarski and Wilczek perform weak-coupling renormalization group studies of SU(N) $\otimes$ SU(N) chiral Lagrangians and find first order chiral restoration transitions for N $\geq$ 3. The transitions are mean field if instanton effects are included but in the absence of instantons remain first order due to fluctuations. The results should be regarded with caution since chiral Lagrangians used in phenonenology have large coupling constants and perturbation theory is unreliable; however, similar models have been studied on a lattice in strong coupling and with Monte Carlo, and the first-order transitions persist[(29)]. It is worth remembering that these calculations show strong dependence on the number of flavors, and suggest that too-naive calculations (like mean-field theory in the Pisarski-Wilczek case) may lead to wrong predictions for the order of the transition.

This picture is supported by the quenched approximation calculations of the SU(2) chiral symmetry transition by the Illinois group, who see a "tumbling" scaling law for quarks in high representations.[(32)]

Chiral symmetry breaking at a short distance scale may actually protect the deconfinement transition from destruction due to too-small m_q, since extra chiral contributions will be added to the quarks' masses.[(33)]

The other class of models work at $g = \infty$, integrate out the gauge fields and get an effective Hamiltonian for mesons which explicitly breaks chiral symmetry[(34-37)]. Then one takes as a partition function

$$Z = \mathrm{Tr}\ \exp(-H_{eff}/T)$$

which can be studied in mean field theory (this is actually not quite what ref.(31) does, but close enough for now) or which can be used to read off the critical behaviour by analogy with known statistical mechanics results. Thus Patel[36] takes Kogut-Susskind fermions and constructs H_{eff} as an O(3) Heisenberg antiferromagnet, with second-order chiral transition. It is amusing (to me) that if one takes SLAC fermions with 3 flavors H_{eff} is an SU(3) ⊗ SU(3) antiferromagnet,[38] which (by analogy with the work of refs(29-30) described in the last paragraph) has a first order transition.

Published Monte Carlo data are contradictory. Two large-m_q calculations (by Hasenfratz, Karsch, and Statemescu[38] and by Engels, Celik, and Satz[9]) have similar data and different interpretations. To quote from thier abstracts, "no transition is observed below a critical mass...of the order of GeV"[38] and "rapid variation of E_G in T, accompanied by strong fluctuations from iteration to iteration, as clear signal of the deconfinement transition."[39]

There are now four groups performing Monte Carlo simulations with light dynamic fermions. Three groups see evidence for a phase transition while one does not. (See Table I.) The groups which see a phase transition all use moderate-sized lattices and small but nonzero quark masses (ma= 0.2 to 0.1) with some attempt to extrapolate to zero quark mass. All groups use some version of fermion latticization which preserves a continuous chiral symmetry in the zero quark mass limit. The fermion

Monte Carlo methods are quite different: Refs.(41) and (43) use pseudo-fermions, ref.(42) uses a microcanonical ensemble, and ref.(44) performs an exact evaluation of the fermion determinant.

Table - SU(3) Dynamic Fermions

Ref.	Method	Flavors	size	ma	$6/g_c^2$
42	microcanonical	4	$8^3 \otimes 4$	.08,.1	5.1
41	pseudofermion	2	$6^3 \otimes 2$	.1	4.7-4.8
43	pseudofermion	3	$6^3 - 8^3 \otimes 4$	.1,.2	5.3±.1
44	exact	4	4^4	0	-

While these results are not in agreement, one can invent a scenario which accounts for their differences. The groups which see a phase transition at small m also see a signal for chiral symmetry restoration at the same value of $6/g^2$. The group which sees no transition works with $m_q = 0$ on a finite lattice. But a finite system with $m_q = 0$ can never undergo chiral symmetry breaking - it always remains in the chirally symmetric phase. So I suggest that the transition which is observed is driven much more strongly by the chiral symmetry restoration transition than by the deconfinement transition - the picture resembles Fig.(4b) or - more likely - (4c).

Although it is premature to quote a number for experiment, asymptotic freedom applied to the results of Table I suggests $T_c \sim 400$ MeV. Clearly more data is needed before any believeable quantitative numbers can be quoted.

V. Conclusion

Perhaps the best conclusion is a wish list: What would I do with a guarantee of 4 hours of free CDC-205 computer time per day for the next year?

a) Measure T_c and ΔE for pure gauge SU(3) matter on large lattices (N_t>10) to see the onset of scaling and get reliable numbers for T_c/Λ_L, $\Delta E/T^4$. Measure plaquette correlation functions for information the behavior of gluon matter (like its mass spectrum) near T_c.

b) Write a fermion-gauge field Monte Carlo using the best presently available fermion algorithm and study the phase structure of an SU(3) theory on a large lattice (but not necessarily in the scaling regieme) as a function of quark mass and possibly number of flavors (since chiral models predict N_F dependence).

I expect both these calculations will be done by somebody during the next year.

There are also several VAX-sized projects which while not relevant to experiment are important in clarifying the nature of the fermion-gauge field transition:

c) What happens in an SU(3) gauge-Higgs system?

d) What happens when the gauge group is SU(2)? Here the deconfinement transition is expected to be completely smoothed by heavy quarks. What happens for light quarks? Is there a chiral transition ?(45)

So far, all lattice Monte Carlo calculations have passed through three stages. The first stage is one of complete ignorance, without even any reliable experimental methods. The second stage is characterized by short simulations on small lattices, with results that appear to be not just qualitatively but quantitatively correct. The third stage involves very long runs on enormous lattices in an attempt to get precise quantitative results. One then finds the earlier results were overstated, and that while the simulations still show the same qualitative behavior, the actual numbers differ from group to group. In my opinion we are in the third state of Monte Carlo for pure boson systems nd in the secnd stage for fermion systems. What is the fourth stage?

Acknowledgements.

I would like to thank P. Damgaard, J. Greensite, P. Hasenfratz, K. Kajantie, F. Karsch, N. Kawamoto, H.I. Miettinen, C. Montonen, R. Raitio, C. Rebbi, H.S.Satz, and B. Svetitsky for conversations and/ or correspondence during the preparation of this talk. I would also like to thank the Research Institute for Theoretical Physics for its hospitality during my visit. This work was also supported by NORDITA and the U.S. Department of Energy through a Young Investigator Grant.

References

* present address

1) L. Susskind, Phys.Rev. D20, 2610 (1979)

2) A. Polyakov, Phys.Lett. 72B, 477 (1978)

3) L. McLerran and B. Svetitsky, Phys.Rev. D24, 450 (1981)

4) J. Polonyi, K. Szlachanyi and J. Kuti, Phys.Lett. 98B, 195 (1981)

5) L. Yaffe and B. Svetitsky, Phys.Rev. D26, 963(1982);
B. Svetitsky and L. Yaffe, Nucl.Phys. B210 [FS6], 423 (1982)

6) K. Kajantie, C, Montonen, and E. Pietarinen, Z.Phys.C9 253 (1981)

7) J. Kogut et al. Phys.Rev.Lett. 51 869 (1983)

8) T. Celik, J. Engels, and H. Satz, Phys.Lett.125B, 411(1983),
29B 383(1983)
Z.Phys.C. 22, 301 (1984).

9) F. Fucito and B. Svetitsky, Phys.Lett. 131B, 165 (1983).

10) F. Karsch and R. Petronzio, CERN-TH 3747 (1983)

11) A.D. Kennedy, J. Kuti, S. Meyer, and B. Pendleton, Santa Barbara preprint NSF-ITP-84-61 (1984).

12) F. Green and F. Karsch, Nucl.Phys. B (in press)

13) M. Gross and J. Wheater, Oxford preprint (1984)

14) G. Batrouni and B. Svetitsky, Cornell preprint CLNS 84/600 (1984)

15) A. Goksch and M. Okava, Phys.Rev.Lett. 52, 1951 (1984)

16) C. Borg and E. Seiler, Nucl.Phys. B215, 125(1983)
E. Tomboulis and L. Yaffe, Phys.Rev. D29, 780(1984)

17) E. Tomboulis, Phys.Rev.Lett. 50, 885(1983)

18) see for instance J. Stack, Phys.Rev. D29, 1213(1984)
and D. Barkai, K. Moriarty, and C. Rebbi, BNL-34462 (1984)

19) M. Fischler and R. Roskies, Pittsburgh preprint PITT-18-84

20) A. Hasenfratz and P. Hasenfratz, Phys.Lett. 104B, 489 (1981),

C. Lang and H. Nicolai, Nucl.Phys. B200 [FS4] 135 (1982)

21) H. Hamber and G. Parisi, Phys.Rev.Lett. 47, 1792 (1981)

22) F. Fucito, E. Marinari, G. Parisi, and C. Rebbi, Nucl.Phys. B180 [FS2] 364 (1981)

23) J. Polonyi and H. Wyld, Phys.Rev.Lett. 51, 2257(1983).

24) T. Banks and A. Ukawa; Nucl.Phys. B225, 145(1983).

25) T. DeGrand and C. DeTar, Nucl.Phys. B225, 590(1983).

26) E. Fradkin and S. Shenker, Phys.Rev. D19, 3682(1979)

27) E. Tomboulis and L. Yaffe, Princeton preprint (1984).

28) S. Raby, S. Dimopoulos, and L. Susskind, Nucl.Phys. B169, 373(1980).

29) J. Kogut, M. Snow, and M. Stone, Nucl.Phys. B200 [FS4] 211(1982)

30) R. Pisarski and F. Wilczek, Phys.Rev. D29, 338(1984).

31) H. Goldberg, Phys.Lett. B127 242(1983)

32) J. Kogut, et.al. Nucl.Phys. B225 [FS9] 326 (1983)

33) cf. V. Alessandrini, LPTHE Orsay 84/14 (1984)

34) P. Damgaard, N. Kawamoto, and K. Shigemoto, Nordita preprint 84/16

35) F. Green, Phys.Lett. B133, 99(1983)

36) A. Patel, Caltech preprint 68-1079(1983)

37) A. Gocksch and M. Ogilvie, BNL-34559 (1984)

38) B. Svetitsky, S. Drell, H. Quinn, M. Weinstein, Phys.Rev. D22, 490(1980).

39) P. Hasenfratz, F. Karsch, and I. Statemescu, Phys.Lett. B133 221(1983)

40) T. Celik, J. Engels, and H. Satz, Phys.Lett. B133, 421 (1983)

41) R. Gavai, M. Lev, and B. Petersson, BI-TP-81/04 (1984)

42) J. Polonyi et al, Illinois preprint (TH)-84-22 (1984)

43) F. Fucito and S. Solomon, Caltech preprint, 1984
F. Fucito, S. Solomon, and C. Rebbi, Caltech preprints CALT-68-1124 and CALT-68-1127(1984)

44) M. Fischler and R. Roskies, Pittsburgh preprint PITT-18-84 (1984)

45) A study of this question was described by Polonyi, in these proceedings.

KINETIC THEORY AND FLUID DYNAMICS IN VERY HIGH ENERGY NUCLEUS-NUCLEUS COLLISIONS†

Gordon Baym
Loomis Laboratory of Physics
University of Illinois
Urbana, Illinois 61801, U.S.A.

The physics of the various regimes and non-equilibrium processes encountered in a central ultra-relativistic heavy ion collision, from the initial kinetic period, through the hydrodynamic phase, hadronization transition and eventual freezeout, are surveyed.

1. INTRODUCTION

The subject of this talk is really how one treats the entire dynamics of ultra-relativistic heavy-ion collisions. The various regimes encountered in a central ultra-relativistic heavy-ion collision are conveniently summarized in the space-time diagram[1] of the collision, Fig. 1, which shows the evolution of the matter along the collision (z) axis as a function of time in the center-of-mass frame. Prior to colliding, the projectile and target nuclei travel essentially along the light cone, until they meet at the origin; the remains of the nuclei, the nuclear fragmentation zones, emerge from the collision again along the forward light cone. The matter between the fragmentation regions is the central rapidity regime, shown in position space in Fig. 2.

The fields in the region between the fragmentation zones are left highly excited by the collision, and the first instants, $\lesssim 1$ fm/c in the local rest frame, are the time of formation of identifiable excitations, i.e., quarks, anti-quarks, and gluons. The detailed description of this particle-production regime lies outside the realm of available kinetic techniques; among the important open questions is that of understanding the distribution of excitations in phase space that emerges.[2] Once excitations have been created, the matter enters the kinetic regime, where one may begin to treat the system in terms of interacting excitations. During this phase, interactions tend to bring the system into local thermodynamic equilibrium, after which one may employ relativistic hydrodynamics to describe the evolution of the system. If, as is predicted, the matter comes into local thermal equilibrium as a quark-gluon plasma, it will undergo the hadronization (and chiral symmetry breaking) transition during the hydrodynamic phase. Then, depending on how strongly coupled the emerging hadronic matter is, the system will either further evolve in the hydrodynamic phase, and then "freezeout," i.e., turn from collision-dominated to free-streaming hadrons, which travel to the awaiting detectors, or else emerge directly from the hadronization transition streaming freely.

† Research supported in part by NSF Grant DMR81-17182.

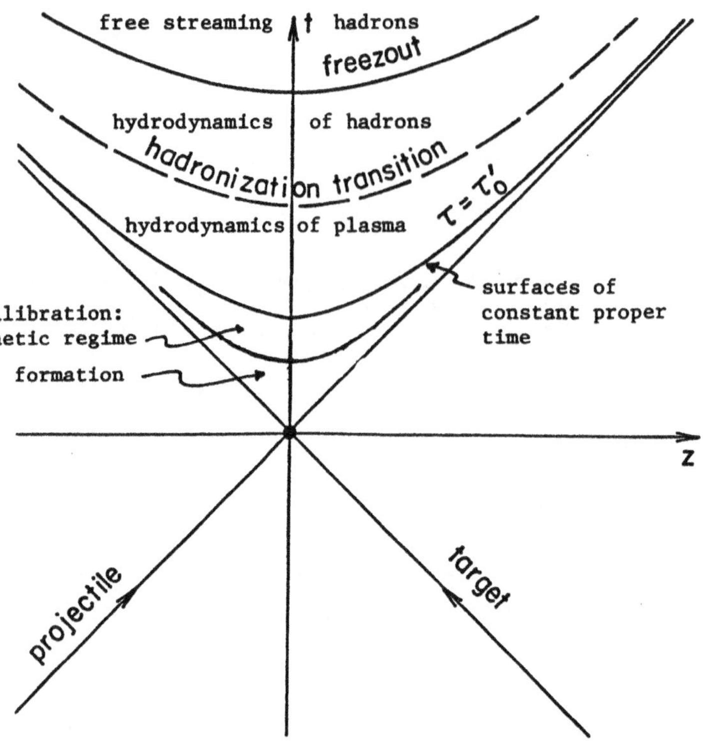

Fig. 1

The length of time that a given element of matter remains in the various phases described above, measured in the local matter rest frame, is approximately the same for all elements in the central rapidity regime. The local fluid velocity there is $\approx z/t$, and the local proper time is $\tau \approx (t^2-z^2)^{1/2}$, so that the surfaces of constant proper time delineating the various dynamic regions in the space-time diagram 1 are approximately hyperbolae as shown.

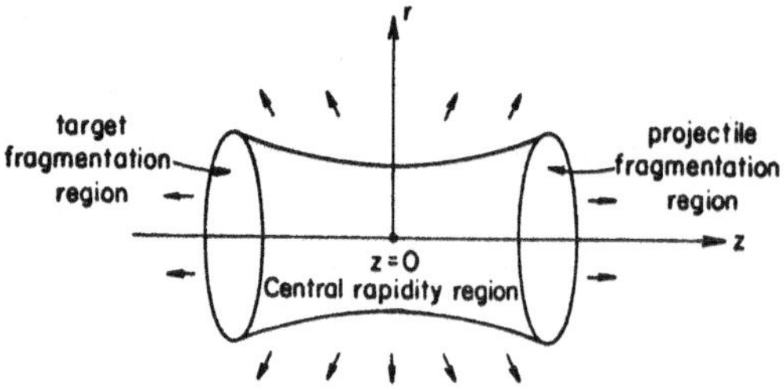

Fig. 2

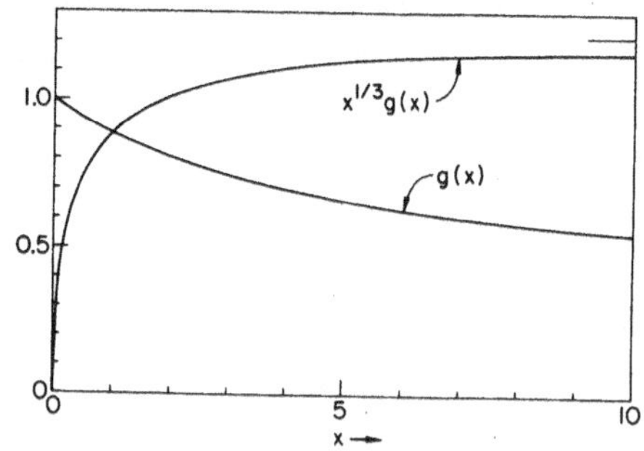

Fig. 3

2. KINETIC REGIME

The dynamics and approach to equilibrium in the kinetic regime can be formulated in terms of the evolution of the distribution function $f_i(p_\mu,x)$, the phase space density of excitations of internal quantum numbers i and 4-momentum p_μ at space-time position x. [The function f, a generalized Wigner distribution function, is definable microscopically in terms of correlation functions of the fields.] The dynamics of f are governed by a quantum kinetic equation[3]; in its simplest form, when the excitations are on shell so that p^o is a function of $|\underset{\sim}{p}|$, and the system varies slowly in space and time compared with characteristic microscopic lengths such as Compton wavelengths of the excitations, the kinetic equation reduces to the familiar semi-classical Boltzmann equation:

$$\frac{\partial f}{\partial t} + \underset{\sim}{\nabla}_p p^o \cdot \underset{\sim}{\nabla}_r f = \left(\frac{\partial f}{\partial t}\right)_{collision}, \tag{2.1}$$

where the right side is the collision operator. Since typical Compton wavelengths in a plasma at temperature $T \sim 200$ MeV are ~ 1 fm, the condition of slow space-time variation is not in fact well-satisfied, and one should more properly employ quantum kinetic equations.

A model solution of (2.1) — assuming spatial variation in the z direction only, Lorentz invariant conditions in the central rapidity region, so that

$$f(p_\perp,p_z,z,t) = f\left(p_\perp,p_z' = (p_z-p^o v)\gamma,\ \tau = (t^2-z^2)^{1/2}\right), \tag{2.2}$$

where $v = z/t$, and $\gamma = (1 - v^2)^{1/2}$, and a single collision time approximation to the (linearized) collision integral — is given in Ref. 4. The features that emerge from this calculation are summarized in Fig. 3, which shows the behavior of the energy density in the central region as a function of proper time τ; in this figure x is the time measured in units of the assumed relaxation time τ_{sc}, and

$$E(t) = \frac{\tau_o E_o}{\tau} g(\tau/\tau_{sc}), \tag{2.3}$$

where E_o is the energy density of the system at an initial time τ_o. Initially the variation of the distribution function is controlled by the longitudinal expansion of the system along the collision axis, and for $\tau \ll \tau_{sc}$, the energy density falls as $1/\tau$. However on longer time scales, collisions bring the system towards local equilibrium, and the energy density falls with the $\tau^{-4/3}$ hydrodynamic behavior,

$$E(t) = 1.22\ \tau_o E_o \tau_{sc}^{1/3}/\tau^{4/3}, \qquad \tau \gg \tau_{sc}; \tag{2.4}$$

as is seen by the curve $x^{1/3}g(x)$ approaching a constant value at large time. One should note that this approach to equilibrium is in fact a slow one, with the deviations falling as $1/\tau$. A surprising feature is the increase with τ_{sc} of $E(\tau)$ at the start of the hydrodynamic regime; the reason is that collisions build up the longitudinal pressure, so that the larger is τ_{sc} the less work any given slice of material (at given z) does on its neighboring slices. As pointed out by Gyulassy and Matsui,[5] the initial energy density in a collision is, as a consequence of this effect, higher in the central region than one would estimate by simple extrapolation backward from the final free streaming collision products. Similarly the entropy density, when thermal equilibrium sets in, is given by

$$s(\tau) = \frac{\tau_o}{\tau} s_{o,eq} (\tau_{sc}/\tau_o)^{1/4} , \qquad (2.5)$$

where $s_{o,eq} \sim E_o^{3/4}$ is the entropy density of an equilibrated system at energy density E_o. Again, the entropy density is larger for larger τ_{sc}; although a slower collision rate equilibrates the system later, the reduction of longitudinal cooling dominates, and the final entropy per co-moving volume falls weakly ($\sim \tau_{sc}^{1/4}$) with increasing collision rate.

Several outstanding problems remain in the description of the kinetic regime. The first, as mentioned, is improvement of the kinetic equations to take into account the fact that the system varies in space and time on scales $\lesssim T^{-1}$, where T is here a characteristic excitation energy. The second is the first principles description of collisions in a many-body system within QCD, taking into proper account the screening of magnetic (transverse) as well as electric (longitudinal) interactions between excitations. On dimensional grounds one expects a mean $\tau_{sc} \sim \alpha^{-2}E^{-1/4}$, where E is the energy density and α the (running) QCD fine structure constant; however detailed calculation must await a firmer understanding of both the screening problem and the infrared structure of QCD. A further question is the role of color degrees of freedom in the dynamics of the kinetic regime; a start on this problem, via generalization of the kinetic equation to include long-ranged color fields, has been carried out by Heinz.[6]

3. HYDRODYNAMICS

Once the system varies slowly in space over distances of order an excitation mean free path, λ, and slowly in time over intervals of order a mean scattering time, τ_{sc}, collisions keep the system in approximate local thermal equilibrium, and one may describe its motion by hydrodynamics. The hydrodynamic equations are simply the five conservation laws of energy, momentum and baryon number:

$$\partial_\mu T^{\mu\nu} = 0 , \qquad (3.1)$$

$$\partial_\mu j_b^\mu = 0 , \qquad (3.2)$$

where $T^{\mu\nu}$ is the stress tensor and j^{μ}_{b} the baryon 4-current, together with relations for $T^{\mu\nu}$ and j^{μ}_{b} in terms of the state variables, e.g., the temperature T, baryon density ρ_b, and fluid velocity v (the velocity of the frame in which the local momentum density vanishes). [It is often useful to regard particle-creation in the initial collision as a source of energy, momentum and baryon number, and represent it by effective sources on the light cone, on the right sides of (3.1) and (3.2).[7,8]] For a non-dissipative fluid,

$$T^{\mu\nu}_{ideal} = (E+P)u^{\mu}u^{\nu} + Pg^{\mu\nu} \tag{3.3}$$

and

$$j^{\mu}_{b,ideal} = \rho_b u^{\mu} , \tag{3.4}$$

where P is the pressure, $u^{\mu} = (1,\underset{\sim}{v})$ is the relativistic 4-velocity, and $g^{oo} = -1$. The dissipative processes of viscosity and thermal conduction lead to respective corrections to (3.3) and (3.4) of order λ/L and τ_{sc}/T_1,[9] where L is a macroscopic length scale and T_1 a macroscopic time scale. In the absence of dissipation, Eqs. (3.1)-(3.4) imply the entropy conservation law:

$$\partial_{\mu}(su^{\mu}) = 0 . \tag{3.5}$$

The hydrodynamic motion following the collision consists of two components: longitudinal expansion along the collision axis, accompanied by transverse expansion, depicted in Fig. 2. As Bjorken showed,[1] the longitudinal motion (neglecting the transverse motion) in the central rapidity region is readily derived if one assumes Lorentz invariance there, i.e., the same conditions at the same proper time $\tau = (t^2 - z^2)^{1/2}$ (an assumption that would also imply a central plateau in the final particle rapidity distributions); the initial conditions are that at an initial proper time t_o', the onset of local equilibrium, the velocity is $v_z = z/t$, while the energy density is a constant $E = E_o$. Then the velocity continues to be given by $v_z = z/t$, while in the absence of dissipation the entropy density falls inversely with the proper time [cf. (2.5)]:

$$s(\tau) = \frac{\tau_o'}{t} s_o . \tag{3.6}$$

The longitudinal flow leads to cooling of the system; for constant sound speed c_s (for which $s \propto T^{c_s^2}$), Eq. (3.6) implies that the temperature in the central region falls with proper time as $\tau^{-c_s^2}$; for a pure "radiation" gas, where $s \propto T^3$, $c_s = \sqrt{3}$.

To go beyond the assumption of Lorentz invariance in the longitudinal expansion, to include, for example, baryon density gradients from the fragmentation regions to the central region, requires numerical solution of the hydrodynamics. These have been given for the longitudinal case in, e.g., Refs. 8 and 10 (see also Ref. 11). Figure 4 shows, for an assumed 50 on 50 GeV per nucleon central collision of two uranium nuclei, a typical result in the ln τ - "rapidity" plane for surfaces of constant energy density (solid curves, labelled in units of GeV/fm^3).[10] The central region is to the left, and the fragmentation region to the right. By contrast, the contours of constant energy density for Bjorken's scaling solution are horizontal in this plot. We also see in this picture the buildup of energy density at early times from excitation formation, included in this calculation by means of a "multi-chain" model.

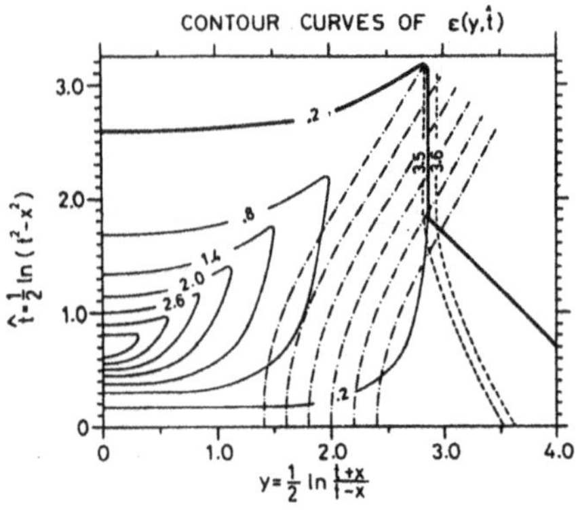

Fig. 4

One important effect of the hydrodynamic coupling between neighboring rapidity slices of the system is the transfer of energy in rapidity space during the evolution. This effect is clearly illustrated in Fig. 5, from Ref. 10, which shows (for one-dimensional motion in a central 50 on 50 GeV per nucleon uranium on uranium collision) (a) the flow of energy per unit rapidity to larger rapidity, compared with (b) the baryon distribution in rapidity which is little modified by the hydrodynamic flow. Such comparison of the energy and baryon rapidity distributions with varying nuclear mass number may be a useful observational indicator of hydrodynamic behavior. More generally, establishment of local equilibrium and passage through a hydrodynamic phase serves to wipe out details of

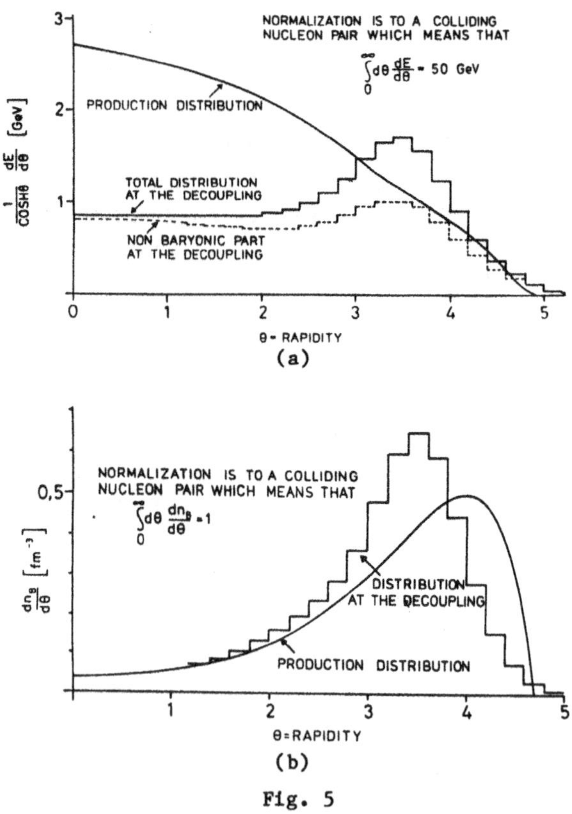

(b)

Fig. 5

the initial distribution of excitations that do not reflect conserved quantities.

Superimposed on the longitudinal expansion is the transverse expansion, which begins with the propagation of a rarefaction wave from the surface inward. To a very good approximation, the transverse motion is just that of a Riemann wave, scaled down in entropy by a factor τ^{-1}, owing to the longitudinal cooling. These features are illustrated in Figs. 6a, for the temperature as a function of transverse radius, and 6b, for the velocity, at various proper times (in fm) after the collision.[12] An important conclusion from these figures is that the temperature of a quark-gluon plasma will be essentially governed by the longitudinal expansion, and that the hadronization transition and final freezeout to free-streaming matter will take place relatively close to the initial collision volume.

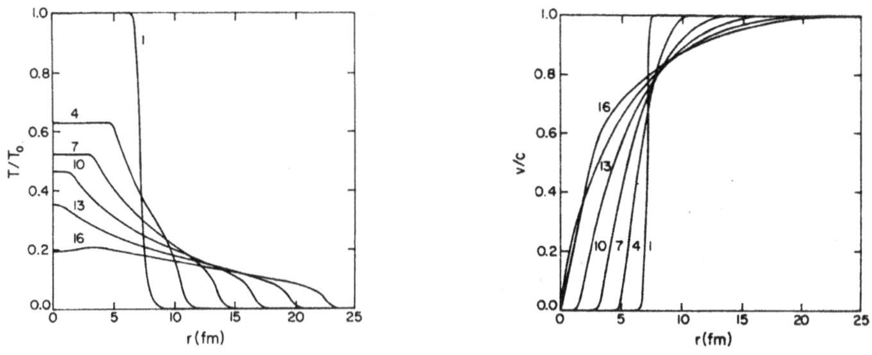

(a) (b)

Fig. 6

4. DISSIPATIVE HYDRODYNAMICS

The various calculations described above were all carried out in the absence of dissipation. Let us briefly consider now dissipative effects, and turn later to calculation of the transport coefficients. Three dissipative processes can occur in the hydrodynamic regime: first or shear viscosity, second or bulk viscosity, and thermal conductivity. For a nearly ideal relativistic gas the first viscosity coefficient, η, is given by

$$\eta = \frac{1}{5} Ts\tau_\eta \qquad (4.1)$$

where T is the temperature, s the entropy density, and τ_η is the collision time for viscosity. For a purely relativistic gas, or an ideal non-relativistic gas of conserved particles, the second viscosity coefficient, ζ, vanishes. The process of thermal conduction is present only in a system with a non-zero baryon number (or other non-zero conserved number, as for the pion gas discussed in sec. 5 below), and represents relative flow of heat with respect to the baryons. All three processes lead to entropy generation (first order in the transport coefficients and second order in the velocity and temperature gradients), and consequently conversion of kinetic into thermal energy.

The heating effects of dissipation are easily seen in the one-dimensional Bjorken scaling solution. For the baryon-free central rapidity region the rate of entropy generation assumes the form in the local rest frame

$$T\dot{s}_{\text{dissipative}} = (\zeta + \frac{4}{3}\eta)(\partial v_z/\partial z)^2 = (\zeta + \frac{4}{3}\eta)/\tau^2 , \tag{4.2}$$

so that the equation of motion of the entropy becomes

$$\frac{\partial s}{\partial t} + \frac{s}{\tau} = \frac{\zeta + \frac{4}{3}\eta}{T\tau^2} . \tag{4.3}$$

The s/τ term arises from the longitudinal expansion and in the absence of dissipation, (3.6) is the solution of (4.3). For constant sound speed and nearly-ideal gas viscosities, (4.1) and $\zeta = 0$, Eq. (4.3) is readily solved and one finds that the temperature with viscosity present compared with what it would be without viscosity, in the large time limit, becomes

$$\frac{T(\tau\to\infty,\text{visc})}{T(\tau\to\infty,\text{no visc})} = 1 + \frac{4}{15}\,\frac{c_s^2}{1-c_s^2}\,\frac{\tau_{\eta,o}}{\tau_o'} \approx 1 + \frac{2}{15}\,\frac{\tau_{\eta,o}}{\tau_o'} , \tag{4.4}$$

where τ_o' is the initial time from which the flow is hydrodynamic, and $\tau_{h,o}$ is the scattering time at τ_o'. For $\tau_{\eta,o}$ and τ_o', of comparable order, the viscous heating is an ~ 10% effect. More detailed solutions of the energy and entropy for Bjorken's scaling solution in the presence of dissipation have been given recently by Danielowicz and Gyulassy.[13]

The importance of dissipation is measured by the effective Reynolds number of the hydrodynamic flow, which is generally given by

$$\text{Re} = \frac{\text{length scale of flow}}{\text{mean free path}} \cdot \frac{\text{flow velocity}}{\text{microscopic velocity}} = \frac{Lv}{\eta\rho} , \tag{4.5}$$

where L is the macroscopic length scale, and η/ρ is the first viscosity per unit mass. When Re >> 1 the effects of viscosity are negligible. For relativistic matter $v \sim c$, and the Reynolds number can be taken to be

$$\text{Re} = \frac{\tau}{s}\left(\frac{\partial s}{\partial \tau}\right)_{\text{diss}} + \frac{15}{4}\frac{\tau_\eta}{\tau} . \tag{4.6}$$

The validity of hydrodynamics itself requires that the mean free path of excitations be small compared with the macrocopic length scales, or relativistically, that Re >> 1; in other words, as dissipative effects become large in high energy nucleus-

nucleus collisions, hydrodynamics begins to become inapplicable.

Simple scaling arguments indicate that the Reynolds number increases with time in the quark-gluon plasma phase, and that in large nuclei one in fact reaches the hydrodynamic regime. Roughly τ_η^{-1} scales in the plasma as $T\alpha^2$, and T as $\tau^{-1/3}$ times a weak power (~ 1/12) of the nuclear mass number, reflecting increased energy deposition (~ $A^{1/3}$) in collisions of larger nuclei. Thus, in the absence of any intrinsic length scales in the dynamics, Re grows as $\tau^{2/3}$. Larger A systems have greater time, ~ $A^{1/3}$, to reach large enough Re to enter a hydrodynamic regime. After the hadronization transition, Re begins to fall, and the system eventually decouples. During its rise and fall, the Reynolds number generally remains well below the critical values for which hydrodynamic flows become turbulent; it may however reach the critical range for onset of turbulence in hydrodynamic jet propagation through a fluid. One might imagine then the possibility that a hadronic jet generated in the initial collision travels through the plasma generating a turbulent wake; such behavior would lead to considerable fluctuations in the observed rapidity distributions in the neighborhood of the emergent jet.

A final remark about the hydrodynamic approach. At the level of smooth initial conditions, as in all the calculations discussed above, hydrodynamics describes only the average event, washing out the fine structure of individual events. An important problem remaining is to understand fluctuations in the initial conditions and to incorporate them into the hydrodynamic description.

5. CALCULATION OF TRANSPORT COEFFICIENTS

There has been much progress recently in calculation of the viscosities and thermal conductivities of the quark-gluon plasma and hadronic phases.[14,15] The basic approach has been to solve the linearized Boltzmann transport equation, writing the excitation distribution in the form $f(\underset{\sim}{p},\underset{\sim}{r},t) = f_{eq}(p) + \delta f(\underset{\sim}{p},\underset{\sim}{r},t)$, where f_{eq} is the (local) equilibrium distribution, and computing the stress tensor $T^{\mu\nu}$ and baryon current j_b^μ in terms of f; the deviation δf yields the dissipative terms in these quantities. Within the single collision time (τ_{sc}) approximation Gavin finds that each light mass quark contributes terms to the first and second viscosities of the plasma:

$$\eta_q = \frac{4}{15}\,\tau_{sc}E_{m=0}(T)\left[1 - \frac{25}{8}\left(\frac{m}{\pi T}\right)^2 + \dots\right] \tag{5.1}$$

and

$$\zeta_q = \frac{5}{21}\,\tau_{sc}E_{m=0}(T)\left(\frac{m}{\pi T}\right)^3 + \dots , \tag{5.2}$$

where m is the mass (assumed << T) and $E_{m=0}(T)$ the energy density, evaluated at m = 0, of the quark species. As we see, finite quark mass leads to a non-vanishing, but small ζ. [Complete results, for general quark mass and net baryon number are given in Ref. 15.] The gluon contributions are [cf. (4.1)]:

$$\eta_G = \frac{4}{15}\tau_{sc}E_{gluon}(T) , \qquad \zeta = 0 . \tag{5.3}$$

The matter in the central rapidity regime after hadronization is, to a good approximation, a pure pionic gas. [Since energy deposition in the nuclear fragmentation regions is expected to be at least several GeV per nucleon, this is not an unreasonable approximation to the matter there as well.] As long as the baryon number is negligible, pions, once well-defined, are effectively conserved in the evolution; one the one hand, maintenance of chemical equilibrium (i.e., zero pion chemical potential) requires the number of pions to decrease as the temperature decreases below the hadronization temperature T_c, while on the other, the lowest order process that can decrease the number of pions is $4\pi \rightarrow 2\pi$, which as a 4-initial body process has rapidly decreasing probability as the pion fluid expands. Conservation of pion number implies that, in addition to viscous diffusion, thermal conduction, the relative flow of energy with respect to the pion number, is a possible transport process in the pion system. Variational solutions of the Boltzmann equation for the transport coefficients, using realistic π-π scattering amplitudes have been given by Gavin.[15] Fig. 7 schematically indicates the form of the first viscosity from the low temperature pionic phase, through the high temperature deconfined phase[14]; the increase of η/T^3 in this latter phase reflects the logarithmic decrease of the running QCD coupling constant with increasing T.

One does not expect these calculations of the transport coefficients to be valid in the neighborhood of the deconfinement transition (indicated by the hatched region in Fig. 7), calculations via semi-classical Boltzmann-like kinetic equation being limited to regimes in which the excitations are well-defined. [As the Landau Fermi-liquid theory shows, however, applicability does not necessarily require the system to be weakly interacting.] Further progress can be made by direct calculation of the correlation functions of the system, which can be approached by means of Monte Carlo lattice gauge theory. For instance, the electromagnetic and lepton pair emission of a plasma is directly related to the Fourier transform of the auto-correlation function of the electromagnetic with itself, $\langle j^\mu_{em}(x) j^\nu_{em}(x')\rangle$.[16] Similarly, the dynamic modes and

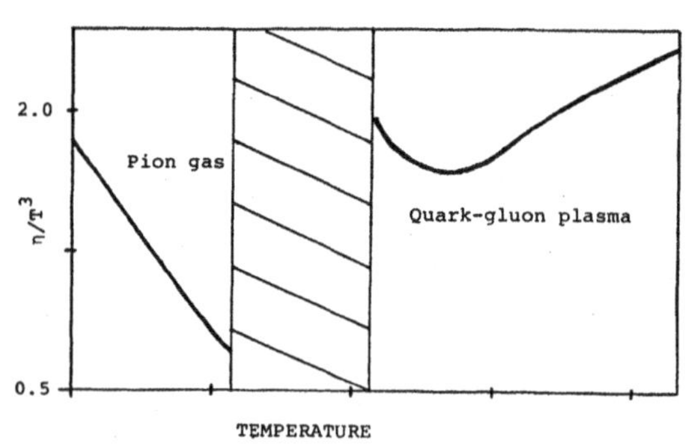

Fig. 7

transport coefficients of a QCD plasma can generally be calculated in terms of the stress-tensor autocorrelation function, $\langle T^{\mu\nu}(x)T^{\lambda\sigma}(x')\rangle$.

For example, the space Fourier transform of the imaginary time ($t = -iu/T$) correlation function, $\langle T_{xz}T_{xz}\rangle$, which is directly calculable by Monte Carlo methods, is given by

$$\int e^{-ikz}d^3r\langle T_{xz}(\underset{\sim}{r},-iu/T)T_{xz}(0,0)\rangle = \int_{-\infty}^{\infty}\frac{d\omega}{2\pi}\frac{\omega e^{-\beta\omega u}}{1-e^{-\beta\omega}}\,b(k,\omega) \ , \tag{5.4}$$

and in the hydrodynamic limit, with k along the z axis,

$$b(k,\omega) \to \frac{2\omega^2\eta}{\omega^2 + (k^2\eta/w)^2} \ , \tag{5.5}$$

where $w = E + P$ is the enthalpy density. Then by use of (5.5) one may try to extract η from calculations of (5.4). While lattice gauge theory calculations that can yield useful results for space and time-dependent correlation functions are still in their infancy,[17] calculation by this means is a potentially important technique, since it allows inclusion of QCD interactions of arbitrary strength, over the entire range from weakly interacting, through the hadronization phase transition, thus going well beyond approaches based on perturbation theory.

6. DYNAMICAL HADRONIZATION TRANSITION

A considerable effort has recently gone into understanding how the first order hadronization transition takes place under the dynamical conditions in a heavy-ion collision. As was shown in Ref. 18, matter in transverse expansion away from the collision axis tends to develop a shock-like discontinuity across the hadronization transition. Friman,[19] and Glendenning and Matsui[20] have studied the transverse motion numerically in terms of expansion of a quark-gluon plasma against an external "bag" pressure and find that as the interior is cooled by longitudinal expansion, the inwardly propagating rarefaction wave becomes accompanied by a compression near the surface, and that the net effect can be a compression of the collision volume in the transverse direction.

The one-dimensional scaling solution for the longitudinal expansion is stable against small perturbations and exhibits no instability against formation of infinitesimal discontinuities across the transition[12]; for smooth initial conditions, the transition would occur in longitudinal expansion by formation of a "two-phase region" consisting, in macroscopic terms, of intermingled regions of hadronic and quark matter. However, as pointed out by Gyulassy et al.,[21] and Van Hove,[2] the sudden release of latent heat in the phase transition can lead to explosive growth in the quark phase of small regions of hadronic phase, i.e., hadronic bubbles, nucleated at various inhomogeneous points in the plasma. Two types of stable "condensation discontinuities" across the bubble wall are possible (see Ref. 9, sec. 124): one analogous to a deflagration burning front in which the incoming matter velocity with respect to the discontinuity is subsonic, as is the matter leaving the discontinuity, and a second, analogous to a detonation, in which both velocities are supersonic. Such explosive bubble growth in the hadronization phase transition leads to certain observable features, including large rapidity density fluctuations, correlations over several units of rapidity, and large nearly-azimuthally distributed transverse energy.

7. FREEZEOUT

As matter expands in the hadronic phase, collisions become less frequent; the matter can no longer maintain itself in thermal equilibrium and it freezes out to a free-streaming gas. The freezeout process is important in that it determines the details of the final distribution of particles that are detected. Freezeout not only introduces non-equilibrium structure in the distribution, analogous to that predicted in the cosmological 3°K microwave background,[22] but it also leads to significant entropy production.

As mean free paths, λ, grow comparable to the system scale size, R, the rate of entropy production, $\sim\lambda$, increases, reaching a maximum when $\lambda \sim R$, and then falling as λ^{-1} for $\lambda >> R$. The net entropy production in this phase is likely weakly dependent on details of the interactions. We may see this by making a simple estimate for a spherically expanding gas of non-relativistic nucleons, for which the total growth of entropy is given in the hydrodynamic phase by

$$\dot{S} \approx \int d^3r \, \frac{\eta}{T} \left(\frac{\partial v}{\partial r}\right)^2 , \tag{7.1}$$

where v is the radial velocity, and in a non-relativistic gas, the viscosity η is $\approx nT\tau$, with n the nucleon density and τ the scattering time for viscosity. If we estimate the velocity gradient as $\approx 1/t$, then $\dot{S} \approx N\tau/t^2$, where N is the total number of nucleons in the gas. Now the scattering time is $\approx 1/n\sigma$, where σ is the nucleon-nucleon cross-section. Since n falls essentially as t^{-3}, we have $\tau \sim bt^3$, where b is a constant. The entropy generation begins to fall off when τ becomes comparable to t, i.e., when $bt^2 \sim 1$; thus we may approximate the total entropy produced in freezeout by

$$\Delta S/N \sim \int_0^{bt^2=1} dt \; bt = \frac{1}{2} \; ; \tag{7.2}$$

the total entropy increase is $\sim 1/2$ unit per nucleon, independent of the size of the cross-section, as long as σ is large enough for the system to be initially collision dominated and thus hydrodynamic.

A detailed calculation[23] of the entropy production of a spherically expanding 80 nucleon gas via the Bertsch-Cugnon Monte Carlo cascade code,[24] exhibits this predicted behavior, as shown in Fig. 8, the calculated final entropy per nucleon as a function of the effective "hard sphere" radius, $a = \sqrt{\sigma/\pi}$. The initial radius of the gas is assumed to be 4.1 fm, for which initially $\lambda/R \sim 1/4a^2_{fm}$, and the initial entropy per nucleon is 3.0 units. The smooth curve is a simple fit to the computed points. For large enough a (corresponding to initial $\lambda/R \gtrsim 1/4$), the system is collision dominated in the first stages of the expansion, and the entropy produced is a constant (coincidentally the same value 1/2 as in the rough estimate above) independent of a.

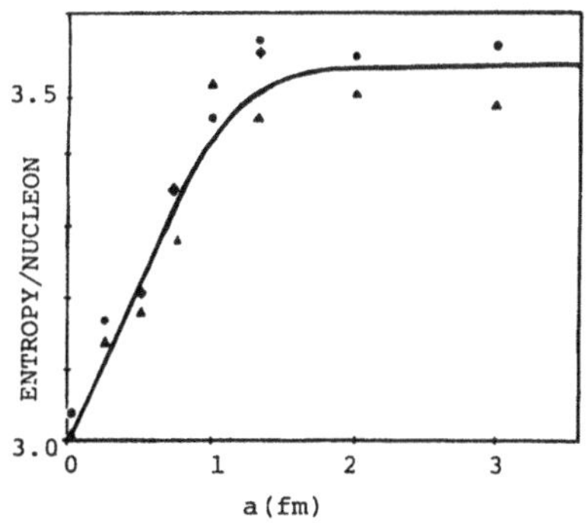

Fig. 8

In a hot relativistic gas the viscosity is proportional to the entropy itself [cf. (4.1)] so that rather than an additive increase in the entropy per particle, as in the non-relativistic example above, one would expect a multiplicative increase in the entropy in freezeout.

In summary, we see that the final state particles emerging from a very high energy nucleus-nucleus collision are formed from matter that has undergone considerable processing from the initial stages of particle production through to the final freezeout of the matter. While one is not yet at the point of predicting accurately final particle distributions, the stages of a central collision are coming into overall focus. The hydrodynamic approach is under good control; however, full three-dimensional modelling of collisions, including dissipative processes and the hadronization phase transition in the equation of state, is an important further step. The area most in need of fundamental studies remains the kinetic regime prior to the onset of the hydrodynamic phase, especially in states far from local equilibrium, including the initial moments of the collision.

REFERENCES

1. J. D. Bjorken, Phys. Rev. D27 (1983) 140.
2. L. Van Hove, Z. Phys. C21 (1983) 93.
3. L. P. Kadanoff and G. Baym, Quantum Statistical Mechanics (W. A. Benjamin, Reading, Mass. 1962).
4. G. Baym, Phys. Letters 138B (1984) 18; Nucl. Phys. A418 (1984) 525c.
5. M. Gyulassy and T. Matsui, Phys. Rev. D29 (1984) 419.
6. U. Heinz, Phys. Rev. Letters 51 (1983) 351; Nucl. Phys. A418 (1984) 603c; Vanderbilt Univ. preprint, 1984.

7. K. Kajantie and L. McLerran, Nucl. Phys. B214 (1983) 261.
8. K. Kajantie, R. Raitio and P. V. Ruuskanen, Nucl. Phys. B222 (1983) 152.
9. L. Landau and E. Lifshitz, Fluid Mechanics (Addison-Wesley, Reading, Mass. 1959), sec. 127.
10. P. V. Ruuskanen, LBL/Jyväskylä preprint, 1984.
11. H. Sumiyoshi, S. Date, N. Suzuki, O. Miyamura and T. Ochiai, Z. Phys. C23, (1984) 391.
12. G. Baym, B. L. Friman, J.-P. Blaizot, M. Soyeur and W. Czyż, Nucl. Phys. A407 (1983) 541.
13. P. Danielowicz and M. Gyulassy, preprint LBL-17278.
14. A. Hosoya and K. Kajantie, Nucl. Phys. B (in press).
15. S. Gavin, Nucl. Phys. A (in press).
16. L. McLerran, Proc. XII Int. Workshop on Gross Properties of Nuclei and Nuclear Reactions (H. Feldmeier, ed., Darmstadt, 1984), 95; L. McLerran and T. Toimela, Fermilab preprint 84/84-T.
17. G. Baym, K. Kajantie, J. Polonyi and H. W. Wyld, to be published.
18. B. L. Friman, G. Baym, and J.-P. Blaizot, Phys. Letters 132B (1983) 291.
19. B. Friman, to be published.
20. N. Glendenning and T. Matsui, preprint LBL-17351.
21. M. Gyulassy, K. Kajantie, H. Kurki-Suonio, and L. McLerran, Nucl. Phys. B237 (1984) 477.
22. R. A. Sunyaev and Ya. B. Zel'dovich, Ann. Rev. Astron. and Astrophys. 18 (1980) 537.
23. G. Baym, G. Bertsch, B. L. Friman and S. Gavin, to be published.
24. G. Bertsch and J. Cugnon, Phys. Rev. C24 (1981) 2514.

HOW WE GOT TO QCD MATTER FROM THE HADRON SIDE BY TRIAL AND ERROR

R. Hagedorn
CERN - Geneva

1. INTRODUCTION

I have 50 minutes and 25 pages of print available to describe the origin and development of a work which took 27 years and was done by perhaps more than 200 people in perhaps more than 300 papers. This leaves me about two minutes of talk per year of work and about three lines of print per paper.

You - and hopefully even those whose work is not reported here - will understand my dilemma: I have to concentrate on the essentials and skip all technical details.

I hope to be able to let you participate in retrospection in the up's and down's of this adventure, which started from a simple statistical model of hadron production without any strive after understanding of the underlying dynamics and without any idea of what finally it would lead to: a phase transition to QCD matter.

This whole history can be squeezed in one sentence coined by H. Satz: "Hadron thermodynamics defines its own limits" [SAT84]; but how and why - to realize and formalize that took a long time, which I shall now describe.

2. PRE-BOOTSTRAP

In 1957, F. Cerulus and I were asked by the then TH Division leader at CERN, B. Ferretti, to do a few calculations ("just a fortnight") of particle production at the future CERN 30 GeV proton synchrotron. We used the Fermi statistical model [FER50] and realized at once that the phase space integrals could not be calculated with the known approximations ($m \to 0$ and $m \to \infty$) since, on the average, the particles (mostly pions) would be neither relativistic nor non-relativistic. We heard of the Monte Carlo method and adopted it to our problem [CER58]. We thus were able to compute reliably hundreds of phase pace integrals with all the then known hadrons and resonances. We even included two-body phase spaces and archived all our results for some reason or another. This proved fortunate when, five years later, L.W. Jones proposed to compute large angle elastic pp scattering and asked us whether we had any possibility to do so. The idea was that there might be a statistical intermediate state decaying into two protons isotropically. The cross-section for this process (visible, if existing, only far outside the diffraction region) should be proportional to the ratio

$$\left.\frac{d\sigma}{dE}\right|_{90^\circ} \sim \frac{\text{2-body phase space}}{\sum_n \text{n-body phase spaces}} \tag{2.1}$$

We dug out our old numerical results and found [FAS63] an energy dependence $\sim\exp(-3.3\ E_{cm}\ [\text{GeV}])$, in agreement with experiments. This numerical result could, in simple analytical models (equal particles) only be reproduced if one suppressed there the factor 1/N! in front of the phase space integrals; otherwise one would obtain $\exp(-\text{const}\ E^{\alpha})$ with $\alpha < 1$ [AUB65, SAT65, VAN65]; but omitting 1/N! amounted to considering the particles to be distinguishable! - a most important observation which triggered all the rest, but which was greeted with pityful smiles by those to whom I talked about it, although I tried to argue that indeed the particles should be effectively distinguishable, since the factor was 1/N! only for like particles; but when there were several different kinds, then

$$\frac{1}{N!}\int\cdots \Longrightarrow \frac{1}{\prod_i N_i!}\int\cdots \tag{2.2}$$

where i goes over all types of particles. We had in our earlier numerical computations used some 40 or 50 different types of particles [including some resonances and counting each state: spin, isospin, antiparticle... as a new kind] and it so happened that on the average, all $\langle N_i\rangle < 1$ so that the main contributions to Eq. (2.1) had come from integrals whose factor $1/\prod_i N_i! = 1$. We then worked out a model of massless distinguishable particles [AUB65] which was so simple that I present it here:

let N masssless, distinguishable particles be enclosed in a box of volume V and let $\varepsilon_1, \varepsilon_2, \ldots, \varepsilon_i, \ldots$ denote the single particle energy levels with occupation numbers $n_1, n_2, \ldots, n_i, \ldots$; then there will be

$$\frac{N!}{n_1!\,n_2!\ldots} \text{ states of total energy } E = \sum n_i\varepsilon_i \;;\; N = \sum n_i \tag{2.3}$$

If particles can be freely created, the partition function will be

$$Z = \sum_{N=0}^{\infty} \sum_{\sum n_i = N} \frac{N!}{\prod_i n_i!} \exp\left(-\frac{1}{T}\sum n_i\varepsilon_i\right) = \sum_{N=0}^{\infty}\left[\sum_i \exp\left(-\frac{\varepsilon_i}{T}\right)\right]^N \tag{2.4}$$

Replacing as usual

$$\sum_i \exp\left(-\frac{\varepsilon_i}{T}\right) \Longrightarrow \frac{V}{2\pi^2}\int p^2 \exp\left(-\frac{p}{T}\right)dp = \frac{VT^3}{\pi^2} \tag{2.5}$$

we find

$$Z = \sum_{N=0}^{\infty}\left(\frac{VT^3}{\pi^2}\right)^N = \frac{1}{1-\frac{VT^3}{\pi^2}} \tag{2.6}$$

which has a pole at $T_c = (\pi^2/V)^{1/3}$, where also the energy density would diverge, implying that T_c is the maximum temperature for this system. Now comes the miracle: putting $V = V_{hadron} = (4\pi/3)(1/m_\pi)^3$ gives $T_c \approx 185$ MeV, very near the "temperature" familiar since many years from cosmic ray physics ($p_\perp$ distribution).

3. EARLY BOOTSTRAP

Quite apart from hadrons not being massless, the above model is pure cheating if you only look at it a little closer. But it had enough cheating power to convince me that something in it must be true. The next move had thus to be to build a better model with real, massive, hadrons while keeping the idea of many different types of them being involved.

For the sake of clarity, I present now a simplified version (Boltzmann statistics, non-relativistic approximation) of the original paper [HAG65].

Consider a mixture of ideal gases, each gas belonging to one species of particle with mass m_i; all particles can freely be created and absorbed. The total partition function is the product of the partition functions of the individual gases (i).

We have

$$\ln Z_i(T,V) = \frac{V}{(2\pi)^3}\int e^{-\sqrt{p^2+m_i^2}/T}\, d^3p = \frac{VT}{2\pi^2}\, m_i^2 K_2\left(\frac{m_i}{T}\right) \approx \approx V\left(\frac{T}{2\pi}\right)^{3/2} m_i^{3/2}\, e^{-m_i/T} \tag{3.1}$$

Thus, in this approximation,

$$\ln Z = \sum_i \ln Z_i = V\left(\frac{T}{2\pi}\right)^{3/2} \sum_i m_i^{3/2}\, e^{-m_i/T} \tag{3.2}$$

where the sum (i) goes over all hadron species which can in principle participate, including resonances and counting each spin and isospin state separately. Assuming most possible particles to be yet unknown, it is convenient to introduce the mass spectrum $\rho(m)$

$$\rho(m)\,dm = \text{Number of different hadron states in } \{m, dm\} \tag{3.3}$$

which, of course, is also unknown (except for a few δ-functions at low lying stable masses and Breit-Wigner functions at the established resonances). We then replace the sum over i by an integral over ρ(m) and obtain

$$Z(T,V) = \exp\left[V\left(\frac{T}{2\pi}\right)^{3/2} \int_0^\infty \rho(m)\, m^{3/2}\, e^{-m/T}\, dm\right] \tag{3.4}$$

For the exact relativistic expression with Fermi and Bose statistics, see [HAG65].

What now? We have a formula which tells us exactly nothing as long as ρ(m) - the complete ρ(m)! - is not given. Here comes the key idea: statistical bootstrap.

3.1 The bootstrap idea

- We are after the description of "fireballs", i.e., highly excited and decaying hadronic systems of hadronic size V_0, composed of hadrons. Our earliest results (phase space, large angle elastic scattering) had suggested that the set of constituent hadrons should also include all resonances, which thererefore must be counted in ρ(m);
- The hadronic system of size V_0 (fireball) is, however, itself a highly excited hadron and is therefore not essentially different from a resonance: there should be just one set of hadrons

  {π, K, resonances, N, Λ, Σ, more resonances ... fireballs ad infinitum}

  and all should appear in ρ(m);
- The partition function (3.4) can be written, using the density of states of the system it describes (a fireball if $V = V_0$ is the hadron volume!)

$$Z(T,V_0) = \int_0^\infty \sigma(V_0, m)\, e^{-m/T}\, dm \tag{3.5}$$

  where $\sigma(V_0, m)$ is the density of states of the fireball at mass m;
- Since ρ(m) should include fireballs and since ρ and σ are both densities of states of essentially the same objects, both functions should be essentially the same at least for large m; a weak requirement ("equal entropy") of this sort was

$$\frac{\log \rho(m)}{\log \sigma(V_0, m)} \xrightarrow[m \to \infty]{} 1 \tag{3.6}$$

On the other hand, one has the identity of Eqs. (3.4) and (3.5):

$$\int_0^\infty \sigma(V_0, m)\, e^{-m/T}\, dm \;=\; \exp\left[V_0\left(\frac{T}{2\pi}\right)^{3/2} \int_0^\infty \rho(m)\, m^{3/2}\, e^{-m/T}\, dm\right] \qquad (3.7)$$

The two equations (3.6) and (3.7) together form the bootstrap condition; in words:

Fireballs

→→ consist of fireballs which→→

↑ ↓

← ← ← ← ← ← ← ← ← ← ← ← ← ← ← ←

3.2 Consequences

From the bootstrap condition, it can be proved that the mass spectrum must grow exponentially (for large m):

$$\rho(m) \xrightarrow[m\to\infty]{} \frac{\text{const}}{m^a}\, e^{+m/T_0} \qquad (3.8)$$

Together with its low mass (empirical) part, this then constituted the so far unknown mass spectrum $\rho(m)$.

Inserting this into Eq. (3.4) shows that the integral does not converge if $T > T_0$: the partition function has, at T_0, a singularity whose nature depends on the value of a. Thus there were three predictions:

- for sufficiently high collision energy, the $p_\perp$ distribution should be (very approximately) $\sim\exp(-p_\perp/T)$ with $T \lesssim T_0$ [often called "$\exp(-6p_\perp)$ law");
- the mass spectrum should grow exponentially;
- T_0 in the $p_\perp$ distribution and in the mass spectrum should be the same.

I presented these predictions and the whole model in a theory seminar in the fall of 1964 – the result was disastrous. Nobody would believe it and I was shouted at: "but the mass spectrum DOES NOT grow exponentially".

True enough I had not even yet checked it. The next morning, I did so with beating heart and found what is depicted in Fig. 3.1 as the October 1964 situation. The $p_\perp$ distribution for 90° elastic pp scattering [ORE64] looked like in Fig. 3.2 where the slope required that T_0 = 158 MeV while for the mass spectrum it was T_0 = 160±10 MeV. After the depressing seminar and an equally depressing night, this was such a relief that I never forgot this experience and henceforth was convinced that, all imperfections notwithstanding, the model was essentially right and that I should continue to work on it no matter what people would say (and they said a lot, most based on the misunderstanding: that in the model a maximal temperature and/or an exponential spectrum were postulated, while these were the results I was so proud

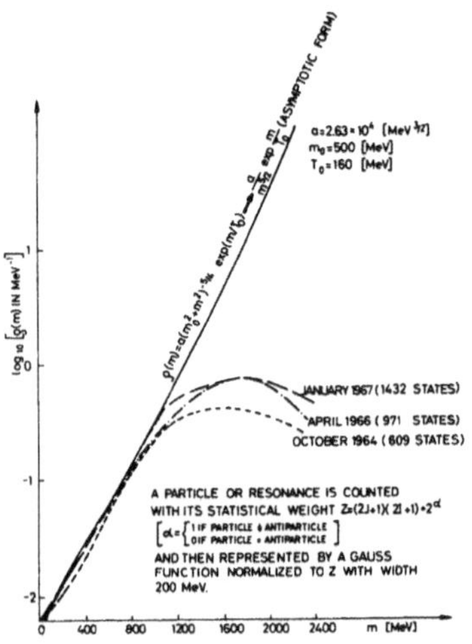

Fig. 3.1: The predicted and the experimental mass spectrum as it evolved from 1964 to 1967.

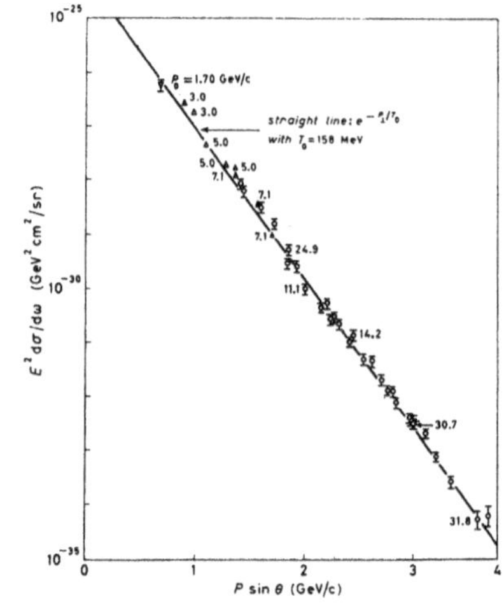

Fig. 3.2: The $p_\perp$ distribution in large angle elastic pp scattering [ORE64] for p_0 from 1.7 GeV/c to 31.8 GeV/c.

of). The general rejection had one immense advantage: those few who worked with me (and myself) could proceed without fear of competition and without hurry.

After this anecdotical digression, consider now the energy density of such a system:

$$\varepsilon = \frac{T^2}{V}\frac{\partial \ln Z}{\partial T} = \left(\frac{T}{2\pi}\right)^{3/2}\left[\frac{3T}{2}\int_0^\infty \rho(m)\, m^{3/2} e^{-m/T}\, dm + \int_0^\infty \rho(m)\, m^{5/2} e^{-m/T}\, dm\right] \tag{3.9}$$

With $\rho(m) \to \text{const.}\; m^{-a} \exp(m/T_0)$ and at $T = T_0$ we find

$$\varepsilon(T_0) = \text{const}\left[\frac{3T_0}{2}\left(m^{5/2-a}\right)_M^\infty + \left(m^{7/2-a}\right)_M^\infty + \text{finite parts}\right] \tag{3.10}$$

where M is assumed so large that the asymptotic expression (3.8) is valid for $m > M$.

The point I wish to make here is that from Eq. (3.10) it follows that

the energy density ε is finite at T_0 if $a > 7/2$ and infinite otherwise,

which could have been interpreted as indicating that since for $a > 7/2$ one could reach the temperature T_0 at finite energy density, one could also pass over it into a region which would need another description; in other words:

for $a > 7/2$ there would be a phase transition,

because all the usual features of a phase transition, in particular a critical temperature T_0, were present.

Unfortunately - and at that time for good reasons! - I discarded this choice of a and preferred a = 5/2, giving infinite energy density at T_0, which therefore should be the highest temperature, approachable only asymptotically. And thus the discovery of the phase transition to the quark-gluon plasma, alas!, escaped me (it was a bit early for that, though).

3.3 Difficulties

The model was still unsatisfactory in two respects.

Inconsistency

While the fireball described by the partition function $Z(V_0,T)$ had a volume V_0, its constituents were treated as pointlike, although they were hadrons and fireballs, too; this inconsistency was not removed until 1978/80; then, however, with striking results (see below).

Difficulty

No reason was given for treating resonances and fireballs as if they were particles; this obstacle was removed a few years later by using the arguments of Beth and Uhlenbeck [BET37] and Belenkij [BEL56]. In fact these arguments put the somewhat shaky bootstrap idea (Section 3.1) on rather firm ground and allowed to state that

a strongly interacting hadron gas can be described alternatively as a mixture of ∞ many ideal gases listed up in the complete mass spectrum

These arguments were given in extenso in three places and cannot be repeated here [HAG70,71,73].

3.4 Early developments

The model could not immediately be applied to hadron collisions, but it was important to do so in order to convince the non-believers (all theoreticians except for a set of measure zero) that it did work. Thus, J. Ranft and myself worked out the so-called "thermodynamical model" in which (see Fig. 3.3) collective velocity distributions were combined with bootstrap thermodynamics in local rest frames [HAG68]. This mammoth paper of 141 pages was too long to be read by anybody - with the consequence that even today people rediscover (independently) results and techniques buried in it. For us, it was a kind of handbook used in further work on the model, in which in particular J. and G. Ranft and their collaborators at Leipzig developed and confronted it with experiment (see later). We produced an "Atlas of Particle Spectra (Fig. 3.4) [GRO70], in which we reproduced measured momentum

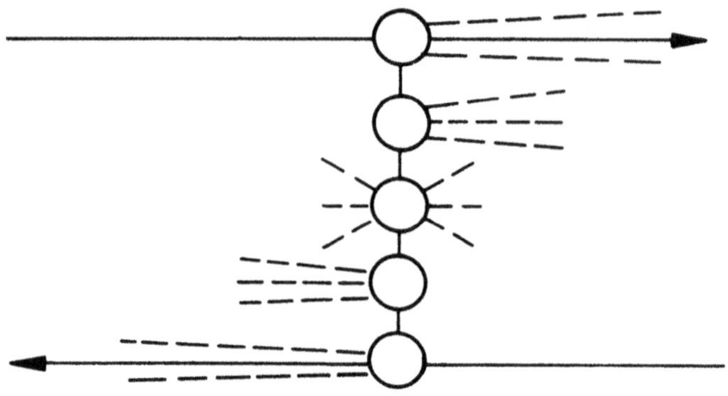

Fig. 3.3: Bootstrap thermodynamics is applied to each bubble in its rest frame. Lorentz transformation to the laboratory (or c.m.) frame and integration over all possible impact parameters and velocities gives realistic predictions of particle momentum spectra.

spectra and predicted unmeasured ones up to p_0 = 800 GeV/c [for p-nucleus (Be, Aℓ, Cu, Pb) up to 70 GeV/c] and discovered that the model obeyed "Feynman scaling" and "limiting fragmentation" [HAG70].

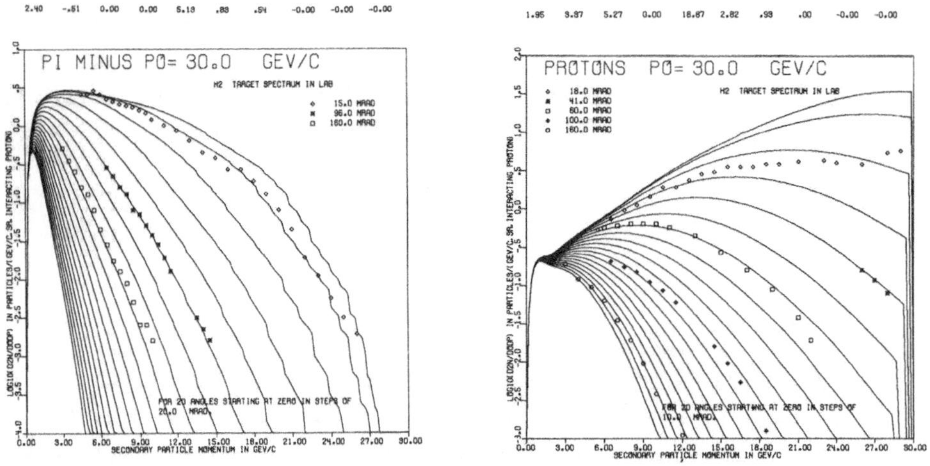

Fig. 3.4: Predicted and measured momentum spectra of π^- and p. Figures taken from the "Atlas of Particle Spectra" [GRO70].

While all these years much talk was about quarks, experimental searches did not reveal any; confinement and QCD were still far in the future. It was speculated that maybe quarks are so heavy that they would be practically unobservable. Bootstrap thermodynamics was presumably the best way to calculate $q\bar{q}$ production rates as a function of their mass (applying perturbation theory to this problem was a then fashionable nonsense). The model was used to predict [HAG68a] $K\bar{K}$, $p\bar{p}$, $d\bar{d}$ and $q\bar{q}$ pair production with the result that the production rates of the first three agreed with experiment and that for quarks a mass of 4-5 nucleon masses would make them practically unobservable (with the techniques known at that time) even if they could, in principle, exist as free particles. If anyone would have suggested to me then a $q\bar{q}$-plasma, I would have declared it impossible (alas!, again).

One often heard objection against the model was that resonances (inside a fireball) lived not long enough to be treated as "particles". G. Matthiae [MAT68] proved, using the principle of detailed balance, that they might just live long enough.

It was tempting to apply the model to astrophysics: big bang and neutron stars [HAG70a]; this triggered a number of papers notably the one by K. Huang and S. Weinberg [HUA70] and by J.C. Wheeler [WHE71]. The then proposed "limited temperature (T_0) big bang" depended entirely on the (at that time reasonable) interpretation of the model as yielding an infinite energy density at T_0.

For an easily readable summary of the situation up to 1972 see [HAG73], where no prior knowledge about statistical bootstrap is required.

3.5 Microcanonical bootstrap

The invention of Veneziano's dual resonance model and the discovery that it also predicted an exponential mass spectrum [FUB69, FUB69a] added to the credibility the statistical bootstrap model had won in the meantime by successful applications.

A number of people became interested and the following few years saw the appearance of some important papers.

S. Frautschi, who also coined the name "Statistical Bootstrap Model" (SBM) [FRA71], made a breakthrough by reformulating the model (grand) microcanonically instead of (grand) canonically (roughly: the original model was the Laplace transform of Frautschi's). His bootstrap equation (BE) reads in invariant notation [YEL73]

$$H\tau(p^2) = H\sum_{m_{in}} \delta_0(p^2 - m_{in}^2) + \sum_{n=2}^{\infty} \frac{1}{n!} \int \delta^4\Big(p - \sum_{i=1}^{n} p_i\Big) \prod_{i=1}^{n} H\tau(p_i^2)\, d^4p_i \tag{3.11}$$

where $\tau(m^2)dm^2 \equiv \rho(m)dm$, H is a universal constant and m_{in} is running over the low mass stable hadrons (often just only the pion). The philosophy was the same as before (with slight differences): $\tau(p^2)$ on the left-hand side of Eq. (3.11) was the density of states of the fireball with mass $m = \sqrt{p_\mu p^\mu}$ to be described while the right-hand side expressed that it was either just one of the input particles or else composed of particles i having mass spectra $\tau(p_i^2)$.

Frautschi, Hamer and Carlitz [FRA71, HAM71, CAR72, FRA72, HAM74, HAM75] drew many important conclusions which would have been impossible to draw in the earlier canonical language. The new language used a vocabulary familiar to particle physicists (phase space) while the canonical one was repellent to them at that time (today, knowledge in thermodynamics is more widespread among high energy physicists). Some of the new results were:

- fireballs would predominantly decay into two fragments, one heavy and one light;
- from iterating their BE with realistic input, they found numerically $T_0 \approx 140$ MeV and $a = 2.9 \pm 0.1$ (which ruled out the until then adopted value $a = 5/2$);
- each imposed conservation law implemented by fixing a quantum number [e.g., baryon number: $\rho(B,m)$] in the mass spectrum, increases the value of a by ½;
- the microcanonical description gives a detailed account of what happens near T_0 (clustering, overshooting T_0,...).

W. Nahm [NAH72] had meanwhile proved analytically that a must be exactly 3. This and the Frautschi-Hamer numerical result strengthened everyone's belief that T_0 should be a limiting temperature, since the energy density $\to \infty$ when $T \to T_0$. Thus: no indication for a phase transition to a new state of matter (alas!, once more).

More checks with experiments were made ($p\bar{p}$-annihilation, pair production, correlations, heavy particle production, ISR secondary spectra, Ericson-fluctuations [FRA78] and so on) which cannot be reviewed here. Instead I refer to the excellent review papers by E.M. Ilgenfritz, J. Kripfganz and H.J. Möhring [ILG77] and by A. Tounsi [TOU73] in which also the current technical improvements of the model (carried out at Leipzig, CERN, Paris) are described.

During these years, a group around L. Sertorio was busy to relate the SBM to the S-matrix theory [BAS73, FRE75, FRE77] and clear up many of the open principal questions. This work is so extensive that I must refer to Sertorio's very complete and important review paper [SER79], which gives a full account of the more theoretical aspects up to 1978.

Other work included attempts - partly successful, partly less - to equip the model with:

- Bose and Fermi statistics (present in the original canonical formulation [HAG65] but extremely difficult to formulate in Frautschi's microcanonical one); Bose statistics was achieved [FAB73, ENG75, KRI75, CHA77], not Fermi;
- internal symmetries [charge, isospin, SU(3),...] using group theoretical methods (based on an early pioneering work by F. Cerulus [CER61]) [ILG73, HAM73, ENG74, ENG74a, CSI74, KRI74, RED80];
- angular momentum conservation [CHI71, HAG77].

Much attention was given to the relation between SBM and dual resonance/string/ Regge pole models with the result that there was indeed a close resemblance but not identity of the models [GOR73, SAT73, DET74, GOR74, STR75]. Related to this question was a reconsideration of large angle elastic scattering and Ericson fluctuations [JOH72, EIL73, MAR73, CHI76, KOG76, FRA78, LOG80].

Nucleon-antinucleon annihilation received new interest [HAM72, MöH75, MöH75a, VAN77, MöH78].

The description of particle production (momentum spectra, production rates,...) was improved [RAN70, HAG72, LET72, LET73, LET73a, KRI75a, RAN75] and extended to unsymmetrical collisions (πN, KN,...) [GER74] and even to p-nucleus collisions [ALA75].

3.6 Exact analytical solutions of the BE

During the same years our analytical understanding of the BE (3.11) advanced and solutions were constructed. The pioneer was J. Yellin, who had proposed Eq. (3.11) and solved it formally by Fourier transform (strictly speaking not existing) and - via Eq. (3.16) - by expansion in terms of phase space integrals [YEL73]. This triggered a series of papers in which the BE was solved by Laplace transform (which does exist); the decay integral equation (sum over all possible decay modes) and its identity with the BE were discovered and a number of formal properties derived [MON72, HAM73, ILG73, HAG73a, FIO77, FEI78, SAT79, FIO80]. Even a second quantized SBM was constructed which led to the strange discovery that the creation and destruction operators of hadron clusters obey a non-associative (and, of course, non-commutative) algebra [ETI77].

3.7 The bootstrap function

The analytical solution of the BE via Laplace transform mentioned above was so important for further development that I present it here. It goes as follows: define the Laplace transforms

$$\Phi(\beta) := \int e^{-\beta_\mu p^\mu} H\tau(p^2)\, d^4p \qquad (3.12)$$

$$\varphi(\beta) := \int e^{-\beta_\mu p^\mu} H\delta_0(p^2 - m_{in}^2)\, d^4p \qquad (3.13)$$

where $(\beta_\mu \beta^\mu)^{-\frac{1}{2}} = T$ = temperature in the rest frame of the system. Laplace transforming Eq. (3.11) then gives

$$\Phi(\beta) = \varphi(\beta) + e^{\Phi(\beta)} - \Phi(\beta) - 1 \quad ; \tag{3.14}$$

writing

$$G(\varphi) := \Phi(\beta) \tag{3.15}$$

this yields the equation

$$\varphi = 2G - e^{G} + 1 \tag{3.16}$$

which has been solved for the "bootstrap function" $G(\varphi)$ by power expansion [YEL73] and by an integral representation exploiting the analytical structure of Eq. (3.16) [HAG82]. Most illuminating is the graphical solution of Eq. (3.16): draw the curve $x = 2y-e^{y}+1$ and interpret it as the graph of the inverse function (Fig. 3.5).

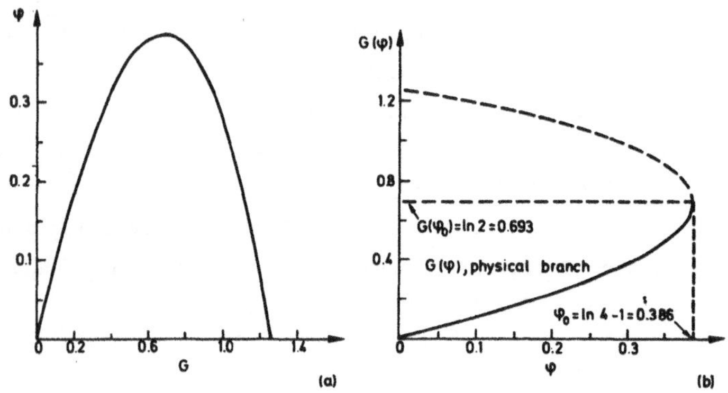

Fig. 3.5: (a) the curve $\varphi(G)$ as given by Eq. (3.16);
(b) its graphical solution by exchanging φ and G.

It follows that for <u>physical</u> systems

$$\varphi \leq \varphi_0 = \ln 4 - 1 \tag{3.17}$$

$$\Phi \leq \Phi_0 = \ln 2 \tag{3.18}$$

In the simple case of one input mass, we have from Eq. (3.13)

$$\varphi(\beta) = 2\pi H m_{in}^2 \frac{K_1(\beta m_{in})}{\beta m_{in}}$$

the condition (3.17) reads

$$2\pi H m_{in}^2 \frac{K_1(\beta_0 m_{in})}{\beta_0 m_{in}} = \ln 4 - 1 \tag{3.19}$$

which fixes a maximal temperature $T_0 = 1/\beta_0 \approx m_{in}$. Thus in this version of the model, T_0 is calculable (depending on H, m_{in}).

Furthermore, one recovers the earlier result of Nahm [NAH72] that the singularity of $G(\varphi)$ at φ_0 is of the square root type: $G(\varphi) = \Phi_0 - \text{const}\sqrt{\varphi_0 - \varphi} + \ldots$ from which it follows by inverse Laplace transformation of Eq. (3.12) that $\tau(m^2) \sim \text{const.}\ m^{-3} \exp(m/T_0)$ [HAG73a] confirming once more that the energy density at T_0 is infinite and hence T_0 a limiting temperature which might be attained or even overpassed only in transient states of superheating, but never in true equilibrium: still no phase transition to a new regime (alas! for the fourth time).

It is important to observe that the same equation (3.16) is obtained independently of:

- the number of space-time dimensions [SAT79],
- the number of input particles (trivial),
- Abelian or non-Abelian symmetry constraints [RED80].

What changes is only the function φ, which may become very complicated depending on various input masses and chemical potentials besides the temperature, while the relation between Φ and φ is always given by Eq. (3.16). Therefore, the "bootstrap function" $G(\varphi)$ has a sort of invariant significance for SBM and it will govern also the discussion of the phase transition to which we will eventually come (note, however, that changing the BE truncating it [CHI72, CHI73, HAG73a] or by making the volume n-dependent [POM51] will change also Eq. (3.16)). This invariance of the Laplace transformed BE suggests that it may appear in other contexts; indeed it does: in renormalization theory, in non-linear differential equations, and in combinatorics (references given in [HAG82]). In this latter field, Eq. (3.16) was discovered already in the last century by E. Schröder [SCH1870] when he solved a combinatorial problem, which he explained in the following words (free translation): "...Given n elements, e.g., material points freely movable in space, one divides the set of n elements arbitrarily into subsets; some possibly containing only one element, some with two, three ... elements; each set of more than one element is to be surrounded by a closed surface, which, together with its content, shall be considered as a new element on an equal footing with all the others. This new set of (obviously less than n) elements is submitted to the same procedure as long as one likes or until it ends because the next step would result in one single element. We wish to find the number of different complexions which can be generated by this procedure..."; this, indeed, is SBM in its most abstract form. He terminates his paper with the prophetic words: "Considerably more difficult would be the solution of the above problems in the case where among the original elements several are identical...", as if he had already known that more than 100 years later we would have difficulties with Bose and Fermi statistics!

3.7 The state of affairs up to 1978

The bootstrap equation (various versions) had been solved and the ensuing thermodynamics had been applied in particle physics, astrophysics and cosmology. For particle physics applications, further assumptions about collective motions had been necessary [HAG68], assumptions which had nothing to do with bootstrap thermodynamics, but contributed to the successes of the model. These and the confirmations coming from the dual models strenthened the belief that the model was basically right. The only challenge: large $p_\perp$, was interpreted as (at least partly) due to pre-equilibrium processes [HAG68, HAG84] and/or angular momentum [HAG77].

The picture was then:

- fireballs consist of fireballs (still pointlike! which is inconsistent, but bothered nobody);
- as a consequence, there is a limiting temperature T_0 which cannot be reached at finite energy density.

This picture seemed to provide a non-perturbative, analytically soluble model of the many-body aspects of strong interactions, running against a "phase transition" whose other side would, however, be a forbidden land. It turned out later that removing the inconsistency of pointlike constituent fireballs would drastically change the situation and make the "phase transition" a true transition with the other side no longer forbidden.

The state of affairs up to 1978 is reviewed in the articles [HAG73]: heuristic introduction, ideology and general results; [TOU73, ILG77]: critical discussion, confrontation with experiments; and [SER79]: more formal questions, relation to S-matrix... Each of these needs the others for complementation.

4. THE PHASE TRANSITION: HADRON MATTER - QUARK MATTER

A phase transition has two sides. In our particular case, only lattice QCD is so far able to handle both sides simultaneously; all other approaches (including ours) treat them separately and join them in hybrid models. That a quark-gluon matter phase might exist has been speculated as early as 1969 [BJO69] (maybe earlier) and the first theoretical paper really consecrated to it seems to be Carruther's in 1973 [CAR73]; from about 1975 on so many papers appeared that I must refer to the articles collected in the Bielefeld proceedings of 1980 and 1982 [SAT80, JAC82] where references are given.

In the present context, the approach from the hadron side is of interest. The obvious argument - who did invent it? - would be: if you compress hadron matter sufficiently, hadrons will overlap and cease to exist as such; you then have a quark-gluon soup. The idea was used in many papers and got much support from the MIT bags [CHO74] which are quark-gluon matter inside, hadrons outside. As many of the properties of bags are strangely resembling the clusters (formerly called fireballs) of the SBM, one might expect that bootstrap thermodynamics would be equivalent to thermodynamics of a bag gas. Indeed: an explicit model constructed by Baacke [BAA77] made a phase transition very likely to happen near T_0. Two years earlier, Cabibbo and Parisi [CAB75] had already proposed that the singularity found in SBM at T_0 should be related to the transition to the quark-gluon phase. These two arguments supported each other beautifully: SBM would provide for the singularity necessary for a true phase transition, while the bag gas model produced the crossing of the pressure curves of the plasma and the hadron gas near T_0.

Unfortunately, SBM was still in an underdeveloped stage in which two arguments spoke against Cabibbo and Parisi's interpretation:

- the energy density $\varepsilon \to \infty$ when $T \to T_0$;
- quarks and gluons did nowhere appear in SBM;

Therefore: how could a singularity, which could not be reached at finite energy density, indicate a transition to a phase whose constituents did not appear in the model? I felt strongly this way.

Some people, however, felt - rightly - that these objections would be overcome by technical development without touching the essential features of SBM. Thus a number of papers, originating mostly in the fertile and critical soil of Bielefeld, investigated the various aspects of a phase transition, in the presence of an exponential spectrum: critical exponents, influence of fugacities, types of transition and others [MIL76, MON77, FIO77,78,79, LET78, SAT78, KAR78, CEL79 - to mention only a few].

At this time, W. Nahm, during a workshop at Erice [CAB78] severely criticized the shortcomings of the model and pointed out that they would forbid its application in astrophysics and to phase transitions, while particle physics applications might still be reasonable [NAH78]. His criticism coincided, however, with a new formulation of SBM, presented just at the same workshop, in which most of the inconsistencies were removed [HAG78] and to which we turn now.

4.1 Hadron volumes

The breakthrough came when J. Rafelski and myself tried to apply SBM to heavy ion collisions. There it was obvious that we could no longer deal with pointlike "constituent clusters" of the extended clusters to be described. It took us considerable time to equip the model with baryon conservation and proper hadron volumes. Our efforts - helped in a later stage by I. Montvay - resulted in the following treatment of the volume question:

- in the fully relativistic formulation we strived for, the volume had to be a four-vector parallel to the system it confined [TOU68]:

$$V \longrightarrow V^{\mu} = A(m, \dots)\, p^{\mu} \tag{4.1}$$

- if a cluster (C) consisted of clusters (i), then:

$$V_C^{\mu} = K \cdot \sum_i V_i^{\mu} \tag{4.2}$$

- with K being some constant. Hence, with Eq. (4.1):

$$A(m_c, \dots)\, p_C^{\mu} = K \sum_i A(m_i, \dots)\, p_i^{\mu} \tag{4.3}$$

- since momentum conservation requires $p_C^{\mu} = \Sigma p_i^{\mu}$, it <u>follows that the volume is proportional to the mass</u>:

$$\begin{aligned} &K = 1 \qquad \text{(dense packing)} \\ &A = \text{const. (independent of } m, \dots) \equiv 1/4B \end{aligned} \tag{4.4}$$

We identify B with the bag constant. These and other results were presented at Erice in the 1978 Workshop on Hadronic Matter at Extreme Energy Density [HAG78]. This paper leaves, however, one question unsolved: after treating the volume correctly on the level of the BE, the hadron gas should take particle volumes also into account (as already done by J. Baacke [BAA77]). We later succeeded in doing so [HAG80] by introducing the notion of the "available volume" defined by subtracting particle (cluster) volumes from the "external" (confining box) volume:

$$\Delta^{\mu} := V^{\mu}_{ext.} - \sum_{\substack{i \\ \text{all part.}}} V_i^{\mu} \tag{4.5}$$

Then, instead of keeping (as usual) the external volume fixed, we required Δ to be constant when summing over particle numbers and integrating over the mass spectrum, whereby V^{μ}_{ext} was pushed to infinity. By this trick, the partition function became <u>formally</u> that of <u>pointlike particles in the available volume Δ</u>, which we knew how to calculate explicitly. Thereafter, Δ had to be eliminated by using

$$\langle V_{ext} \rangle = \Delta + \frac{\langle E \rangle}{4B} \tag{4.6}$$

Then the energy density (and all other physical quantities) could be calculated. It turned out that

$$\frac{\langle E\rangle}{\langle V_{ext}\rangle} = \varepsilon(T,\mu) = \frac{\varepsilon_{pt}(T,\mu)}{1+\varepsilon_{pt}(T,\mu)/4B} \qquad (4.7)$$

(and similarly for other quantities). Here ε_{pt} is the fictitious "point particle energy density" which we can explicitly calculate and which diverges on the "critical curve" [given by an equation like (3.19) with the baryon chemical potential μ incorporated]. Thus on the critical curve <u>the energy density is now finite</u> [Eq. (4.7) with $\varepsilon_{pt}(T,\mu) \to \infty$]:

$$\varepsilon(T_{crit},\mu_{crit}) = 4B = \text{bag energy density} \qquad (4.8)$$

The baryon number density is also finite along this curve while the average cluster mass and volume tend to infinity, the pressure to zero: <u>everything coalesces into one infinite cluster with the energy density of a bag</u>. Since inside highly excited (large) bags we have a quark-gluon gas, we should identify the single infinite cluster with quark-gluon matter, the transition to which is achieved just on the critical curve.

We finally have arrived at a sophisticated, relativistic Van der Waals-type gas of strongly interacting particles: finite (mass proportional) cluster volumes represent repulsive forces, while the exponential mass spectrum resulting from bootstrap represents the attractive forces. The thus completed SBM leads to a phase transition on a critical curve (Fig. 4.1) qualitatively common to many models

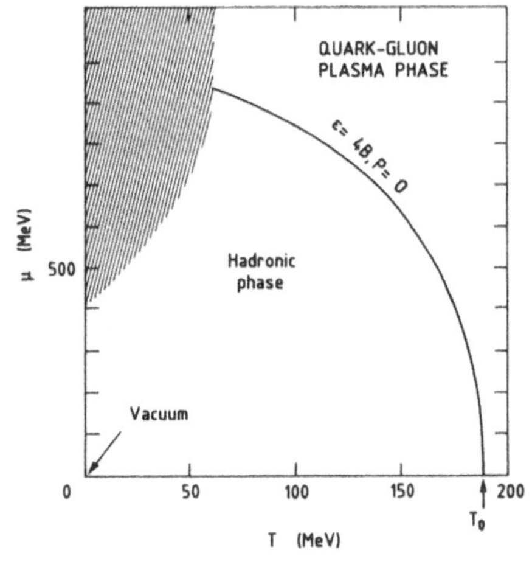

Fig. 4.1: The singular curve of the statistical bootstrap model. In the shaded region the model is unreliable, because there the (otherwise negligible) effects of Bose-Einstein and Fermi-Dirac statistics become important.

[CHI78, CEL79, KAP79, KAL79, SHU80, KAJ81, KUT81, JAC82] to mention only a few of the many references. The results of QCD lattice calculations (see [JAC82]) support our conclusions.

Yet, one might argue, how do you know that the singularity indicates a phase transition to quark-gluon matter: nowhere in the SBM quarks do appear as input particles! Short of a proof, I can at least propose a chain of heuristic arguments:

- the identical properties of QCD bags and SBM clusters: both have the same mass volume relation and the same mass spectrum up to finest details with respect to conservation laws [KAP82, RED83] - these identities lead to the conclusion that QCD bags of interacting fields and SBM clusters are complementary descriptions of the same objects below the transition;
- if we thus are free to claim that inside our clusters we indeed have quark-gluon matter (though correlated by colour neutrality of constituent subclusters), then we indeed have implicitly quarks and gluons in the model;
- the very nature of the bootstrap approach makes it rather irrelevant which are the input particles, provided the basic quantum numbers are represented. Instead of starting with quarks and gluons and imposing their interaction and confinement to colourless bound states we have accepted Nature's own solution of the QCD bound state problem and used the π and the nucleon (others could be added) as input (after all: to describe a helium gas, you do not start with neutrons, protons and electrons!).

Thus without having a formal proof, I believe to be justified in asserting that SBM predicts a phase transition (of which not yet all details are known) to a quark-gluon matter phase as J. Rafelski and I have proposed it at the 1980 Bielefeld Symposium [HAG80a, RAF80].

A further criticism was brought up by V.V. Dixit and H. Satz: in the standard thermodynamical limit a hard sphere gas, even with an exponential mass spectrum, cannot produce a singularity at any finite temperature [DIX81]; therefore our method with the available volume violates the standard rules of taking the thermodynamic limit. The question was discussed at length [HAG83] and it was shown that since hadron matter at the regime near transition cannot be enclosed in boxes of fixed volume, the standard thermodynamical limit is not suitable. Using the "grand canonical pressure ensemble" (without recourse to the available volume technique) the result of Dixit and Satz was reproduced but also our old results were recovered, which settled the question.

Here I stop. Work goes on. For a review of present developments, applications, problems, and progress, see J. Rafelski and M. Danos [RAF82,83], also [JAC82], and a rather daring and controversial paper of mine [HAG84].

And if you ask me now why it took 27 years to arrive at the present (still problematic) state, let me answer with Shakespeare [SHA1601]:

There are more things in Heaven and Earth,
Horatio, than are dreamt of in your philosophy.

ACKNOLWEDGEMENTS

My sincere thanks and admiration go to all who have worked in this field and brought it into the present state, but in particular to those with whom I had, over the years, the pleasure of close contact, discussions and collaboration: A. Auberson, R.D. Carlitz, F. Cerulus, M. Chaichian, J. Engels, B. Escoubès, E. Etim, G. Fast, R. Fiore, S. Frautschi, H. Grote, L.W. Jones, J. Kripfganz, J. Letessier, I. Montvay, W. Nahm, J. and G. Ranft, J. Rafelski, K. Redlich, H. Satz, K. Schilling, L. Sertorio, A. Tounsi, L. Turko, J. Vandermeulen and U. Wambach.

REFERENCES

ALA 75 J.P. Alard, A. Baldit, R. Brun, J.P. Costilhes, J. Dhermain, J. Fargaix, L. Fraysse, J. Pellet, G. Roche, J.C. Tamain, A. Cordaillat and A. Pasinetti, Nuovo Cimento A30 (1975) 320.

AUB 65 G. Auberson and B. Escoubès, Nuovo Cimento 36 (1965) 628.

BAA 77 J. Baacke, Acta Phys.Polon. B8 (1977) 625.

BAS 73 A. Bassetto and L. Sertorio, Nuovo Cimento 14A (1973) 548.

BEL 56 S.Z. Belenkij, Nucl.Phys. 2 (1956) 259.

BET 37 E. Beth and G.E. Uhlenbeck, Physica 4 (1937) 915.

BJO 69 J.D. Bjorken and E. Paschos, Phys.Rev. 185 (1969) 1975.

CAB 75 N. Cabibbo and G. Parisi, Phys.Lett. 59B (1975) 67.

CAB 78 N. Cabibbo and L. Sertorio, Eds., "Hadronic Matter at Extreme Energy Density" Workshop, Erice, Italy, 1978, Plenum Press (1980).

CAR 72 R.D. Carlitz, Phys.Rev. D5 (1972) 3231.

CAR 73 P. Carruthers, Collective Phenomena 1 (1973) 147.

CEL 79 T. Celik and H. Satz, Z.Phys. C1 (1979) 163.

CER 58 F. Cerulus and R. Hagedorn, Supp. Nuovo Cimento 9 (1958) 646, 659.

CER 61 F. Cerulus, Nuovo Cimento 19 (1961) 528

CHA 77 M. Chaichian and M. Hayashi, Phys.Rev. D15 (1977) 402.

CHI 78 S.A. Chin, Phys.Lett. 78B (1978) 552.

CHI 71 C.B. Chiu and R.L. Heimann, Phys.Rev. D4 (1971) 3184.

CHI 72 C.B. Chiu and J. Johnson, Phys. Lett. B42 (1972) 475.

CHI 73 C.B. Chiu, Nucl.Phys. B54 (1973) 170.

CHI 76 C.B. Chiu, Phys.Rev. D13 (1976) 2091.

CHO 74 A. Chodos, R.L. Jaffe, K. Johnson, C.B. Thorn and V.F. Weisskopf, Phys.Rev. D19 (1974) 3471.

CSI 74 F. Csikor, I. Farkas, Z. Katona and I. Montvay, Nucl.Phys. B74 (1974) 343.

DET 74 J. Dethlefsen, H.B. Nielsen and H.C. Tze, Phys.Lett. 48B (1974) 48.

DIX 81 V.V. Dixit, F. Karsch and H. Satz, Phys.Lett. 101B (1981) 412.

EIL 73 G. Eilam, Y. Gell, B. Margolis and W.J. Meggs, Phys.Rev. D8 (1973) 2871.

ENG 74 J. Engels, K. Fabricius and K. Schilling, Phys,Lett. 53B (1974) 65.

ENG 74a J. Engels, K. Fabricius and K. Schilling, Nuovo Cimento 23A (1974) 581.

ENG 75 J. Engels, K. Fabricius and K. Schilling, Phys.Lett. 59B (1975) 477.

ETI 77 E. Etim and R. Hagedorn, Nucl.Phys. B131 (1977) 429.

FAB 73 K. Fabricius and U. Wambach, Nucl.Phys. B62 (173) 212.

FAS 63 G. Fast and R. Hagedorn, Nuovo Cimento 27 (1963) 208;
G. Fast, R. Hagedorn and L.W. Jones, Nuovo Cimento 27 (1963) 856.

FEI 78 M. Feigenbaum and L. Sertorio, Nuovo Cimento, 43A (1978) 31.

FER 50 E. Fermi, Progr.Theor.Phys. 5 (1950) 570.

FIO 77 R. Fiore, R. Page and L. Sertorio, Nuovo Cimento 37A (1977) 45.

FIO 78 R. Fiore, R. Page and L. Sertorio, Nuovo Cimento 44A (1978) 531.

FIO 79 R. Fiore, Z.Phys. C1 (1979) 305.

FIO 80 R. Fiore, Z.Phys. C4 (1980) 281.

FRA 71 S. Frautschi, Phys.Rev. D3 (1971) 2821.

FRA 72 S. Frautschi, Nuovo Cimento 12A (1972) 133.

FRA 78 S.C. Frautschi, in [CAB78], p. 1.

FRE 75 P. Fré and L. Sertorio, Nuovo Cimento 28A (1975) 538.

FRE 77 P. Fré, Fortschr.Phys. 25 (1977) 579.

FUB 69 S. Fubini, D. Gordon and G. Veneziano, Phys.Lett. 29B (1969) 679.

FUB 69a S. Fubini and G. Veneziano, Nuovo Cimento 64A (1969) 811.

GER 74 H.R. Gerhold and B. Buschbeck, Nuovo Cimento A22 (1974) 39.

GOR 73 M.I. Gorenstein, V.A. Miransky, V.P. Shelest, B.V. Struminsky and G.M. Zinovjev, Phys.Lett. 43B (1973) 73.

GOR 74 M.I. Gorenstein, V.A. Miransky, V.P. Shelest, G.M. Zinovjev and H. Satz, Nucl.Phys. B76 (1974) 453.

GRO 70 H. Grote, R. Hagedorn and J. Ranft, CERN Report (Black) (1970).

HAG 65 R. Hagedorn, Suppl. Nuovo Cimento 3 (1965) 147.

HAG 68 R. Hagedorn and J. Ranft, Suppl. Nuovo Cimento 6 (1968) 169.

HAG 68a R. Hagedorn, Suppl. Nuovo Cimento 6 (1968) 311.

HAG 70 R. Hagedorn, Nucl.Phys. B24 (1970) 93.

HAG 70a R. Hagedorn, Astron. & Astrophys. 5 (1970) 184.

HAG 71 R. Hagedorn, CERN Yellow Report 71-12.

HAG 72 R. Hagedorn and J. Ranft, Nucl.Phys. B48 (1972) 157.

HAG 73 R. Hagedorn, in Cargèse Lectures in Physics, Vol. 6, Gordon and Breach (1973).

HAG 73a R. Hagedorn and I. Montvay, Nucl.Phys. B59 (1973) 45.

HAG 77 R. Hagedorn and U. Wambach, Nucl.Phys. B123 (1977) 382.

HAG 78 R. Hagedorn, I. Montvay and J. Rafelski, in [CAB78], p. 49.

HAG 80 R. Hagedorn and J. Rafelski, Phys.Lett. 97B (1980) 136.

HAG 80a R. Hagedorn and J. Rafelski, in [SAT80], p. 237.

HAG 82 R. Hagedorn and J. Rafelski, Comm.Math.Phys. 83 (1982) 563.

HAG 83 R. Hagedorn, Z.Phys. C17 (1983) 265.

HAG 84 R. Hagedorn, CERN Preprint TH. 3684 (1983), to be published in Rivista Nuovo Cimento (1984).

HAM 71 C.J. Hamer and S. Frautschi, Phys.Rev. D4 (1971) 2125.

HAM 72 C.J. Hamer, Nuovo Cimento 12A (1972) 162.

HAM 73 C.J. Hamer, Phys.Rev. D8 (1973) 3558.

HAM 74 C.J. Hamer, Phys.Rev. D10 (1974) 1458.

HAM 75 C.J. Hamer, Nucl.Phys. B95 (1975) 173.

HUA 70 K. Huang and S. Weinberg, Phys.Rev.Lett. 25 (1970) 895.

ILG 73 E.M. Ilgenfritz and J. Kripfganz, Nucl.Phys. B56 (1973) 241.

ILG 73 E.M. Ilgenfritz and J. Kripfganz, Nucl.Phys. B62 (1973) 141.

ILG 77 E.M. Ilgenfritz, J. Kripfganz and H.J. Möhring, Fortschr.Phys. 25 (1977) 123.

JAC 82 M. Jacob and H. Satz, Eds., "Quark Matter Formation and Heavy Ion Collisions", Bielefeld Workshop 1982 (World Scientific, 1982).

JAC 82a M. Jacob and J. Tran Thanh Van, Eds., Physics Reports 88 (1982) 321.

JOH 72 F. Johns, B. Margolis, W.J. Meggs and R.K. Logan, Phys.Rev.Let. 29 (1972) 756.

KAJ 81 K. Kajantie and H. Miettinen, Z.Phys. C9 (1981) 341.

KAL 79 O.K. Kalashnikov and V.V. Klimov, Phys.Lett. 88B (1979) 328.

KAP 79 J.I. Kapusta, Nucl.Phys. B148 (1979) 461.

KAP 82 J.I. Kapusta, Nucl.Phys. B196 (1982) 1.

KAR 78 F. Karsch and D.E. Miller, Phys.Lett. 79B (1978) 259.

KOG 76 S. Kogitz, R.K. Logan and S. Tanaka, Phys.Rev. D14 (1976) 1420.

KRI 74 J. Kripfganz and E.M. Ilgenfritz, Phys.Lett. 48B (1974) 329.

KRI 75 J. Kripfganz, Nucl.Phys. B100 (1075) 302.

KRI 75a J. Kripfganz and J. Ranft, Phys.Lett. 55B (1975) 301.

KUT 81 J. Kuti, B. Lukacs, J. Polonyi and K. Szlachanyi, Phys.Lett. 95B (1980) 75.

LET 72 J. Letessier and A. Tounsi, Nuovo Cimento 11A (1972) 353.

LET 73 J. Letessier and A. Tounsi, Nuovo Cimento 13A (1973) 557.

LET 73a J. Letessier and A. Tounsi, Nuovo Cimento 15A (1973) 358.

LET 78 J. Letessier, A. Tounsi and H. Satz, Nuovo Cimento 45A (1978) 580.

LOG 80 R.K. Logan, M. Bates and S. Kogitz, Phys.Rev. D25 (1982) 2441.

MAR 73 B. Margolis, W.J. Meggs and R.K. Logan, Phys.Rev. D8 (1973) 199.

MAT 68 G. Matthiae, Nucl.Phys. B7 (1968) 142.

MIL 76 D.E. Miller, Phys.Lett. 64B (1976) 193.

MöH 75 H.J. Möhring, J. Kripfganz, E.M. Ilgenfritz and J. Ranft, Nucl.Phys. B85 (1975) 221.

MöH 75a H.J. Möhring, Nucl.Phys. B87 (1975) 509.

MöH 78 H.J. Möhring, Acta Phys.Polon. B9 (1978) 125.

MON 72 I. Montvay, Phys.Lett. 42B (1972) 466.

MON 77 I. Montvay and H. Satz, Nuovo Cimento 39A (1977) 425.

NAH 72 W. Nahm, Nucl.Phys. B45 (1972) 525.

NAH 78 W. Nahm, in [CAB78], p. 219.

ORE 64 J. Orear, Phys.Lett. 13 (1964) 190.

POM 51 I.Ya. Pomeranchuk, Dokl.Akad.Nauk 78 (1951) 889.

RAF 80 J. Rafelski and R. Hagedorn, in [SAT80], p. 253.

RAF 82 J. Rafelski, in [JAC82a], p. 331.

RAF 83 J. Rafelski and M. Danos, "Perspectives in High Energy Nuclear Collisions", Inst.f.Theor.Phys.Frankfurt Univ. and US Nat.Bureau of Standards: NBSIR-83-2725.

RAN 70 J. Ranft, Phys.Lett. 31B (1970) 529.

RAN 75 J. Ranft and G. Ranft, Nucl.Phys. B92 (1975) 207.

RED 80 K. Redlich and L. Turko, Z.Phys. C5 (1980) 201.

RED 83 K. Redlich, Z.Phys. C21 (1983) 69.

SAT 65 H. Satz, Nuovo Cimento 37 (1965) 1407.

SAT 73 H. Satz, Phys.Lett. 44B (1973) 373.

SAT 78 H. Satz, in [CAB78], p. 269.

SAT 79 H. Satz, Phys.Rev. D20 (1979) 582.

SAT 80 H. Satz, ed., "Statistical Mechanics of Quarks and Hadrons", Bielefeld Symposium 1980 (North Holland 1981).

SAT 84 H. Satz, "The Limits of Hadron Thermodynamics", Bielefeld Preprint (May 1984), to be published in Fortschr.Phys. 1984.

SCH 70 E. Schröder, Z.Math.Phys. 15 (1870) 361. (For a modern representation of this work, see: L. Comtet, Advanced Combinatorics, Reidel Pub.Co., 1970).

SER 79 L. Sertorio, Rivista Nuovo Cimento 2 (1979) 1.

SHA 1601 W. Shakespeare, Hamlet (1601).

SHU 80 E.V. Shuriak, Physics Reports 61 (1980) 71.

STR 75 A. Strumia and G. Venturi, Lett. Nuovo Cimento 13 (1975) 337.

TOU 68 B. Touschek, Nuovo Cimento B58 (1968) 295.

TOU 73 A. Tounsi, XVI Scottish Universities Summer School 1973, London 1974.

VAN 65 J. Vandermeulen, Bull.Soc.Roy.Sci, Liège, Belgium, 34e année, NO 1/2 (1965), p. 34.

VAN 77 J. Vandermeulen, Bull.Soc.Roy.Sci, Liège, Belgium, 46 (1977) 57.

WHE 71 J.C. Wheeler, Astrophys.J. 169 (1971) 105.

YEL 73 J. Yellin, Nucl.Phys. B52 (1973) 583.

NUCLEAR STOPPING POWER

by

Wit Busza
Massachusetts Institute of Technology
Cambridge, Massachusetts 02139
USA

The most important input needed for estimating the optimum energy of relativistic heavy ion colliders is the loss in rapidity of the nucleons as two heavy ions collide, together with information on how the lost energy is distributed amongst the produced particles. In principle there is no way of obtaining this information without actually colliding heavy ions. However, knowledge of proton-proton and proton-nucleus collisions may be a useful guide as to what happens in a nucleus-nucleus collision. In this brief review I plan to summarize the relevant pp and pA data and discuss the question "can one infer from pA interactions what happens in AA interactions?"

At the outset let me state that since the Quark Matter '83 Conference there has been little change in our understanding of this subject.

There is no consensus on the mechanism by which the baryon loses momentum in a pp collision and how, in detail, particles are produced. The experimental situation at high energies is as follows.[1] In more than 90% of inelastic collisions the outgoing baryon is a proton or neutron and thus the baryon momentum loss can be fairly well inferred from the $pp \to pX$ and $pp \to nX$ data without information on other channels. Approximately twice as often the leading baryon is a proton rather than a neutron. For energies greater than a few tens of GeV the data approximately scales and can be factorized. i.e. the invariant inclusive differential cross-sections at a given value of Feynmann x and of P_T are approximately independent of incident energy, and the P_T dependence is only weakly dependent on x. See for example figs. 1-3.

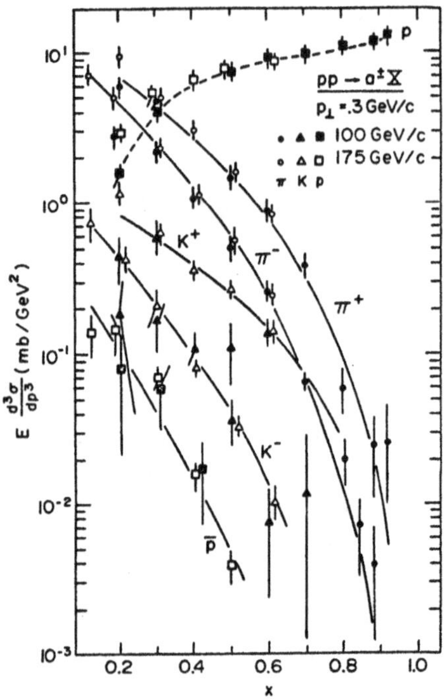

Fig. 1. Data obtained by Brenner et al.[2] which illustrate that the invariant differential cross-sections are approximately independent of incident energy.

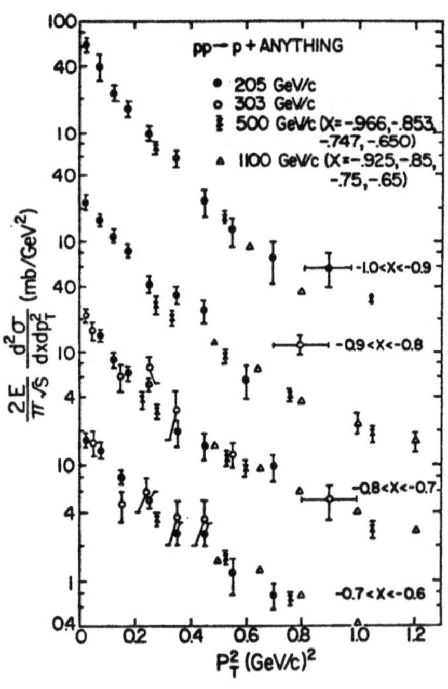

Fig. 2. Data compiled by Whitmore[3] which illustrate that the P_T-distributions in $pp \to pX$ are approximately independent of energy and Feynmann x.

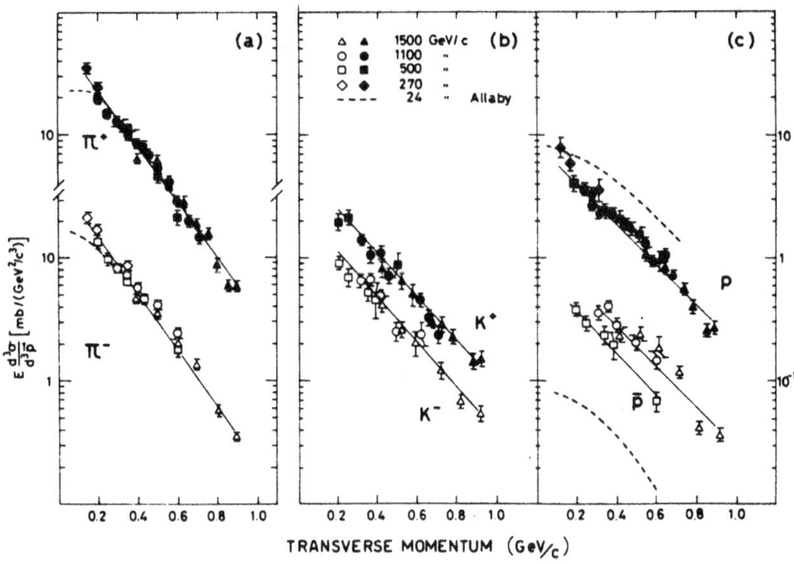

Fig. 3. Data compiled by Bertin et al.[4] which illustrate that in $pp \rightarrow hX$ the P_T-distributions are not very sensitive to the produced particle type.

Integrating over P_T one obtains for $pp \rightarrow pX$ an x distribution which is approximately flat, as can be seen in figs. 4 and 5. In fig. 5 the difference in absolute magnitude of the cross-sections at different energies may be partly due to a breakdown of scaling but more likely is indicative of experimental difficulties with normalization.

A flat distribution in x means that the rapidity loss has an exponential distribution, as shown in fig. 6. It is tempting to speculate that such a simple distribution is a reflection of a simple mechanism of slowing down. For example it would follow if in the rest frame of the proton an approximately constant force acted for a time which had a Poisson distribution, the force and mean time being given by the standard string tension and energy density. As can be seen in fig. 4, the preliminary data of Forrest et al.[5] shows that the $pp \rightarrow nX$ spectrum is considerably softer than that for $pp \rightarrow pX$. This result is consistent with the earlier ISR results.[6]

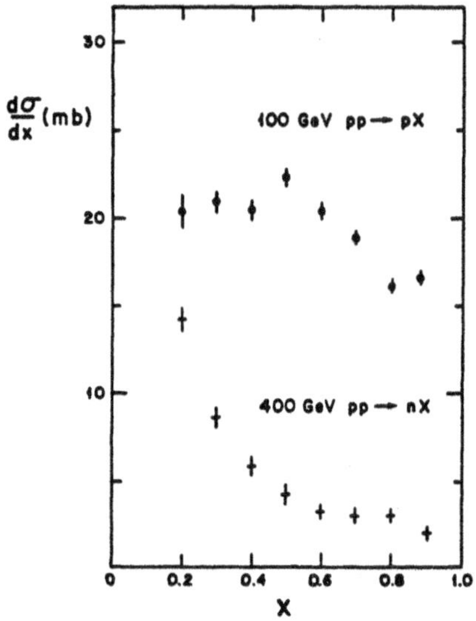

Fig. 4. Comparison of the x-dependence of the cross-sections for pp→ pX[2] and pp→ nX (preliminary data of Forrest *et al.*[5]).

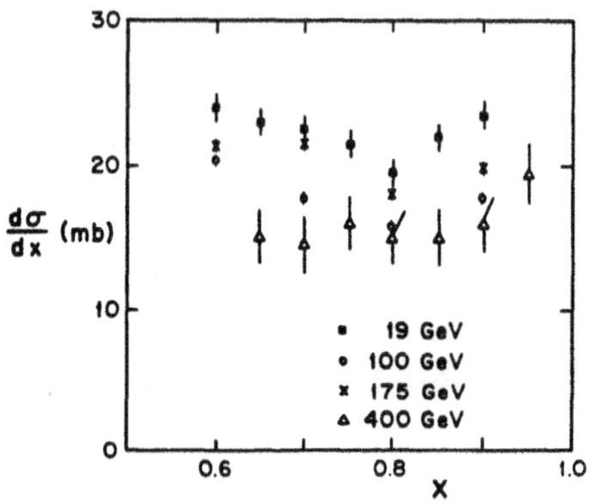

Fig. 5 Data on the x-dependence of dσ/dx at various energies. The 19 GeV and 400 GeV data are from the compilation by Whitmore[3], the 100 GeV and 175 GeV data are from Brenner *et al.*[2].

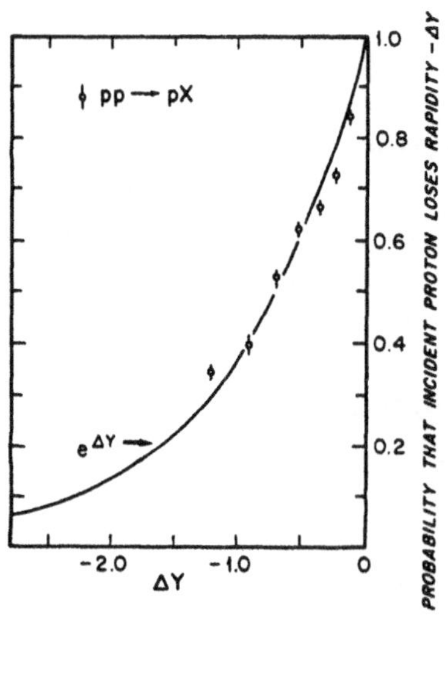

Fig. 6. Probability distribution for protons to lose rapidity − Δy in pp→ pX. The data are from ref. 2.

From the $p \to p$ and $p \to n$ data one can calculate the probability that the incident baryon in pp collisions loses rapidity, $-\Delta y$. The probability distribution is independent of absolute normalization and thus not very sensitive to the choice of data. Using the data shown in fig. 4 one obtains the result shown in fig. 7. Although the measurements cover a relatively small range in rapidity loss, since the area under the curve has to be unity, there is little flexibility in the extrapolation of the data to higher rapidity losses and thus the average rapidity loss $\langle \Delta y \rangle$ or the median rapidity loss $\Delta y_{\frac{1}{2}}$ can be reasonably well determined. The results are given in fig. 7. As can be seen, even in pp collisions the fractional energy loss of the baryon on the average is very significant, of the order of 60%.

There is little information, in particular for highly inelastic collisions, on the correlation between the energy loss of the baryon and the momentum distribution of the produced particles, i.e. on the question "where does the energy go?". Brenner et al.[2] have looked at this question in $pp \to pX$ collisions for energy losses of up to 70%. See for example fig. 8. Their data suggest that the rapidity distribution of the produced particles is simply determined by the available energy. i.e. as less and less energy is taken up by the baryon, higher and higher rapidity particles are produced. If the

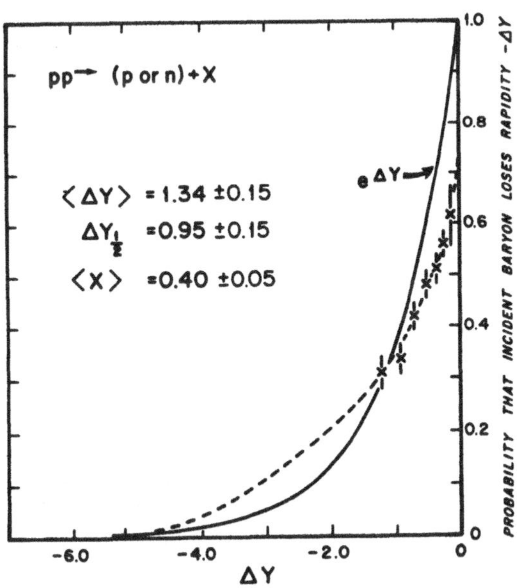

Fig. 7. Probability distribution for the baryon to lose rapidity $-\Delta y$ in $pp \to$ baryon + X The experimental points are based on the data shown in fig. 4. The broken curve is a crude extrapolation of the data using the constraint that the area under the curve is unity. See text.

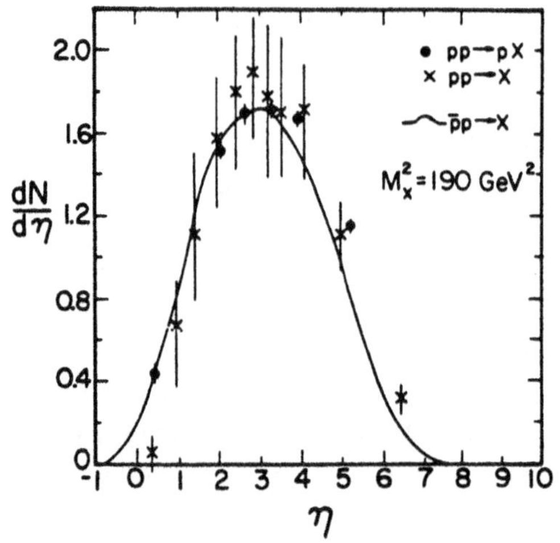

Fig. 8. Data from Brenner *et al.*[2] comparing the pseudo rapidity distribution of charged particles for the two processes pp → pX and pp → X at a common $s = M_x^2 = 190\ \text{GeV}^2$. In the pp → pX data the proton's incident and outgoing momentum are 175 GeV/c and 70 GeV/c respectively.

same trend were to be observed in collisions where the baryon loses most of its energy, and if this correctly reflected what happens, in heavy ion collisions, it would indicate that current estimates of energy densities achievable in heavy ion collisions is optimistic. However, it should be pointed out that the range of baryon rapidity loss covered by the Brenner *et al.*[2] data is too small for alarm at this time.

Let me now discuss the data on energy loss in baryon-nucleus collisions.

There have been several high energy experiments which have made fairly extensive studies of the A-dependence of the inclusive production of particles. These are summarized in Table I.

Table I: Summary of data on pA→hX

Energy of incident proton (GeV)	Produced particle studied	Range of P_T (GeV/c)	Reference
24	$\pi^\pm$, $K^\pm$, p, $\bar{p}$	0.05-1.5	Eichten *et al.*[7]
100	$\pi^\pm$, $K^\pm$, p, $\bar{p}$	0.18,0.3,0.5	Barton *et al.*[8]
300	Λ, Λ, K_s°	0-2	Skubic *et al.*[9]
400	Ξ°	0-2	Pondrom *et al.*[10]
400	n	0-2	Forrest *et al.*[5]

In figs. 9 and 10 examples are shown of how the invariant differential cross-sections for $pA \rightarrow hX$ at a fixed $P_T \sim 0.3$ GeV/c depend on A. As expected from scaling, the A-dependence of the data is approximately independent of incident energy. However, if one compares directly the cross-sections of Eichten et al.[7] at 24 GeV with those of Barton et al.[8] at 100 GeV, clear differences are seen (fig. 11). Whether this is a manifestation of a violation of scaling or an outright experimental discrepancy is not clear at present. This discrepancy has serious implications as to stopping in heavy ion collisions and it is important that it be resolved. Some light may be thrown on this issue by the experiment of Remsberg et al.[11] using 18 GeV protons at BNL, and the analysis of data taken by the European hybrid spectrometer at CERN and by the Hybrid Spectrometer at Fermilab.

The fact that in fig. 10 the A-dependence seems to be also independent of the particle produced is a curious phenomenon, presently not understood[8].

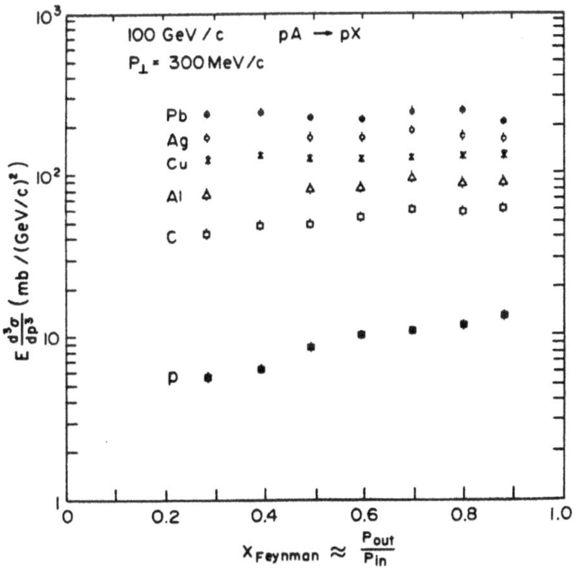

Fig. 9. Example of the A-dependence of the invariant differential cross sections as measured by Barton et al.[8].

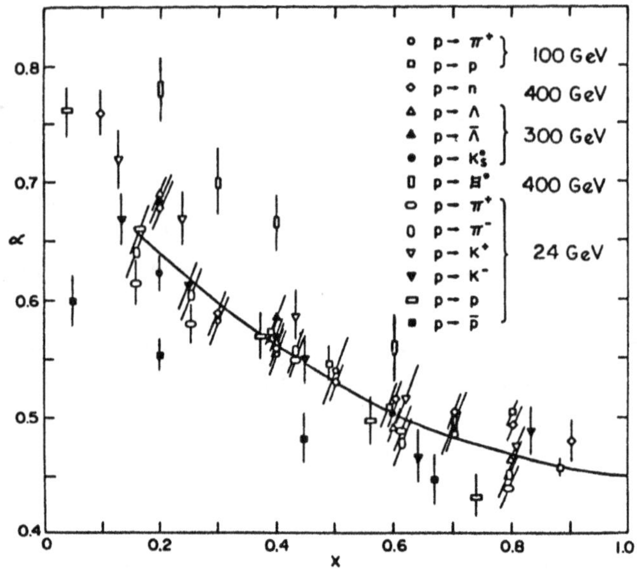

Fig. 10. Compilation made by Barton et al.[8] of the world data on the A-dependence of projectile fragments in high energy proton collisions. α is the parameter obtained in a fit of the measured invariant cross-section at P_T = 300 MeV/c to the form const. A^{α}. α = 0.72 corresponds to $E\frac{d\sigma}{dp^3}$ α σ inelastic.

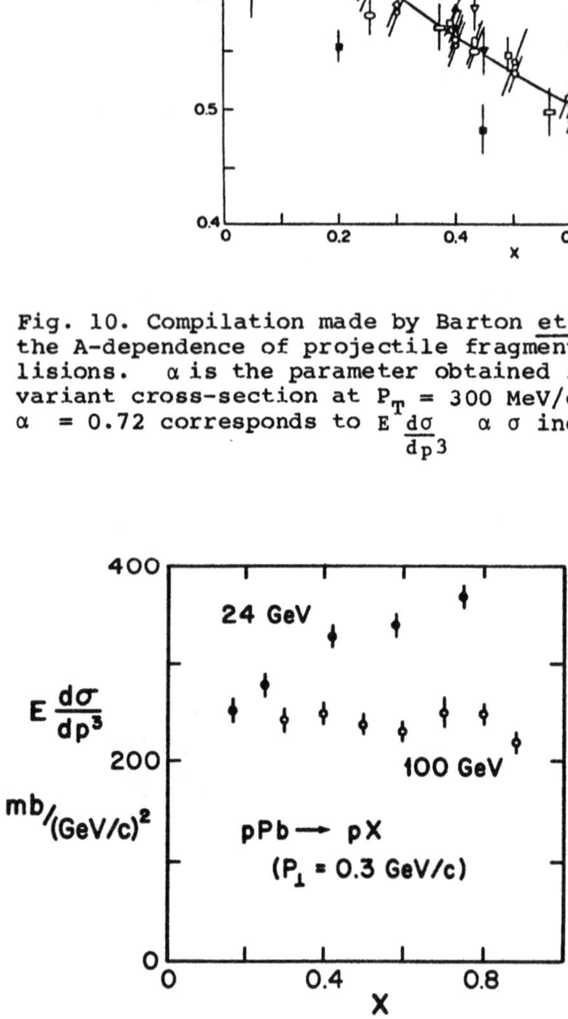

Fig. 11. Direct comparison of the 24 GeV (ref. 7) and 100 GeV (ref. 8) differential cross-sections for the process pPb → pX at P_T = 300 MeV/c.

The results in fig. 9 and 10 immediately suggest that on the average the baryon loses more energy in a collision with a nucleus than with a nucleon, and that the increase in energy loss is not compensated by an increase of particle prediction with $x \gtrsim 0.15$ (upper two units of rapidity).

The measurements listed in Table I are not comprehensive, i.e. they do not cover the complete range of P_T and x. However, since in most cases the P_T distributions of particles are not very dependent on the particle types and incident or outgoing momentum, and since the reaction cross-sections are known,[12] enough data exist to allow useful extrapolations to all regions of phase space. For example, extrapolation and integration of the $pA \to pX$ and $pA \to nX$ data yields the results shown in fig. 12, for the dominant channel of the type $pA \to$ baryon + X.

In a manner similar to the analysis of energy loss in pp collisions, discussed earlier, together with A.S. Goldhaber[13], I have used the $pA \to pX$ data* to determine baryon energy loss in p-lead collisions. Our analysis is summarized in figs. 13 and 14. In the extrapolation of data to regions of phase space where there are no measurements we made the following assumption:

i) the inclusive P_T distributions are independent of x and A,

ii) the ratio of the production of various baryons is independent of A.

----

*.) At the time we did this analysis the $pA \to nX$ data were not available. This new information does not significantly alter our results.

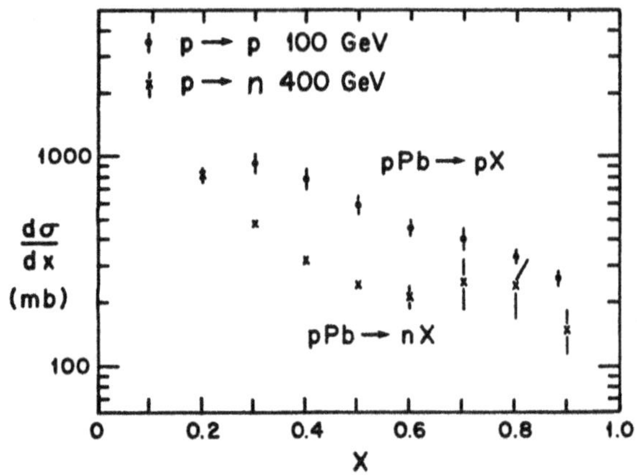

Fig. 12. dσ/dx as a function of x for pPb → pX (ref. 8) and pPb → nX (preliminary data of Forrest et al.[5]). The pPb → pX was obtained from that shown in fig. 9 assuming that the P_T dependence is the same as in pp → pX. The main contribution to the enhancement around x = 0.8 in the p → n data is due to production at 0° . i.e. from some coherent channel.

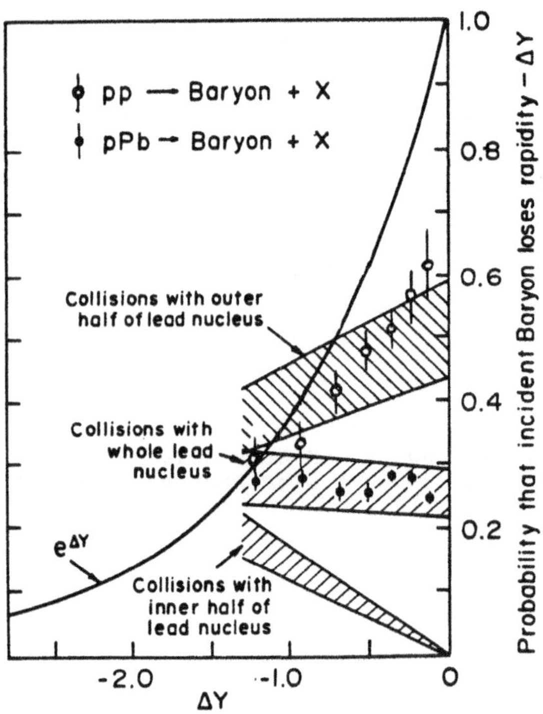

Fig. 13. Probability distributions for protons to lose rapidity − Δ y in collisions with protons and lead nuclei. The distributions were calculated from the data shown in figs. 4 and 12. The shaded regions reflect experimental errors and uncertainties in analysis of data. See text and ref. 13.

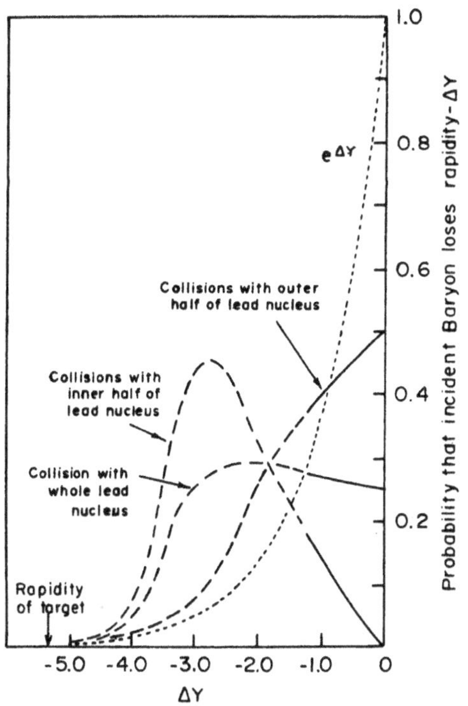

Fig. 14. Extrapolated probability distributions for rapidity loss of protons striking lead nuclei. The dashed lines are the extrapolated portions. The constraint that the normalized total, central, and peripheral inclusive cross-sections should all be smooth makes the extrapolations nearly unique. Figure is taken from ref. 13.

In order to have some idea about energy loss in central collisions with the lead nucleus we assumed the data can be divided into two separate regions: (i) periphery of nucleus equivalent to 3.3 smaller nuclei with A = 15, having in total half the lead reaction cross-section, and (ii) central part of nucleus which contributes no events at x = 1. The results of our analysis are given in Table II.

Table II

Target With Which Baryon Collides	$\bar{\nu}$	Arithmetic Mean Shift of Rapidity	Median Shift of Rapidity $\Delta y_{\frac{1}{2}}$	Median Percentage Energy Loss	Momentum Corresponding To $\Delta y_{\frac{1}{2}}$
Proton	1	1.34 ± 0.15	0.95 ± .15	60 ± 5%	1.0 ± 0.2
Lead Nucleus	3.8	1.9 ± 0.3	1.9 ± 0.3	85 ± 5%	$3.1 ^{+1.0}_{-0.8}$
Central Half of Lead Nucleus	5.8	2.4 ± 0.5	2.5 ± 0.5	90 ± 5%	$5.7 ^{+1.2}_{-0.6}$

These results are very encouraging from the point of view of heavy ion collisions. They show that a baryon loses a significantly larger fraction of its energy in a collision with a large nucleus than with a proton. If they correctly reflect the energy loss that will occur in heavy ion collisions, they suggest that baryon densities 5-10 times greater than in nuclear matter should be achieved in head on collisions of heavy ions provided that each beam has ~ 5 GeV/nucleon. They do suggest that in order to produce in heavy ion colliders a relatively baryon free central plateau one needs beams, with energies $\gtrsim$ 75 GeV/nucleon.

These are important inferences and therefore it is now necessary to address the question "can one infer from pA interactions what happens in AA interactions?".

There are many uncertainties that enter into the process of inferring from pA results what happens in AA collisions. For example, there could be significant explicitly collective effects of unknown character, or the stopping power of a target nucleus already struck by one "layer" of projectile nucleons could be modified by the time a second layer came along. Aside from such possibilities, we do not know the distribution in space-time of the baryons following a heavy ion collision. The slowing down could take place over larger distances than present estimates, based on our understanding of confinement, would suggest. Alternatively, there could be strong correlations in how much various nucleons slowed down. Both of these effects could significantly reduce the baryon density.

At present, too little is understood about soft hadronic processes[1] to find theoretical answers to these uncertainties. Lacking theoretical understanding, the next best is to look for guidance in phenomenology. In principle such guidance could be very reliable since there is a reasonable measure of how seriously to take a given model's predictions about nucleus-nucleus interactions. It is the degree to which the model can explain, without too many ad hac assumptions the various features of hadron-nucleus collisions[14] from hadron-hadron data.

Unfortunately, a straight forward multiple scattering model, with no ad hac assumptions in which interactions are sequential and independent of each other and assumed to give instantaneously the hadron-hadron result, cannot be a realistic model of a hadron-nucleus collision. This is immediately apparent if one considers the particle production process in various frames of reference[15]. In particular, consider a p-A collision in the rest frame of the incident proton as illustrated below.

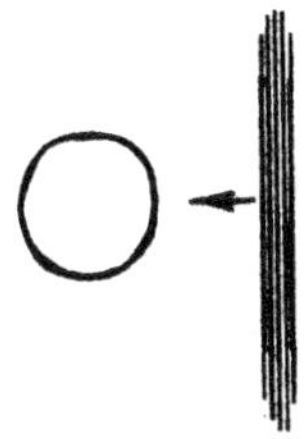

A highly Lorentz contracted nucleus collides with the stationary proton. The produced particles which have a rapidity close to that of the proton have, in this frame, a low velocity and a size comparable to that of the proton. They have to be "produced" at a time after the incident nucleus has passed the interaction region. Clearly it is nonsense to talk as if the process consisted of particles produced in a collision of the proton with one of the nucleons in the nucleus followed by reinteractions of these particles with subsequent nucleons in the nucleus.

As the above discussion illustrates, modelling hadron-nucleus interactions at high energies is non trivial. As a result and because such modelling may be illuminating as to the mechanism of particle production, since the early seventies it has attracted a great deal of attention[16]. Despite the effort, in my opinion, the present status of the hadron-nucleus theoretical and experimental phenomenology is inadequate to throw significant light on the question of deriving A-A from p-A results[15]. Briefly, the reasons for this are:

i) on the experimental side the data have not been sufficiently precise, in particular there is no good way of separating nuclear fragments from the produced low rapidity particles. As a result, for example at low rapidity, experiments disagree with each other and it is never clear, for comparison with models, which particles are included in various measured distributions.

ii) on the theoretical side there are equally serious problems. On the one hand predictions tend not to be very sensitive to model assumptions (main features dominated by variations in impact parameter and peripheral collisions) and on the other there is great flexibility in the predictions. One has, for example, a choice of the exact shape of nuclei used in computation and can improve agreement by including or excluding cascading of the slow particles.

As far as I can see, the only general conclusions that can be drawn from these efforts at modelling are:

i) there are no obvious reasons why baryon stopping in nucleus-nucleus collisions should be less than in baryon-nucleus collisions.

ii) the results of heavy ion collisions should, as a by-product, provide valuable information about the mechanism of particle production and soft hadronic interactions in general.

iii) it will be non trivial to extract new physics from the mass of relatively uninteresting data[17].

To conclude, in my opinion, the loss of rapidity suffered by a baryon as it collides with a nucleon or a nucleus is experimentally fairly well established. More extensive and precise measurements in particular over a very wide energy range are needed to clarify discrepancies such as that between the 24 GeV and 100GeV $pPb \rightarrow pX$ data, and check that the present results hold at higher energies. I do not believe they will fundamentally alter the picture as to nuclear stopping power. The same cannot be said about knowledge of where the energy goes when a baryon loses a significant fraction of its energy. Little is know about this topic, in particular for hadron-nucleus collisions. It is an important experimental question with serious implications for quark-gluon plasma searches and should be answered as soon as possible.

The question "can we infer from pA interactions what happens in AA interactions?" is still an open question and will most likely remain so until we actually start studying heavy ion collisions at high energies. My guess is that, since the stopping in pA collisions is less than what one would have expected from the most naive extrapolation of pp data, so in AA collisions it will be less than the most naive extrapolation of pA data.

REFERENCES

1. For discussions and reviews of various aspects of soft hadronic physics see the proceedings of the annual Symposia on Multiparticle Dynamics. e.g. XV Symposium, Lund, June 1984; see also Fiałkowski, K. and Kittel, W., Reports on Progress in Physics 46 (1983) p. 1283.

2. Brenner, A.E., et al., Phys. Rev. D26 (1982) p. 1497

3. From the compilation by Whitmore, J. Physics Reports 10C (1974) p. 273

4. From the compilation by Bertin, A. et al., Phys. Lett. 42B, (1972) p. 493

5. Forrest, P., et al., private communication from Heller, K.

6. Engler, J., et al., Nucl. Phys. B84 (1975) p. 70

7. Eichten, T., et al., Nucl. Phys. B44 (1978) p. 333

8. Barton, D.S., et al., Phys. Rev. D27 (1983) p. 2580

9. Skubic, P., et al., Phys. Rev. D18 (1978) p. 3115

10. Pondrom, L., private communication, to be published in Physics Reports.

11. Remsberg, L.P., et al., AGS Proposal 790, 1984

12. Carroll, A.S., et. al., Phys. Lett. 80B (1979) p. 319

13. Busza, W. and Goldhaber, A.S., Phys. Lett. 139B (1984) p. 235; Goldhaber, A.S., Proceedings of Conference on the Intersections of Particle and Nuclear Physics, Steamboat Springs, May 1984.

14. For a review of hadron-nucleus data see, for example: Otterlund, I., Nucl. Phys. A418 (1984) p. 87c; Fredrikson, S., et al., R.I.T.A., Stockholm report TRITA-TFY-84-06; Busza, W., Acta. Phys. Pol. B8 (1977) p. 333

15. For a discussion of why to-date experimental data in hadron-nucleus collisions have not been too successful in discriminating between various models of particle production, see Busza, W., Proceedings of the XV Symposium on Multiparticle Dynamics, Lund, June 1984.

16. For reviews of theoretical models see, for example, Nikolaev, N.N., Sov. J. Part. Nucl. 12 (1981) p. 63; Chao, W.Q., et al., Heidelberg preprint 1983. Examples of more recent attempts at modelling hadron-nucleus or nucleus-nucleus physics and other references can be found in: Capella, A., et al., CERN preprint TH.3700-CERN (1983); Białas, A., Munich preprint MPI-PAE/PTh 33/84 (1984); Hwa, R.C., et al., Phys. Rev. Lett. 52 (1984) p. 492 and Oregon preprint OITS-256 (1984); Hufner, J. and Klar, A, to be published Phys. Lett. B (1984); Csernai, L.P., and Kapusta, J.I., Minnesota preprint UMTNP-70 (1984); Wong, C.Y., Phys. Rev. Lett. 52 (1984) p. 1393 and Phys. Rev. D30 (1984).

17. See, for example, Busza, W., Nucl. Phys. A418 (1984) p. 635c.

STOCHASTIC EVOLUTIONS AND HADRONIZATION OF HIGHLY EXCITED HADRONIC MATTER

P. Carruthers
Theoretical Division
Los Alamos National Laboratory
Los Alamos, NM 87545

ABSTRACT

Stochastic ingredients of high energy hadronic collisions are analyzed, with emphasis on multiplicity distributions. The conceptual simplicity of the k-cell negative binomial distribution is related to the evolution of probability distributions via the Fokker-Planck and related equations. The connection to underlying field theory ideas is sketched.

I. INTRODUCTION

As we have heard in the session on lattice QCD, it is no easy matter to calculate equilbrium properties of uniform thermalized hadronic matter. Work is only beginning to determine the low-lying excitations (quasi particles) of such a medium. In the absence of such knowledge, it is hard to make a complete list of "signals" of the legendary QCD plasma.

As an example consider the energy loss of a fast electron passing through a metal. $dN/d\Delta E$ exhibits sharp peaks at multiples of the plasma frequency $\Delta E = n\hbar\omega_{pl}$ giving clean evidence for the degenerate electron plasma. The analogous experiment for the QCD plasma is very hard to carry out, since a single controllable quark probe is not available. Suppose that in a nucleus-nucleus collision the initial overlap produces a plasma which is then "probed" by the last nucleons to enter the interaction regions. Although we expect plasmons to be excited, it is hard to tell what the "signal" might be. Likewise it could happen that a single fast nucleon entering the plasma could create a shock wave (due to the existence of sound waves in the medium[1]) whose Mach cone would lead to an excess of transverse hadrons. Again the superposition of many distributed vertices probably smears out any clear signal.

It is clear that a nucleus-nucleus collision provides a very messy laboratory. To complicate matters further, despite years of work there is no compelling theory of multiparticle production in hadron-hadron collisions.

A fundamental and detailed description of the evolution of a high energy collision is indeed formidable. Although we have recast field theory in a form exposing its kinetic and transport aspects,[2] mathematical obstacles have prevented any real progress. In QCD the demands of gauge invariance pose further problems. If we insist on manifest gauge invariance, then the Wigner phase space distributions for quarks is

$$F_{\alpha\beta}(p,R) = \int d^4r\, e^{ip\cdot r}\mathrm{Tr}\left[\rho\bar{q}_\alpha(R - \tfrac{1}{2}r)\ (P \exp i \int_c A\cdot dl)\ q_\beta(R + \tfrac{1}{2}r)\right] \quad (1)$$

Here α,β are spinor indices, ρ the density matrix, c the path running between R + ½ r to R - ½ r and P the ordering symbol for the noncommuting matrixes. $A\cdot dl = A^\mu_a\lambda^a dl_\mu$. For gluons one cannot work locally with the vector potential but is led to objects like

$$F_{\mu\nu,\rho\sigma}(p,R) = \int d^4r\, e^{ip\cdot r}\mathrm{Tr}\left[\rho F_{\mu\nu}(R - \tfrac{1}{2}r)\ P \exp(i \int_c A\cdot dl)\ F_{\rho\sigma}(R + \tfrac{1}{2}r)\right] \quad (2)$$

The unruly properties of the gauge "connector" $P \exp(i \int A\cdot dl)$ makes it very hard to formulate QCD kinetic theory in a gauge invariant way.

II. PHENOMENOLOGY OF MULTIPLICITY DISTRIBUTIONS IN HADRONIC COLLISIONS

The situation becomes more optimistic if we adopt more modest goals and look to experiment to learn what simplifications nature has chosen to make. Experience shows that gross features of highly excited systems can often be described by a small number of variables, the simplest example being that of thermodynamic equilibrium. For nonequilibrium systems there exists a powerful and flexible framework for the description of complex nonlinear systems, associated with the names of Langevin, Fokker, Planck and Kolmogorov, to name a few. From the study of hadronic number distributions (especially KNO scaling[3]), we were led (as was Biyajima independently) to the view that many aspects of the hadronization process could be well described as stochastic number evolutions of this type.

In the following we shall focus on hadronization in hadron-hadron collisions. We shall need to emphasize at one time or another the dynamical, geometrical or statistical aspects of a collision, although, of course, each is interwoven with the other. Similarly we shall categorize collisions as elastic, diffractive, or inelastic or nondiffractive inelastic.

Experiment tells us that not all the kinetic energy of colliding particles is available for hadronization. To a large extent the colliding particles retain their integrity, losing an <u>average</u> of one-half their initial energy. Event by event the mass M available for hadronization will vary,[4-6] as emphasized by Weiner at this conference.[4] For the nondiffractive inelastic events with initial particles of momentum P_1 and P_2, final leading particles P_R and P_L the available mass is given by

$$M^2 = (P_1 + P_2 - P_R - P_L)^2 \tag{3}$$

$$0 \leq M \leq W - 2m$$

for pp collisions.

The relevant stochastic ingredients of a typical hadronic collision are

(a) hadronization mass M^2,

(b) impact parameter b, and

(c) number of produced particles, broken down further by charge or flavor, etc.

Correlations typically exist among these quantities. We can ask (theoretically, at least) what is the probability distribution of charged particles for a given b. Experimentally we can measure P(M,W), the probability of producing a hadronization mass M at c.m. energy W as well as $P_n(M)$, the probability of producing n particles from a given initial mass M. Only by knowing $P_n(M)$ can we disentangle the stochastic aspects of the hadronization process with given M from the variability of M itself. The quantity measured is

$$P_n(W) = \int_0^\infty dM\ P_n(M,W)\ P(M,W) \tag{4}$$

While theory should predict $P_n(M,W)$ and P(M,W), these important quantities are accessible experimentally and should provide useful guidance to the theorist. Actually such information lies unpublished in existing data tapes!

The observed multiplicity distribution $P_n(W)$ is broad, with a long tail. Is $P_n(M,W)$ also broad or is it narrow, being smeared by a broad distribution of hadronization mass, a point of view advocated by Goulianos et al and Minh Duong-Van? Another instructive mathematical example is suggested by the Poisson transform

$$P_n = \int_0^\infty dx\ f(x)\ \frac{(x\bar{n})^n\ e^{-x\bar{n}}}{n!} \tag{5}$$

where $\bar{n}$ is the average multiplicity for the distribution P_n. Eq. (5) represents a broad class of distributions. For example, in the case that $f(x) = k^k x^{k-1} e^{-kx}/(k-1)!$ we obtain the negative binomial distribution

$$P_n^k = \frac{(n + k - 1)!}{n!(k - 1)!}\ \frac{(\bar{n}/k\ )^n}{(\bar{n}/k + 1)^{n+k}} \tag{6}$$

which has been shown[9] to give a good description of the hadronic KNO plot.

Comparison of (4) with (5) suggests the possibility that for given M the hadron counting distribution is simply Poisson, the observed $P_n(W)$ resulting from a P(M,W) having the shape $(M/\bar{M})^{k-1}\ \exp(-kM/\bar{M})$ with $\bar{M}$ some fixed reference mass. Other possibilities exist, of course.

Before leaving this topic we stress once again the necessity of employing M rather than W. For example, for such a simple quantity as the bulk multiplicity, errors continue to propagate, although Basile[5] et al have shown that

$$\bar{n}(M,W) = \sum_n n\, P_n(M,W) \tag{7}$$

depends on M rather than W, a result in complete accord with the physical picture described above. Rejections of the Fermi-Landau power law has been based on just this confusion: the use of $W^{\frac{1}{4}}$ instead of $M^{\frac{1}{4}}$. By the same reasoning one would have to reject the more fashionable QCD leading log multiplicity formula.

Eq. (5) possesses the remarkable property of form invariance[10] under elimination of an unobserved component. For clarity consider partitioning n into charged and neutral contributions: $n = n_{ch} + n_0$. If we measure only charged particles, then a range of n (or n_0) will contribute for a fixed n_{ch}. If the probability of a particle being charged (neutral) is p(q), with p + q = 1, the probability of the pair (n_{ch}, n_0) is $n!p^{n_{ch}}q^{n_0}/n_{ch}!n_0!$ and compounding with (5) reproduces (5) with $n \to n_{ch}$, $\bar{n} \to \bar{n}_{ch}$. It should be stressed that the binomial distribution used here does not imply the absence of charged-neutral correlations. For example the negative binomial (6) leads to a linear dependence of $\bar{n}_0$ on n_{ch}, as observed. This result and the similar forward/backward correlation,[11] lends further support to the validity of the negative binomial distributions for the description of hadronic multiplicity distributions.

Details of the application of (6) and its refinements to pp and p$\bar{p}$ multiparticle data may be found in Ref. (9, 11, 12).

III. STOCHASTIC DYNAMICS: EVOLUTION OF PROBABILITY DISTRIBUTIONS

The observed counting distribution is an S-matrix quantity, the result of a complex dynamical evolution. One is naturally led to ask how the probabilities for the occurrence of various quanta evolve in time. Alternatively one can ask for the rate change dN_i/dt or better yet the flows $p\cdot\partial f(p,R)$ of phase space distributions. Before exploring these possibilities, we wish to focus on some purely statistical features of Eq. (6).

The simplest meaning of (6) is simply a generalized Bose-Einstein distribution in which n particles are distributed among k boxes of equal average occupancy $\bar{n}/k$. Another interpretation arises in the semiclassical theory of photoelectron counts. Here k sources emit radiation fields in a gaussian field ensemble and it is unnecessary to know that $\bar{n}$ is related to k. From this, one can reasonably assert that the hadronization process acts as if k separate cells (superclusters?) emit "out" fields behaving as gaussian random variables. In this case the only remaining dynamical quantity is the number of cells (phenomenologically $k \approx 4$). Clearly k does not have to be integral and probably represents a statistical average.

Next we shall discuss another possible meaning of the parameter k, in terms of the parameters appearing in the Fokker-Planck equation which described the time dependence of probabilities. These parameters are in turn related to kinetic equations (e.g., Langevin equations) which show the relation to the microscopic parameters such as couplings and cross sections.

Rate equations for number populations are familiar in chemistry, solid state physics and jet calculus. The treatment of fluctuations is tricky if done casually but often a phenomenological noise term gives good results. The noisy number flow is (for gaussian white noise) equivalent to a (linear) differential equation for the probabilities (generalized Fokker--Planck equation: FPE). For application to hadronic collisions we can further imagine that there is a regime in which quarks and glue are the best variables, so that it is necessary to evolve the joint distribution $P(n_q, n_G, t)$ for the various flavors, colors, etc. Near and at hadronization it may be essential to use hadron number coordinates. No one knows how to join the two regimes. In the following we shall ignore this problem since it is difficult even to explain the conceptual framework. We shall write equations as if there were only one species of Bose particle.

Consider a Boson field theory with discrete modes labeled ℓ. The basis of oscillator coherent states

$$|\alpha\rangle = e^{-\frac{1}{2}|\alpha|^2} \sum_{n=0}^{\infty} \frac{\alpha^n}{\sqrt{n!}} \, |n\rangle \tag{8}$$

α a complex number and $|n\rangle$ an oscillator number state are convenient for analyzing the kinetic (and classical limit) features of the time evolution. The density matrix ρ, represented by $\langle\alpha|\rho|\beta\rangle$ can be reconstructed from its diagonal element $\langle\alpha|\rho|\alpha\rangle$ if $\langle\alpha|\rho|\beta\rangle$ is an analytic function of the two complex variables α^* and β. Alternatively one can adopt the "P-representation"

$$\rho(t) = \int d^2\alpha \; \Phi(\alpha,t) \; |\alpha\rangle\langle\alpha| \tag{9}$$

which represents an enormous class of useful states.[13-14] Here α is $\{\alpha_\ell\}$ representing whatever modes enter the problem.

In this basis, the equation of motion for ρ is given by[15]

$$i \, \frac{\partial\rho(J,\phi)}{\partial t} = \mathscr{L}\!\left(J, \phi, \frac{\partial}{\partial J}, \frac{\partial}{\partial\phi}\right) \; \rho(J,\phi) \tag{10}$$

where α is related to action angle variables by $\alpha = \sqrt{J}\, e^{i\phi}$. $\mathscr{L}$ is given by a finite power series of order $(\hbar/J)^n$ with the classical term coinciding[16] with the formulation of phase space evolutions due to Brout and Prigogive. By assuming random initial phases, the number rate equation can be rigorously derived. Alternatively, Eq. (10) can be written for the weight Φ of Eq. (9):

$$i \frac{\partial\Phi(\alpha,t)}{\partial t} = \tilde{\mathscr{L}}\left(\alpha,\alpha^*, \frac{\partial}{\partial\alpha}, \frac{\partial}{\partial a^*}\right) \Phi(\alpha,t) \tag{11}$$

when phase averaged equations of type (10-11) are of the FPE type, though with many variables.

The connection with the Poisson transform is easily seen by taking the diagonal element of (9):

$$P_n(t) = \langle n|\rho|n\rangle = \int d^2\alpha \; \Phi(\alpha,t) \frac{|\alpha|^{2n} e^{-|\alpha|^2}}{n!} \tag{12}$$

We remark that when Φ is gaussian, one recovers Eq. (6). Comparing with Eq. (5), we find the correspondence[10]

$$f(x,t) = \int \frac{d\phi}{2\pi} \Phi(e^{i\phi}\sqrt{x\bar{n}},t) \tag{13}$$

Thus we expect the kernel of the Poisson transform (which is also the KNO scaling form) to obey an FPE. At some "late" time (or rather hypersurface) the evolution is terminated by physical separation of the system (i.e., hadronization).

A nice mathematical example was given by Biyajima, where an exactly soluble FPE evolves to a stationary solution

$$f(x,t) \sim \psi_k(x) = \frac{k^k}{(k-1)!} \; x^{k-1} \; e^{-kx} \tag{14}$$

which happens to be the KNO form of (6) for $x = n/\bar{n}$, and which reproduces the discrete P_n^k of Eq. (6) through use of the Poisson transform, Eq. (5).

We have noticed yet another FPE, equivalent to the Langevin equation with multiplicative noise

$$\frac{dx}{dt} = dx - bx^2 + xF(t) \tag{15}$$

which gives the same stationary solution. So differing dynamical schemes (even for one variable) can give the same long-time probability distribution. From (15) we can show that k is 2d/Q or 2b/Q where Q is the strength of the noise $\langle F(t)\, F(t')\rangle = 2Q\delta(t - t')$.

Since we are running out of space, time and results, we close by giving a qualitative outline of the logical connection among the concepts mentioned above. By application of these classical techniques of statistical physics, we believe that a successful phenomenology of hadronization can be constructed.

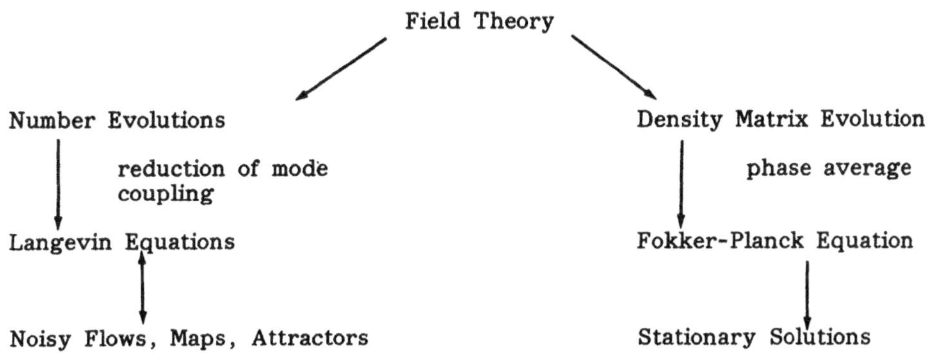

1. The hadronization process is an example of a driven (by the kinetic energy of the colliding particles) self-damped (dissipative) system. The effective number of degrees of freedom is thus reduced.
2. For high excitations ($Q^2 \gg \Lambda^2$), QCD is scale invariant. The realization of this scale invariance is through cascade (Cantor set) structures (as in turbulence).
3. The proper description of the dissipative evolution is by means of noisy flows (coupled Langevin equations) and Fokker-Planck equations.
4. Hadronization is, or is near, an out of equilibrium phase transition, described by the FPE in close analogy to the σ model or Ginzburg-Landau theory of Type II superconductors. Hence universality and scaling behavior (described by fractal dimensions of the evolution) will make the results better than the accuracy of the approximations.
5. The nonlinearities of the evolution can automatically self-organize into the emitting cells. It is possible that the cells are in phase space rather than ordinary space.
6. Fractal structures of the hadronization hypersurface should be sought both in xyzt space and in the effective dynamical phase space.

ACKNOWLEDGEMENTS

The ideas of this paper owe much to conversations with my collaborators Minh Duong-Van, Chia Chang Shih and Fredrik Zachariasen. This research was supported by the United States Department of Energy.

REFERENCES

1. P. Carruthers, Phys. Rev. Lett. 50 (1983) 1179.
2. P. Carruthers and F. Zachariasen, Phys. Rev. D13 (1976) 950. Rev. Mod. Phys. 55 (1983) 245.
3. Z. Koba, H. B. Nielsen and P. Olesen, Nucl. Phys. B40 (1972) 317.
4. R. Weiner (this proceedings). See also G. N. Fowler, E. M. Friedlander, M. Plümer and R. M. Weiner (to be published).
5. M. Basile et al, Nuovo Cimento 47A (1982) 244.
6. P. Carruthers, Nucl. Phys. A418 (1984) 501c.
7. K. Goulianos, H. Sticker and S. N. White, Phys. Rev. Lett. 48 (1982) 1454.
8. Minh Duong-Van, presented at the "Chaos Workshop, Dynamics Days," Center for Non-Linear Dynamics (La Jolla, CA, Jan. 1984).
9. P. Carruthers and C. C. Shih, Phys. Lett. 127B (1983) 242, LA-UR-84-2505, to appear in the proceedings of the XV Multiparticle Dynamics Symposium (Lund, Sweden, June 1984).
10. P. Carruthers and C. C. Shih, Phys. Lett. 137B (1984) 425.
11. P. Carruthers and C. C. Shih (to be published).
12. M. Biyajima, Prog. Theor. Phys. 69 (1983) 966.
13. R. Glauber, Phys. Rev. 130 (1963) 2529; ibid 131 (1963) 2766.
14. J. Klauder and E. C. G. Sudarshan, "Quantum Optics" (W. A. Benjamin, Inc., N.Y., 1968).
15. P. Carruthers and K. S. Dy, Phys. Rev. 147 (1966) 214.
16. R. Brout and I. Prigogine, Physica 22 (1956) 621.
17. P. Carruthers in Los Alamos preprint LA-UR-84-1009 (proceedings of the Workshop on $\bar{p}p$ Options for the Super Collider (University of Chicago, Feb. 1984).

COLLECTIVE FLOW AND INTRANUCLEAR CASCADE DYNAMICS

J. CUGNON,
Université de Liège,
Institut de Physique au Sart Tilman,
Bâtiment B.5
B-4000 LIEGE 1, Belgique

The prospect of having very soon ion beams of high energy in order to test the current ideas about the phase transition between hadronic matter and quark-gluon plasma is very fascinating. However, the question of whether the nuclei are large enough systems to exhibit bulk dynamics, embodied by a quantity like the equation of state (EOS) is worth to be adressed. The answer (or at least a hint to it) can be searched for in the wealth of experimental data accumulated during the past ten years by the Bevalac people as well as in the theoretical effort that it has provoked. However, despite of this considerable effort, it has not yet been possible to determine clearly whether the collision between two heavy ions in the GeV per nucleon energy range proceeds from hydrodynamics or from the intranuclear cascade (INC) dynamics. The former, which can be considered as the short mean free path limit, may also generate EOS effects, whereas the INC model embodies a rather long mean free path situation and corresponds more or less to an ideal gas EOS. It may be more correct to say that in the hydrodynamical approach, all the spatial (and also temporal) fluctuations have been averaged out on all scales smaller than a given length [1], which may be considered as the mean free path. Therefore, the dynamics is thus governed by quantities which are associated with "cells" of linear dimension equal (at least) to the mean free path. A cell usually contains several particles and therefore the hydrodynamics can be considered as being sensitive to the bulk properties of the matter : EOS and transport coefficients.

The inclusive measurements are not very helpful in distinguishing between the two approaches. So the attention has turned to exclusive measurements. The hope is that the so-called global analysis of these measurements, very familar to high-energy physicists, will help to discriminate [2-7]. The basic idea is that bulk properties lead to a correlated flow, namely that if a nucleon is going outward in a given direction, there is a good chance to see other ones in approximately the same direction. The correlated flow will expectedly show up in the orientation of the sphericity ellipsoid, one of the most commonly used global variables constructed with the momenta of the ejectiles. However, as it has been shown by Danielewicz and Gyulassy [6], the correlated flow may be strongly masked by fluctuations due to the finite number of particles. Recently, the plastic ball group published their measurements of Ca+Ca and Nb+Nb at 400 MeV/A [8]. The so-called $dN/d\cos\theta$ plot (which express more or less the frequency of having an ellipsoid with a major axis pointing at an angle θ from the beam direction in the c.m. frame) clearly shows a peak for Nb+Nb high multiplicity events around $\theta \approx 23°$, whereas the maximum always arises at 0° in the Ca+Ca system.

The Nb+Nb results are more or less reproduced by the hydrodynamical calculations of ref. [9], although few attention is paid to the bias introduced by the detector (see below). Note that hydrodynamics should also predict a peak in the $dN/d\cos\theta$ plot for Ca+Ca, because in this model, the collective flow comes from the large incompressibility of the matter. Both the Ca+Ca and Nb+Nb data are reproduced by the INC calculation of ref. [10]. The detail of the calculation will not be discussed here, but it is necessary to quote its principal features in order to understand the relationship between the collective flow and the observables. The coalescence-invariant [4,5] sphericity tensor (i,j = cartesian indices)

$$Q_{ij} = \sum_{\nu} \frac{1}{2m_\nu} p_i^{(\nu)} p_j^{(\nu)} \quad , \tag{1}$$

where $\vec{p}^{(\nu)}$ is the momentum of nucleon ν, is constructed for each event after a filter is applied to simulate in the most accurate way the acceptance of the detector : this essentially amounts to discarding the target spectators with a lab energy of less than 20 MeV and the participant free neutrons (see ref. [10] for the detail). Furthermore the number of proton participants M_ν is calculated and finally the $dN/d\cos\theta$ plot is constructed for various bins in multiplicity M_ν. The results are shown in Fig. 1, along with the data as given by ref. [8] for the largest multiplicity bin. The qualitative features of the results are preserved when changing the filter and the definition of M_ν in a reasonable way.

It is interesting to have a closer look at the detail of the results. Fig. 2 gives for the most interesting impact parameters the distribution of the projection of the unit vector $\vec{e}_1$ along the major axis of the full ellipsoid (without filter) on the impact parameter axis. For central collisions (b=0) the ellipsoid points in the ave-

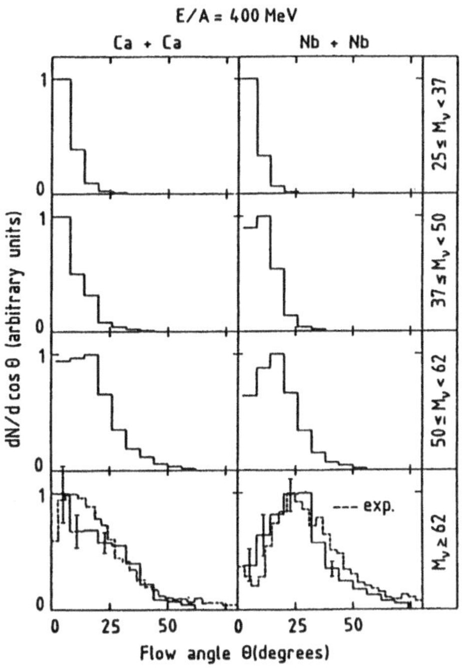

Fig. 1 : Full lines : dN/d cos θ plot as calculated in ref. [10] for various charge multiplicity bins (see text for detail). Dotted lines : experimental data of ref. [8] for the largest measured multiplicity bin. The error bars give the precision of the calculation.

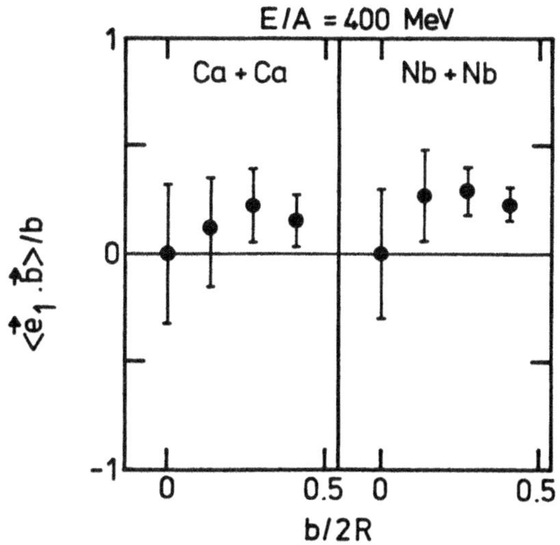

Fig. 2 : Average values (dots) of the cosine of the angle between the major axis of the ellipsoid and the impact parameter vector (see text) as calculated in ref. [10]. The error bars give the event to event fluctuations of the same quantity. The quantity R is the nuclear radius.

rage towards the beam axis, with a large fluctuation however. For intermediate impact parameters, the ellipsoid points towards finite angles indicating an intrinsic sideward collective flow. It is worthwhile to note that the fluctuations are smaller for the heavier system, which according to ref. [6], indicates that the fluctuations are largely dominated by finite particle effects and are probably insensitive to the dynamics itself. The intrinsic collective flow is thus largely dominated by the intermediate impact parameters. The flow angle for Nb+Nb is at most $\sim 15°$ whereas the observed angle is 23°. From Fig. 2, it is also evident that there is an intrinsic flow in Ca+Ca which does not yield a peak in the dN/d cos θ plot at nonzero angle. These apparent contradictions are due to the filter and to the fluctuations which act in an intricate way. Fig. 3 illustrates this effect for the most significant impact parameter. The removal of the target spectators shifts the average angle $\bar{\theta}$ towards a larger value,

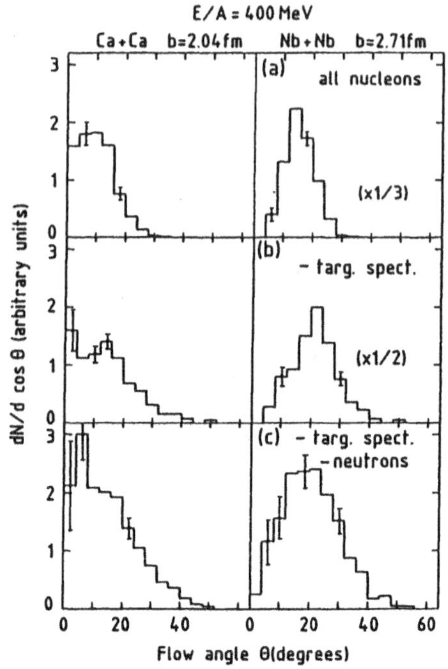

Fig. 3 : dN/d cos θ plot for a typical impact parameter, as calculated in ref. [10]. Part (a) corresponds to the full sphericity tensor. Part (b) is obtained by removing the target spectators. Part (c) is obtained by further removing the free neutrons.

essentially because the spectators are more or less aligned with the beam axis, whereas the participants take the largest contribution to the sideward flow. Furthermore, the elimination of the free neutrons slightly decreases $\bar{\theta}$. Both operations increase the fluctuations because the number of particles diminishes. In Nb+Nb, a net peak remains. On the contrary, in Ca+Ca, the larger fluctuations are sufficient to wash out the manifestation of the flow, owing to a smaller intrinsic flow angle.

At this stage, it is not possible to say that INC and hydrodynamics are equally good to reproduce the experimental Nb+Nb data, because the acceptance of the detector is treated in a very crude way in ref. [9]. Furthermore, hydrodynamics is unable to predict fluctuations without further assumptions. But, let us consider for the moment that the two approaches are equally good. What are the consequences of that ? In our opinion, they are numerous and important :
(1) There is a definite collective flow in heavy ion collisions, even in Ca+Ca (this is rather a consequence of the good success of the INC); (2) With Nb+Nb, the size of the system is large enough compared to the

mean free path for hydrodynamics to be valid, at least for the description of the expansion of the system (this would be in keeping with the result of ref. [11] which shows that the expansion conserves entropy). The collision process starts to be dominated by the bulk dynamics ; (3) The properties of the collective flow would be determined by the maximum pressure, which is more or less the same in both approaches [12] ; (4) The details of the equation of state, if the latter is different in this region from a free gas equation of state do not manifest themselves in a sufficient manner to be observed on the flow angle, which is calculated with an accuracy of a few degrees only. In this perspective, the Au+Au data and calculations are waited for, since all the features would be magnified in this system, according to the above discussion.

References

[1] K.G. Wilson, Rev.Mod.Phys. 55(1983)583

[2] H. Stöcker, L.P. Csernai, G. Graebner, G. Buchwald, H. Kruse, R.Y. Cusson, J.A. Maruhn and W. Greiner, Phys.Rev. C25(1982)1873

[3] J. Kapusta and D. Strottman, Phys.Lett. 106B(1981)33

[4] J. Cugnon, J. Knoll, C. Riedel and Y. Yariv, Phys.Lett. 109B(1982) 167

[5] M. Gyulassy, K.A. Frankel and H. Stöcker, Phys.Lett. 110B(1982)185

[6] P. Danielewicz and M. Gyulassy, Phys.Lett. 129B(1983)283

[7] J. Cugnon and D. L'Hôte, Nucl.Phys. A397(1983)159

[8] H.A. Gustafsson et al., Phys.Rev.Lett. 52(1984)1590

[9] G. Buchwald, G. Graebner, J. Theis, J. Maruhn, W. Greiner and H. Stöcker, Phys.Rev.Lett. 52(1984)1594

[10] J. Cugnon and D. L'Hôte, to be published

[11] G. Bertsch and J. Cugnon, Phys.Rev. C24(1981)2514

[12] M. Gyulassy, private communication.

PHASE TRANSITIONS IN THE EARLY UNIVERSE

L. F. Abbott
Physics Department
Brandeis University
Waltham, MA 02254

I. Introduction

Phase transitions in the early universe may have played a crucial role in determining the large-scale density and distribution of the matter we see today. I discuss here two aspects of this role for phase transitions. First, I review in a general way the beautiful ideas of inflationary cosmology[1] and then discuss how the confinement/chiral symmetry breaking transition in QCD may have affected cosmology and in particular the relevance of this transition to one of the great puzzles in cosmology--the nature of the dark matter in the universe.

II. Inflation

In cosmology a distinction is made between the coordinate distance between two objects and the physical distance. The relation between these is given in terms of a time-dependent scale factor, R(t), by

$$L_{physical} = R(t)L_{coordinate} \tag{1.1}$$

Objects such as galaxies are generally at fixed coordinate sites. Since the Hubble constant defined by

$$H = \frac{\dot{R}}{R} \tag{1.2}$$

is positive the physical distance between galaxies increases with time. This is the expansion of the universe.

The evolution of the scale factor, R(t), with time is determined by the Friedmann equations,

$$\left(\frac{\dot{R}}{R}\right)^2 = \frac{8\pi G}{3}\rho \pm \frac{1}{R^2} \tag{1.3}$$

and

$$\frac{\ddot{R}}{R} = -\frac{4\pi G}{3}(\rho + 3p) \qquad (1.4)$$

where ρ is the matter energy density, p its pressure and G is Newton's constant. The $\pm$ sign in Eq. (1.3) depends on whether the energy density of the universe is less than or greater than the critical density, $\rho_c = 3H^2/8\pi G$.

Inflation is motivated by the flatness problem of standard big-bang cosmology. Consider the quantity

$$\Omega = \frac{\rho}{\rho_c} \qquad (1.5)$$

which is the ratio of the energy density to the critical energy density. Ω can be measured and although the measurements are very difficult we know that Ω is somewhere between about 0.1 and 1. Now from Eq. (1.3)

$$\Omega = 1 \mp \left(\frac{1}{\dot{R}^2}\right) \quad . \qquad (1.6)$$

Under normal conditions, the pressure and energy density are positive so Eq. (1.4) tells us that $\ddot{R} < 0$ or equivalently that $\dot{R}$ decreases with time. This is because the gravitational attraction of the matter in the universe slows down the rate of expansion. Since $\dot{R}$ decreases with time, from Eq. (1.6) we see that Ω gets further and further away from one as time goes on. The universe is now about ten billion years old which is an enormously long time relative to any time scale in particle physics and yet Ω is still remarkably close to one. The only way we can explain this is to assume that in the early universe $\dot{R}$ was gigantic so that even today after decreasing for ten billion years it is still large enough to make the term $1/\dot{R}^2$ in Eq. (1.6) small.

How did $\dot{R}$ ever get so large? In the standard cosmology the huge value of $\dot{R}$ must be put in by hand as a very unusual and unnatural initial condition. This is the flatness problem. However, suppose there existed an epoch in cosmology during which $\dot{R}$ increased with time. If this period lasted long enough then $\dot{R}$ could grow to an enormous value and the problem would be solved. This is the solution proposed by inflation. The inflationary epoch is characterized by increasing $\dot{R}$ or equivalently $\ddot{R} > 0$ which from Eq. (1.4) implies an unusual equation of state, since

$$\rho + 3p < 0 \quad . \qquad (1.7)$$

Due to the fact the $\rho + 3p$ is negative during inflation we have a sort of antigravity situation--gravity actually accelerates the rate of expansion. During inflation $\dot{R}$ increases and becomes enormous, typically so enormous that we expect that the $1/\dot{R}^2$ term in Eq. (1.6) is still tiny today. Inflation therefore predicts that $\Omega = 1$ to a high degree of accuracy.

The condition for inflation, $\rho + 3p < 0$, is satisfied in all known models by dominating the energy density of the universe with the vacuum energy of a metastable false vacuum state. For this $p = -\rho$ so $\rho + 3p = -2\rho < 0$ resulting in an exponentially increasing $\dot{R}$. Phase transitions are crucial for achieving the unusual equation of state required to produce inflation.

Inflation solves another great problem in standard cosmology--the horizon problem. The distance scale over which causal processes can operate at time t in cosmology is of order $1/H(t)$. In a normal cosmological epoch this distance increases more rapidly than the distance between objects at fixed coordinate sites--that is $1/H$ increases more rapidly than R since as we have seen $R/(1/H) = \dot{R}$ decreases with time. This means that as time goes on we can detect and interact with more and more of the universe, and just reflects the fact that it takes the light from distant objects a finite time to reach us, so the longer we wait the further we see. The horizon problem arises because today, by detecting the microwave background radiation, we can observe regions of the universe which were separated by a distance greater than $1/H(t_E)$ when they emitted this radiation at time t_E. Yet when we measure the the blackbody temperatures of these regions they agree to better than 0.01%. In the standard cosmology we are left with the mystery of how regions which were not in causal contact could have come to exactly the same temperature.

Inflation solves this mystery. As we have seen, during an inflationary period $\dot{R}$ increases with time. This means that the distance between objects, which is proportional to R, grows more rapidly than the causal horizon size, $1/H$. Objects initially inside one horizon (i.e. with separations less than $1/H$), get pushed outside the horizon to separations larger than $1/H$. When we later detect regions separated by a distance greater than $1/H(t_E)$ at the same temperature this is no longer surprising. These regions were once in causal contact (at an early time t they had separation less the $1/H(t)$) before inflation thrust them apart.

There is another important consequence of the unusual relation between the horizon growth and expansion rate during inflation. Consider a quantum fluctuation in the energy density with a wavelength which is smaller than the horizon size, 1/H. Of course we expect such a wavelength to be very small since this is a quantum effect. The wavelength of this fluctuation will get red-shifted and grow like R. During inflation this grows faster than 1/H so the fluctuation can become much larger than the horizon size. To an observer inside the horizon the fluctuation will seem to disappear. However, when inflation ends and the horizon starts to grow with respect to the flucuation wavelength, eventually there will come a time when the fluctuation will re-enter the horizon. Our observer will then re-discover the flucuation, only now instead of being elementary particle sized it will be huge--perhaps of galactic scale. This is how inflation provides the seeds for the formation of the large-scale structure in our universe. Because of inflation we can now not only understand the origin of primordial fluctuations but also compute the spectrum of these fluctuations and relate its magnitude to parameters of an underlying field theory.[2] This is a spectacular achievement--it enables us to make definite predictions about the very large-scale structure of our universe.[3]

III. The QCD Confinement/Chiral Symmetry Breaking Transition

The confinement/chiral symmetry breaking transition in QCD which is the topic of this conference almost certainly played no role in the inflationary process itself. However, it still may be extremely relevant to inflationary cosmology. As we have discussed, inflation predicts that, to a very high degree of accuracy,

$$\rho = \rho_c = \frac{3}{8\pi G}H^2 \qquad (2.1)$$

in today's universe. We know from studies of nucleosynthesis that the density of ordinary baryons in the universe cannot be greater than about 20% of ρ_c. Therefore, if inflation is correct at least 80% of the energy density of the universe must be in the form of dark matter of unknown composition. One of the great challenges in the application of particle physics to cosmology is to figure out what this dark matter is. I will discuss three possibilities, all related to the QCD transition.

The confinement/chiral symmetry breaking transition took place at a temperature of around 100 MeV, about 10^{-5} sec after inflation ended. The first two proposals I will discuss both assume that this transition is first order. The first proposal was discussed at last year's Quark Matter meeting[4] so I will just mention it briefly. It assumes that the first order transition occurred explosively producing violent fluctuations. These fluctuations might have formed black holes which are now the dark matter. Witten has noted[5] that if such a violent process occurred, gravitational waves produced during the phase transition should be detectable today through accurate timing measurements of pulsars--in particular the extremely stable millisecond pulsar. Thus, in principle, this proposal can be tested.

Witten has recently considered[5] the possibility that the QCD phase transition occurred very close to equilibrium rather than violently. This is certainly a strong possibility because the cosmological cooling rate of order 10^{-5} seconds is much slower than the typical strong interaction time scale, 10^{-24} seconds. By making a series of rather bold assumptions Witten suggests that dark matter could be produced by such a gentle transition in the early universe.

If the phase transition occurs near equilibrium then regions of quark matter coexist with regions of hadronic matter for a time of order 10^{-5} seconds while the expansion of the universe removes the latent heat released during the transition. At some point, we have a universe which is mostly hadronic matter but with scattered bubbles of quark matter which have not yet been converted. Outside these bubbles, in the ordinary hadronic phase, baryon number is carried by baryons the lightest of which has mass m_p. Inside the quark matter bubbles, baryon number is carried by quarks. If the phase transition takes place at $T_c \approx 100$ MeV then we expect the ratio of baryon number density outside the bubbles to that inside the bubbles to be proportional to $\exp(-m_p/T_c)$. In other words, most of the baryon number is inside the bubbles.

These quark bubbles are, of course, not stable at this point and so they continue to shrink. There are two major mechanisms for conducting the latent heat released by the contraction of the bubbles away from them. This must occur if they are to continue to shrink. First, hadrons can evaporate off the surface of the bubbles. Second, neutrinos can carry away the heat released. Although the neutrino process involves the weak interaction it is enhanced because it does not have to occur exclusively on the surface of the bubble like hadron evaporation. Witten in fact estimates that these two processes are probably competative. If hadron evaporation is appreciable then the bubbles eventually disappear and there are no long-term cosmological effects of the QCD transition. However, if the neutrino process is the dominant heat transport mechanism, then as the bubbles cool they will not lose any baryon number. Thus, they end up as chunks of quark matter with fairly high baryon number densities. Witten then conjectures that such objects might be stable. Thus, they remain until the present and Witten suggests therefore that the dark matter may be chunks of quark matter which he estimates have radii of order

$$r = (10^{-2} \text{ to } 10)\text{cm} \tag{2.2}$$

and masses

$$m = (10^{9} \text{ to } 10^{18})\text{g} . \tag{2.3}$$

Is it reasonable to assume that quark matter is stable? We know, of course, that the energy per baryon in quark matter consisting of only up and down quarks must be greater than the energy per baryon in nuclear matter, otherwise nuclear matter would be unstable. Thus,

$$(E/B)_{\substack{\text{quark matter}\\ \text{with u and d}}} > (E/B)_{\text{nuclear matter}} \tag{2.4}$$

so quark matter consisting of up and down quarks cannot be stable. However, the cosmologically produced quark matter will have a substantial fraction of strange quarks in it. Above some critical baryon number, $B > B_{crit}$, it seems likely that the energy per baryon in quark matter consisting of up, down and strange quarks is less than that in quark matter without strangeness,

$$(E/B)_{\substack{\text{quark matter}\\ \text{with u, d and s}}} < (E/B)_{\substack{\text{quark matter}\\ \text{with u and d}}} \quad \text{for } B > B_{crit} . \tag{2.5}$$

This is because of the exclusion principle. If we have a large enough piece of quark matter consisting of up and down quarks then it will be energetically favorable to turn some of the up and down quarks in states near the Fermi surface into strange quarks which can occupy states with much lower energies. For a high enough Fermi energy this will more than compensate for the larger strange quark mass. Comparing the inequalities (2.4) and (2.5) we see that it is possible that

$$(E/B)_{\substack{\text{quark matter}\\ \text{with u, d and s}}} < (E/B)_{\substack{\text{nuclear}\\ \text{matter}}} \tag{2.6}$$

for $B > \tilde{B}_{crit} > B_{crit}$. Despite this, ordinary nuclear matter would be effectively stable because its decay would be extremely high order in the weak interactions due to the fact that so many strange quarks must be produced. Stable strange quark matter would presumably be made from roughly equal numbers of up, down and strange quarks to minimize the energy penalty caused by the exclusion principle.

Strange quark matter has recently been discussed in detail by Farhi and Jaffe.[6] They conclude that within the uncertainties of the strange quark mass, the bag constant and the QCD coupling constant it is possible that strange matter is stable, and they discuss some of its properties. They doubt that $\tilde{B}_{crit}$ could be less than six and suggest it is probably closer to 100 or 1000 if indeed strange matter is stable. Chunks of quark matter have relatively low Z values for their baryon number. We would naively expect quark matter (in the absence of an accompanying electron cloud) to be positively charged because the larger strange quark mass should cause a slight shortage of strange quarks compared with u's and d's. If strange matter is stable and positively charged then it eats up neutrons but not protons or other charged nucleii at ordinary energies because of Coulomb repulsion. This type of quark matter would not be harmful to ordinary nuclear matter, but if strange quark matter is stable then neutron stars are actually made out of it. If strange matter is negatively charged then, as Farhi and Jaffe point out, it is extremely dangerous since it will eat up all ordinary nuclear matter. Various methods for detecting a cosmic influx of strange quark matter are currently being considered.[6]

Finally, I would like to describe a third way in which QCD could have produced the dark matter. This production mechanism is not actually a result of the confinement/chiral symmetry breaking transition, but rather of the SU(2) x U(1) breaking transition and the fact that QCD gets strong at energies of order 100 MeV.

This proposal suggests that the dark matter of the universe is made up of non-relativistic, extremely light axions.[8] Axions were first introduced to explain the absence of CP violation in the strong interactions. The QCD Lagrangian

$$L = -\frac{1}{4}F_{\mu\nu}^{2} + i\bar{q}\gamma\cdot Dq - m\bar{q}q + (g^2/32\pi^2)\theta F_{\mu\nu}\tilde{F}_{\mu\nu} \qquad (2.7)$$

where $F_{\mu\nu}$ is the QCD field strength tensor, $\tilde{F}_{\mu\nu} = \varepsilon_{\mu\nu\alpha\beta}F_{\alpha\beta}$ and q is the quark field. In order to agree with limits on the electric dipole moment of the neutron we must have[9] $\theta < 10^{-9}$. To explain why this parameter is so small the axion was introduced.[10] This modifies the Lagrangian (2.7) to

$$L = -\frac{1}{4}F_{\mu\nu}^{2} + i\bar{q}\gamma\cdot Dq - m\bar{q}q + (g^2/32\pi^2)(\frac{a}{f_a} + \theta)F_{\mu\nu}\tilde{F}_{\mu\nu} + \frac{1}{2}(\partial_\mu a)^2 \quad (2.8)$$

where a is the axion field. Now the parameter θ can be eliminated by redefining

$$\tilde{a} = a - f_a\theta \quad (2.9)$$

and the absence of strong CP violation becomes a dynamical question--namely why is the vacuum expectation value of $\tilde{a}$ zero,

$$\langle\tilde{a}\rangle = 0 \quad ? \quad (2.10)$$

Peccei and Quinn[10] in fact showed that QCD instantons produce a potential for the axion field with a minimum at $\tilde{a} = 0$, which solves the strong CP problem.

The potential which determines the expectation value of $\tilde{a}$ is formed by instantons. However, at temperatures much above 100 MeV, instanton effects are irrelevant. Therefore, in cosmology we have the situation that in the early universe when T >> 100 MeV, the axion field had no potential and was free to establish any value whatsoever for $\langle\tilde{a}\rangle$. The value of $\langle\tilde{a}\rangle$ was presumably determined by chance, and it seems highly unlikely that the value $\langle\tilde{a}\rangle = 0$ could have been choosen randomly. Therefore, as the universe cools to near 100 MeV and instantons become relevant the axion field finds itself with a non-zero value of $\langle\tilde{a}\rangle$ but now in a potential with a minimum at $\langle\tilde{a}\rangle = 0$. It therefore starts to oscillate around the minimum of the potential. This axion field oscillation is completely equivalent to a coherent state of axions at rest. If f_a is around 10^{12}GeV then enough axions are produced by this mechanism to give the universe a critical energy density and the axions have lifetimes longer than the present age of the universe so they are still around today. The mass of the axion is roughly $m_\pi f_\pi/f_a$. Thus, if $f_a \approx 10^{12}$GeV, the dark matter is composed of non-relativistic axions of mass 10^{-5}eV.

Although the QCD phase transition did not cause inflation, through one of these mechanisms it may have played an essential role in producing the dark matter needed to reconcile inflation with our observations of the universe.

References

1. A. Guth, Phys. Rev. D23 (1981) 347.
2. A. Guth and S.-Y. Pi, Phys. Rev. Lett. 49 (1982) 1110; J. Bardeen, P. Steinhardt and M. Turner, Phys. Rev. D28 (1983) 679; A. Starobinskii, Phys. Lett. 117B (1982) 175; S. Hawking, Phys. Lett. 115B (1982) 295.
3. P.J.E. Peebles, Astrophys. J. 263 (1982) L1; L. Abbott and M. Wise, Phys. Lett. 135B (1984) 279; Astrophys. J. 282 (1984) L47.
4. D. Schramm and K. Olive, Proceedings of Quark Matter '83. See also: C. J. Hogan, Phys. Lett. 133B (1983) 172; L. Van Hove, CERN preprint TH.3623 (1983); M. Crawford and D. Schramm, Nature 298 (1982) 538; T. DeGrand and K. Kajantie, University of Helsinki preprint HU-TFT-84-34 (1984).
5. E. Witten, Phys. Rev. D30 (1984) 272.
6. E. Farhi and R. Jaffe, MIT preprint CTP#1160 (1984).
7. E. Farhi and R. Jaffe (in preparation); A. DeRujula and S. Glashow (in preparation).
8. J. Preskill, M. Wise and F. Wilczek, Phys. Lett. 120B (1983) 127; L. Abbott and P. Sikivie, Phys. Lett. 120B (1983) 133; M. Dine and W. Fischler, Phys. Lett. 120B (1983) 137.
9. V. Baluni, Phys. Rev. D19 (1979) 2227; R. Crewther, P. DiVecchia, G. Veneziano and E. Witten, Phys. Lett. 88B (1979) 123.
10. R. Peccei and H. Quinn, Phys. Rev. Lett 38 (1977) 1440; S. Weinberg, Phys. Rev. Lett. 40 (1978) 223; F. Wilczek, Phys. Rev. Lett. 40 (1978) 279.

MESSAGE FROM LIGHT IONS: AN EXPERIMENTAL REVIEW OF THE SEARCH FOR ABNORMAL PHENOMENA IN HIGH-ENERGY NUCLEUS-NUCLEUS COLLISIONS

R. Szwed

Institute of Experimental Physics
University of Warsaw, Warsaw, Poland

Abstract: The experimental data from the CERN Intersecting Storage Rings (ISR) and from Dubna and Berkeley on the search for abnormal nuclear phenomena in high-energy nucleus-nucleus collisions is reviewed.

1. INTRODUCTION

In a year and half from now, the first experiments at the CERN Super Proton Synchrotron (SPS) using light ion beams (up to ^{16}O) should take place [1]. For the first time, both the powerful accelerator ($E > 10$ GeV/n) [2] and the detectors will be used. Since the ion program at the SPS is not far ahead, there is much theoretical speculation about the possible observation of abnormal nuclear phenomena in ^{16}O interactions with nuclear targets [3,4].

The aim of the present paper is to present a brief review of the experimental results from the CERN Intersecting Storage Rings (ISR) and from Dubna and Berkeley in order to illustrate the present experimental status in the search for abnormal nuclear phenomena in the nucleus-nucleus collisions.

A straightforward extrapolation from the existing data to the interactions of ^{16}O with nuclear targets at the 200 GeV/n (maximum SPS energy) will be attempted. The weak points of the present experimental approaches to the search for abnormal nuclear phenomena will be pointed out.

The paper is organized as follows: In Section 2 the pion production is discussed and the experimental data on the temperature of produced pions are reviewed. In Section 3 the new experimental data on strangeness production from Dubna and the ISR are presented. The pion and baryon density as a function of the rapidity and multiplicity is discussed in Sections 4 and 5, respectively. Section 6 contains a brief comment on the study of the particle production with large transverse momenta in nucleus-nucleus collisions. In Section 7 some technical remarks on the central collision trigger are presented.

After each section there will be a summary, divided into three parts:

S (summary): A brief summary of the pure experimental facts reviewed in the section.

E (extrapolation): An extrapolation from the known experimental results to ^{16}O-^{207}Pb collisions at 200 GeV/n.

D (drawbacks): The weak points of the experimental approach discussed in the section.

The conclusions are given in the last section.

2. PION PRODUCTION AND TEMPERATURE

2.1 PION YIELDS

Pions are the most frequently produced particles in high-energy collisions; therefore, knowledge of the pion production mechanism is essential for the understanding of the main features of high-energy nucleus-nucleus collisions.

The pion multiplicities from the collisions of various projectile nuclei A_P (d, α, ..., Ne) with various targets A_T (Li, ..., Pb) were measured at the Dubna energy [momentum: p = 4.2 (4.5) GeV/c per nucleon or kinetic energy: E_K = 3.36 (3.66) GeV per nucleon] by track detectors [5-8]. They were well reproduced by the models assuming independent multiple nucleon scattering [9]. The ISR αα and αp study [10-12] led to similar conclusions. The models [13-16] assuming independent parton-parton collisions and using an eikonal approximation to calculate different multiple scattering terms, predict rather well the multiplicities and rapidity distributions observed in αα ($\sqrt{s}$ = 30.4 GeV/n corresponding to a laboratory momentum p = 516 GeV/c per nucleon) and αp ($\sqrt{s}$ = 44 GeV/n corresponding to a laboratory momentum p = 1031 GeV/c per nucleon). The best agreement with the data is achieved within the dual-parton model [16,13]. For more information, the reader is refered to the existing reviews on the subject [17,18].

2.2 PION TEMPERATURES

The other characteristic of the production mechanism is the temperature of the pion source. A large increase in the temperature is necessary to pass from the normal hadronic phase to the quark-gluon phase [3,4]. Following the arguments of Hagedorn and Rafelski [19,20], it can be said that 'the temperature is most reliably measured through the measurement of the mean transverse momenta of the particles'. In fact, within a given thermodynamical model (see, for example, ref. [19]), the average transverse momentum of produced pions can be related to the temperature of the collision. Figure 1 shows a compilation of the experimental data from nucleon-nucleon [12,21,22] and nucleus-nucleus [6,8] collisions. The average transverse momenta of produced negative pions, $\langle p_T^{\pi} \rangle$, are plotted as a function of the initial projectile momentum. A strong increase of the average transverse momentum measured in nucleon-nucleon collisions is seen in the range from 1 to 10 GeV/c (the energy threshold effects), whereas for the higher energies the rise is logarithmic, which is equivalent to the well-known fact of the limited transverse momenta of particles produced in the high energy hadron collisions. Owing to the rather drastic increase of $\langle p_T^{\pi} \rangle$ in the few-GeV energy range, more can be learnt about the production mechanism by studying the average p_T (≡ temperature) dependence on the energy and centrality. The centrality dependence of the $\langle p_T^{\pi} \rangle$ was extensively studied by the Dubna track detectors [6,8] by colliding many different pairs of nuclei and using different triggers, namely: 'minimum bias trigger' (all inelastic events); 'semi-central trigger' (charged forward-going fragments suppressed); 'central trigger' (charged and neutral forward-going fragments suppressed) [9]. As an example, some measurements are displayed in Fig. 1. No rise of the $\langle p_T^{\pi} \rangle$ with the increasing centrality and size of colliding nuclei is seen within the experimental errors, contrary to what is expected within the simple thermodynamical models. The quantitative agreement with the data is achieved within a multiple scattering model [23] which uses an extension of the Glauber model and the theoretical ideas coming from the particle-nucleus interactions. As seen in Fig. 1, the $\langle p_T^{\pi} \rangle$ measured for the different pairs of the colliding nuclei is slightely lower than that extrapolated from the nucleon-nucleon data.

At Berkeley, the slope factor E_0, when inclusive cross-sections of the produced particles are parametrized by an exponential formula $\exp(-E_K^*/E_0)$ (E_K^* is a kinetic energy in the c.m. frame), is usually taken as a measure of the temperature. In Fig. 2 the energy dependence of the temperature of pions and protons at $\Theta_{cm} = 90^o$ is plotted for collisions of ^{20}Ne-^{23}Na^{19}F at E_K = 0.4, 0.8, and 2.1 GeV/n (p = 0.95, 1.46, 2.89 GeV/c per nucleon) [24]. The shape of the energy dependence of the temperature is quite similar to the one from Fig. 1, but a direct comparison is impossible. Nevertheless, using the model [19], it is possible to relate T and $\langle p_T^{\pi} \rangle$, and the result obtained in such a way[1] is plotted in Fig. 1. The Berkeley data-points follow rather closely the energy dependence of $\langle p_T^{\pi} \rangle$ measured in the nucleon-nucleon interactions.

The temperatures and the average transverse momenta of pions were obtained for the αα, αp, and pp scatterings at the ISR energies [12] by the fitting procedure using Hagedorn's prescription [20]. The results are listed below and are plotted in Fig. 1:

	$\sqrt{s}$ [GeV]	$\langle p_T^h \rangle$ [MeV/c]	T [MeV]
αα:	31.2	363 ± 20	145 ± 8
αp:	44	357 ± 20	143 ± 8
pp:	30.4	364 ± 19	147 ± 7
pp:	44	366 ± 20	145 ± 7

---

[1] The temperature values T evaluated from the $\langle p_T^{\pi} \rangle$ (or vice versa) are only approximately correct owing to the ambiguities in the calculation connected to the inclusion of pions from the Δ decays and other effects [19].

As one can see, the temperatures or average transverse momenta of the negative hadrons produced in αα and αp collisions are similar to those measured in the pp collisions at the corresponding c.m. energies.

S: The multiplicity and rapidity results from nucleus-nucleus collisions at low (p = 4.5 GeV/c per nucleon) and high (p = 516 GeV/c per nucleon) energies, are in good agreement with predictions of the theoretical models, which assume additivity of the independent parton-parton (nucleon-nucleon) interactions.

No increase of the temperature ($\langle p_T^{\pi}\rangle$) is seen (within experimental errors) with increasing centrality and size of the colliding nuclei. The energy dependence of $\langle p_T^{\pi}\rangle$ is close to the one observed for nucleon-nucleon collisions.

E: The simple extrapolation from the existing data to the case of ^{16}O colliding with nuclear targets at 200 GeV/n gives the following temperature and average p_T of produced pions: $\langle p_T^{\pi}\rangle \approx 335$ MeV/c , $T \approx 140$ MeV.

D: There is no systematic study of the temperature of the produced pions using different colliding ion pairs and the central collision trigger at Berkeley energies.

3. STRANGENESS PRODUCTION AND TEMPERATURE

3.1 STRANGENESS YIELDS

There is a common belief that the abundant strangeness production will serve as evidence of the phase transition to the quark-gluon plasma.

Two experiments using the streamer chambers [25-27] and one using the bubble chamber [28] measured strange-particle yields by identifying the neutral strange decays (K_S^0 and Λ). At Berkeley the semi-central collisions of ^{40}Ar with $^{39}K^{33}Cl$ as a target were investigated [25], and the following value of the ratio of the average number of produced Λ's, $\langle n^{\Lambda}\rangle$, to the average number of produced negative pions, $\langle n^{\pi}\rangle$, was found: $\langle n^{\Lambda}\rangle/\langle n^{\pi}\rangle = 0.04/6 \approx 0.7 \times 10^{-2}$. The topology for Λ-producing events was similar to that for the rest of the events. The Dubna experiments observe a strong increase of the strange particle yield when, say, the 'central' ^{16}O-^{207}Pb interactions are compared with 'inelastic' ^{4}He-^{6}Li, but a strong increase of the pion production is also observed and the relative ratio stays constant, as one can see in Fig. 3 [27]. The amount of produced K_S^0's and Λ's is proportional to the number of nucleons participating in the collision [27]. Taking into account a rather large energy difference (threshold effects) between the Berkeley and Dubna experiments, the $\langle n^{\Lambda}\rangle/\langle n_-\rangle$ ratios measured at Dubna and Berkeley are astoundingly close.

The only available information at higher energies comes from the R807 experiment at the CERN ISR [29]. The multiplicity dependence of the ratio of strange charged kaons to the charged pions is presented in Fig. 4 for αα, αp, and pp collisions at the corresponding c.m. energies. The ratio does not depend on the multiplicity (centrality).

3.2 TEMPERATURE

A compilation of the Λ's average transverse momenta measured in high-energy nucleus-nucleus collisions is presented in Fig. 5. The data come from two Dubna experiments performed at p = 4.2 GeV/c per nucleon [28] and p = 4.5 GeV/c per nucleon [26,27], and the Berkeley experiment[2] at p = 2.89 GeV/c per nucleon [25]. In order to obtain statistics, some data are summed, as indicated in Fig. 5. It is worth noticing that the inelastic ^{4}He-^{6}Li interactions can be treated, to a first approximation, as the ideal average nucleon-nucleon collisions, owing to the smallness of the colliding objects and inclusion of the averaging over the Fermi momentum. No significant dependence of $\langle p_T^{\pi}\rangle$ on the size of the colliding nuclei and centrality is seen; nevertheless, the $\langle p_T^{\Lambda}\rangle$ for He-Li inelastic collisions is the smallest.

In Fig. 6 the compilation of the proton-proton [30] and nucleus-nucleus [25,27,28] data on the K_S° and Λ average transverse momenta is shown in a way that is similar to that for the pions in Fig. 1. The solid line recalls the behaviour of the π data. It can be seen from Fig. 6 that the $\langle p_T\rangle$ of K_S°'s are smaller than those for Λ's. The energy dependence of $\langle p_T^{K_S^{\circ},\Lambda}\rangle_{pp}$ seems to be the same as for $\langle p_T^{\pi}\rangle_{pp}$. Unfortunately, all the nuclear collision data have very large experimental errors, so no final conclusions are possible. Still, the $\langle p_T^{\Lambda}\rangle$ for the centrally colliding symmetric nuclei (^{12}C-^{12}C, ^{16}O-^{20}Ne, and ^{20}Ne-^{20}Ne) is noticeably larger. This particular sample was investigated in more detail by the streamer chamber group [31], and the following very promising observations were made. In a certain fraction of the investigated sample, the Λ's were produced outside the kinematical limits. The authors called this sub-sample 'Λ^{out}'. Whereas the temperature of produced Λ's in the inelastic He-Li ('average nucleon-nucleon' collisions) is around 107 ± 20 MeV, for the 'Λ^{out}' sample it was 150 ± 19 MeV. Also, the pion temperature was much larger for those events, T^{π}('Λ^{out}') = 114 ± 11 MeV, whereas for all other central or non-central collisions it was around 95 ± 5 MeV. Moreover, pions were produced isotropically in the nucleon-nucleon c.m. frame (flat distribution in $\cos\Theta^*$) in the 'Λ^{out}' sample, whilst for the other samples they were forward-backward peaked. The selection of the 'Λ^{out}' sample corresponds to a cross-section $\sigma = 10^{-3}\times\sigma_{inel}$, therefore the phenomenon could not be observed in the inclusive measurements. Assuming that the data are not affected by statistical fluctuations, these observations can be interpreted as evidence that the Λ's and π's are emitted from a single hot source, perhaps already from the other phase.

S: There is no evidence of copious strange-particle production in high-energy nucleus-nucleus interactions; no dependence on centrality and size of colliding nuclei is observed. The temperatures of strange particles produced in central nucleus-nucleus collisions are higher than the ones observed in the average nucleon-nucleon collisions.

---

[2] The error of the $\langle p_T^{\pi}\rangle$ was not given in the original publication, but must be slightly larger than the one of the Dubna data because of smaller statistics.

The recent Dubna streamer chamber observation [31] shows that in the selected sub-sample of central collision events of symmetric nuclei, the Λ's and π's are produced from a single hot spot.

E: The extrapolation from existing data gives the following predictions for T^{Λ} and $\langle p_T^{\Lambda}\rangle$ to be observed in the central collision events at 200 GeV/n: $T^{\Lambda} \approx 150$ MeV and $\langle p_T^{\Lambda}\rangle \approx 600$ MeV/c. If collisions of the symmetric nuclei (^{16}O-^{16}O) were much better sources of the 'hot' Λ's, the largest experimental problem -- how to deal with high multiplicity -- will be less serious at the SPS experiments.

D: The main drawback of the experimental study discussed in this section is the very low statistics. The small statistics is a result of the fact that no dedicated experiment measuring only strange particles has yet been done.

4. PION DENSITIES

Most of the predicted exotic phenomena in the high-energy nuclear collisions is strictly correlated with the high densities of the hadronic matter. At the energy $E_K = 3.66$ GeV ($p = 4.5$ GeV/c per nucleon) the average multiplicity of pions produced in nucleon-nucleon collisions is, $\langle n^{\pi}\rangle < 1$, whereas at $p = 200$ GeV/c, $\langle n^{\pi}\rangle \approx 8.5$, so the situation is completely different in these two energy regions. At the SPS energies the pions will contribute to the bulk of the particle density.

The αα experiment at the ISR gives information about the pion densities at even higher energies [32]. Figure 7 gives the multiplicity dependence of semi-inclusive rapidity distributions, for secondaries emitted into different rapidity regions in the αα and pp collisions [12]. In the central rapidity region (Fig. 7a,d) we observe a linear growth of the rapidity density with multiplicity for both negative and positive secondaries. It is interesting to note that αα data obey the same linear relation as that obeyed by the pp data, but obviously extended to much larger multiplicities. In the intermediate region of rapidity $1 < y < 3$ (Fig. 7b) the particle density in αα and pp collisions still increases approximately linearly with the multiplicity. The growth is faster in the case of the αα scattering than in the pp case. The same effect is more clearly manifested for the fragmentation region $3 < y < 6$ (Fig. 7c), where almost no increase of the particle density is seen in the pp case. Comparing the particle densities per rapidity unit in the average case [12,18] and for the largest multiplicity sample in the central rapidity region (Fig. 7a), it can be seen that the particle density increases by a factor of ≈ 5. Adding up positive and negative secondaries in Fig. 7 at $n_{ch} = 50$, we obtain $dn/dy \approx 15$ charged particles produced per rapidity unit. On top of this, fluctuations of the particle densities in an individual event can occur. Figure 8 shows such a fluctuation. The event was chosen out of 30,000 events as one of a few largest multiplicity events. In this particular event, 26 particles produced in one rapidity unit are seen. So the particle density can be quite large within one event because of the possible event-by-event fluctuations.

S: The highest pion densities are obtained for highest-multiplicity events and in the central rapidity region. This suggests that the central rapidity region may well be the most promising, when looking for exotic phenomena connected with the high local energy or hadron density.

E: Although, it is rather difficult to make any predictions about the pion densities for the ^{16}O-^{207}Pb collisions by extrapolating from the αα data, some indirect ways do exist. The multi chain dual-parton model introduced by the Orsay group [33] has been able to describe the multiparticle production in hadron-nucleus and recent αα data with considerable success [16]. Therefore, it is relatively safe to use this model to predict pion densities at the new SPS experiments. The model predicts that even collisions of oxygen with heavy targets will produce an energy deposition slightly smaller than 1.5 GeV/fm^3 [34], the magic number believed to be a threshold density between the hadronic and the quark-gluon phase. The only hope, as the authors claim [34], is in the selection of very central events, by increasing the veto angle for the forward going projectile fragments.

D: Within the present experimental configuration it was impossible to set up the central collision trigger at the ISR. The subject will be discussed in more detail in Section 7.

5. BARYON DENSITIES

The baryon density achievable in central collisions of heavy nuclei at the high energies is a topic of current interest. The reason is twofold: firstly, the baryon density can contribute, for a large part, to the total hadron density when two large nuclei collide; secondly, the only way the baryons are stopped is via emission of the pions and other particles. Therefore, the stopping of nucleons is an equivalence of the abundant particle production. The stopping power or nuclear transparency is really a very important problem, because it is the 'to be or not to be' of high nuclear densities and all subsequent effects [35]. Recently, Busza and Goldhaber have interpreted hadron-nucleus data claiming stopping of the protons traversing the lead nucleus [36]. There is also evidence for isotropic flow of the energy and particles in ^{93}Nb-^{93}Nb high-multiplicity collisions at 0.4 GeV/n kinetic energy [37]. However, the incident energy is very small and the cascade development is well justified even in the forward region [35], which is certainly not the case at 200 GeV/n energy.

Therefore it is interesting to look for the baryon densities at the high-energy nucleus-nucleus collision. The only candidate is again the αα ISR data. Unfortunately, as a rule in the high-energy experiments, there is no particle identification in the full phase space, so even the proton rapidity distributions cannot be extracted from the data in a direct way. But, in the case of isoscalar beam and target collisions (αα or dd scattering), fortunately it is possible to extract the proton rapidity spectra by playing a trick [12,32], which is the following: for both negative and positive secondaries the proton mass is assigned. Since the rapidity spectra for negative and positive pions are the same -- as there is the same amount of u and d quarks in the initial state -- their contributions cancel out in the difference. The influence on the result of the difference in positive and negative K production and of heavier fragments (registered with very small efficiency) is neglected. The result of such a procedure is shown in Fig. 9 for three different multiplicity bins. In the smallest-multiplicity bin (most peripheral collisions) it can be seen that the proton rapidity distribution is peaked at 3.5, exactly coinciding with the beam rapidity, and that it falls rapidly when the central rapidity region is approached. Large errors at central rapidity are the result of subtracting two large numbers. The mid-rapidity gap is clearly visible. When more central events (larger-multiplicity bin) are selected a shift of the proton peak towards the central rapidity region is seen. We can assume that the neutron density is the same as the proton one in the $\alpha\alpha$ collision. Therefore we can risk claiming that baryons can be stopped by nuclear matter when two large nuclei collide.

However, a warning seems to be necessary. At least part of the effect can be understood within a very simple phenomenological picture, namely: in the $\alpha\alpha$ collision with a large centrality, many nucleons have to interact with large inelasticity, so on the average the particle production is enhanced. Thus triggering on the high multiplicity, events can be selected, in which nucleons have undergone collisions with a large inelasticity and therefore have shifted towards the smaller rapidities. The disentangling of how much of the stopping of the nucleons is due to the secondary nucleon-nucleon collisions is the real non-trivial point. A comparison of the $\alpha\alpha$ and dd data can possibly shed light on the problem, and such a study will be undertaken by the R418 ISR Collaboration using existing experimental data.

S: The $\alpha\alpha$ data demonstrate that baryons can be slowed down by nuclear matter.

E: The baryon mid-rapidity gap can presumably be filled up, at least partially, in the oxygen collisions with large centrality, even at 200 GeV/n.

D: The real central collision trigger was not used in the ISR $\alpha\alpha$ experiments.

6. PARTICLE PRODUCTION WITH LARGE TRANSVERSE MOMENTA

At this conference there are a few papers [38] on particle production with large transverse momenta (or energy flow), presenting new interesting results from the $\alpha\alpha$ study at the ISR [39,40]. Therefore, this section will be limited to just a short remark.

Figure 10 brings to mind the well-known results obtained at the ISR, demonstrating 'Abnormal Nuclear Enhancement' (ANE) (discovered previously at Fermilab [41]) observed in the $\alpha\alpha$ collisions [42,17]. The lines plotted in Fig. 10 correspond to the theoretical predictions based on the multiple parton scattering mechanism [43-45]. As can be seen in this figure, only the model assuming multiple scattering and large Fermi momenta [44] gives satisfactory agreement with the data. The large Fermi momenta (p_F > 500 MeV) may appear from a non-trivial source. In fact, usually decays of the N^* or Δ resonances, or the existence of such exotic objects as multiquark states, enhance the large Fermi momentum tail [46].

It may well be that the study of the particle production with large transverse momenta is the best tool for searching for new physics phenomena in nucleus-nucleus collisions at high energies [35].

S: The ANE was observed in the h-nucleus and in the αα interactions.

E: The cross-section for the particle production with large p_T will be amplified in the case of oxygen collisions, hence it grows with A_P and A_T at least as:

$$\sigma_{A_P A_T} / \sigma_{nn} \approx A_P \times A_T = 16 \times 207 = 3312.$$

Presumably $\sigma_{A_P A_T}$ will be much higher because of the ANE, but this is still to be measured. Much better experimental conditions for tagging projectile fragments exist at fixed-target experiments.

D: In spite of the fact that the ANE was observed for both the h-nucleus and the αα experiments, no single experiment using heavier colliding nuclei was done at Berkeley or Dubna.

7. CENTRAL COLLISION TRIGGER

It is fairly clear that the highest hadron densities are achieved in events with the highest centrality. So events in which as many nucleons as possible deposit as much energy as possible should be selected. The way to do this is by using a high-resolution forward hadron calorimeter and, eventually, a tracking device for charged fragments as the main components of the central collision trigger.

The typical experimental configuration (fixed-target experiment) is shown in Fig. 11. The ^{4}He beam was chosen to simplify the picture. The projectile nucleus fragments in a target, and the nuclear fragments (except neutrons) are bent differently because of the different rigidities. The circles plotted in the plane placed perpendicularly to the beam direction 5 m away from a target, correspond to the transverse Fermi momentum of 100 MeV/c in the ^{4}He rest system. The detector is placed in a typical 1.5 T magnetic field. The distance d between centres of the outgoing neutrons and protons will change drastically with the incoming beam momentum, as indicated in Fig. 11. This sets rather serious constraints on the calorimeter device.

Whilst a central collision trigger is rather easy to build at fixed-target experiments, it is still only 'half a central trigger', because the information about the target nucleus is lost. At the colliders the geometry of the colliding nuclei is symmetric, therefore a central collision trigger can be applied by vetoing nuclear fragments of both colliding nuclei. Unfortunately, it was impossible to apply such a trigger at the ISR, because it requires a special intersection region, described below. An attempt to veto the events with fast nuclear fragments registered at the forward regions of the detector, led to the selection of peripheral events with fragments lost in the beam tube.

At the moment, the construction of a few heavy-ion colliders is proposed [47]. The new colliders should have very special intersection regions (at least some of them) suitable for a heavy-ion study. A simplified sketch of the intersection layout is presented in Fig. 12. The outgoing vacuum pipes should have a sharp perpendicular wall at the edge to minimize the amount of the material to be traversed by the nuclear fragments. The prolongation of the vacuum tube should have a diameter as small as possible (limited only by current acceleration technology).

All stripped nucleons, even those having very small perpendicular Fermi momentum, should traverse the wall and be registered in the calorimeter device surrounding the beam pipe in order to tag nuclear fragments.

Figure 9, which demonstrates the stopping power in the $\alpha\alpha$ collisions, also shows that even for the largest multiplicity bin the protons are peaked at the beam rapidity. This shows that the off-line multiplicity cut is only a poor-man's central collision trigger, and the chance to investigate the real central $\alpha\alpha$ collisions at the ISR is lost for ever.

D: No real central collision trigger (dedicated calorimeter device) was used up to now in the nucleus-nucleus study[3] at any energy.

8. CONCLUSIONS

Even the present brief review shows that a lot of study has been done experimentally in the field of the high-energy nucleus-nucleus collisions, and a great deal of the phenomena occurring when two nuclei collide are now understood. So far, no evident features or signals of the phase transition have been observed.

But a lot of experimental information is still missing. The best examples of the 'should be done' experiments in the few-GeV energy range are perhaps: a high statistics (preferably electronic) experiment measuring, in an exclusive way, strange particle production; and a similar experiment measuring particle production with large transverse momenta -- both using a real central collision trigger.

As far as high energies are concerned, we will have to wait until the appropriate beams are available. When preparing the ion adventure at the SPS we shall keep in mind that the ISR missed the charm and beauty not because of the lack of energy or proper beams but rather because of the lack of proper equipment at the right time.

Acknowledgements

I would like to thank all my colleagues in the ISR $\alpha\alpha$ adventure -- in particular M.A. Faessler, H.G. Fischer, S. Frankel, W. Geist, and I. Otterlund -- for enlightening discussions and close collaboration.

It gives me a great pleasure to express my appreciation to H. Bialkowska, M. Gazdzicki, E. Skrzypczak, A. Sandoval and R. Stock, who provided me with much of the indispensable information on the current high-energy nucleus-nucleus experiments.

The many discussions with B. Badelek, A. Capella, W.Q. Chao, L. Lesniak, H.J. Pirner, G. Wilk, and specially with R. Hagedorn on the theoretical aspects of the high-energy nucleus-nucleus collisions, are acknowledged with gratitude.

---

[3] The only device that I am aware of was the 'neutron veto counter' used by the streamer chamber group at Dubna [48]. Unfortunately, it did not have any granularity or energy measurement capability.

I would also like to thank the Dubna SKM-200 Collaboration and the CERN ISR R807 and R418 Collaborations for providing me with their unpublished results.

The comments of M.A. Faessler, K. Rybicki, and J. Zakrzewski and their careful reading of this manuscript were of very great help.

References

[1] R. Stock, 'Quark matter searches using ^{16}O beams at the CERN SPS', talk given at this conference.
[2] H. Haseroth, report CERN/PS/LI/82-12 (1982), and in Proc. Workshop on SPS Fixed-Target Physics in the Years 1984-1989, Geneva, 1982 (CERN 83-02, Geneva, 1983), Vol.II, p. 443.
[3] See articles by J. Rafelski, H. Satz, J. Kapusta and L. McLerran, in Phys. Rep. 88 (1982) 321.
[4] M. Jacob, CERN report Ref.TH.3728 (1983), and in Proc. Third Int. Conf. on Ultra-Relativistic Nucleus-Nucleus Collisions (Quark Matter '83), BNL Upton, 1983 (Nord Holland, Amsterdam, 1984), p. 7c.
[5] V.D. Aksinenko et al., Nucl. Phys. A324 (1979) 266.
V.D. Aksinenko et al., Nucl. Phys. A348 (1980) 518.
M.Kh. Anikina et al., Z. Phys. C9 (1981) 105.
[6] A.Kh. Abdurakhimov et al., Nucl. Phys. A362 (1981) 376.
[7] E.A. Abdrakhmanov et al., Z. Phys. C5 (1980) 1.
H.N. Agakishev et al., Z. Phys. C16 (1983) 307.
[8] H.H. Agakishev et al., Z. Phys. C12 (1982) 283.
[9] R. Szwed, Warsaw Univ. report IFD/2/81 (1981), and in Proc. 5th High Energy Heavy Ion Study, Berkeley, 1981 (LBL-12652, Berkeley, 1981), p. 371.
[10] T. Akesson et al., Phys. Lett. 119B (1982) 464.
[11] W. Bell et al., Phys. Lett. 128B (1983) 349.
[12] W. Bell et al., 'Charged particle spectra in $\alpha\alpha$ and αp collisions at the CERN ISR', CERN ISR R418 experiment, paper in preparation.
[13] W.Q. Chao and H.J. Pirner, Z. Phys. C14 (1982) 165.
[14] S.J. Brodsky et al., Phys. Rev. Lett. 39 (1977) 1120.
[15] A. Bialas et al., Z. Phys. C13 (1982) 147.
L. Lesniak, private communications.
[16] A. Capella et al., Phys. Lett. 108B (1982) 347.
A. Capella, private communication.
[17] M.A. Faessler, 'Experiments with alpha particles at the CERN Intersecting Storage Rings', preprint CERN-EP/84-64 (1984), to be published in Phys. Reports.
[18] R. Szwed, in Proc. Int. Conf. on High Energy Nuclear Physics, Balatonfured, 1983 (KFKI, Budapest, 1983), p. 197.
[19] R. Hagedorn and J. Rafelski, Phys. Lett. 97B (1980) 136.
[20] R. Hagedorn, CERN preprint Ref.TH.3684 (1983).
[21] A. Abdivaliev et al., JINR report BI-1-122181 (1979).
[22] A.M. Rossi et al., Nucl. Phys. B84 (1975) 269.

H. Boggild et al., Nucl. Phys. B57 (1973) 77.
J.W. Chapman et al., Phys. Lett. 47B (1973) 465.
T. Kafka et al., Phys. Rev. D16 (1977) 1261.

[23] C. Pajeras and A.V. Ramallo, Phys. Lett. 120B (1983) 441.

[24] S. Nagamiya et al., Phys. Rev. C24 (1981) 971.

[25] J.W. Harris et al., Phys. Rev. Lett. 47 (1981) 229.

[26] M. Anikina et al., Phys. Rev. Lett. 50 (1983) 1971.
M. Anikina et al., in Proc. Int. Conf. on High Energy Nuclear Physics, Balatonfured, 1983 (KFKI, Budapest, 1983), p. 433.

[27] M. Anikina et al., 'Characteristics of Λ and K^0 particles produced in central nucleus-nucleus collisions at a 4.5 GeV/c momentum per incident nucleon', to be published in Z. Phys. C.

[28] N. Akhababian et al., JINR report D1-82-455 (1982).
S. Arakelian et al., JINR report P1-83-354 (1983).

[29] S. Frankel, private communication, CERN ISR R807 experiment, paper in preparation.

[30] K. Jaeger et al., Phys. Rev. D11 (1975) 1756.
P. Aahlin et al., Phys. Scripta 21 (1980) 12.
H. Boggild et al., Nucl. Phys. B57 (1973) 77.
V.V. Ammosov et al., Nucl. Phys. B115 (1976) 269.
J.W. Chapman et al., Phys. Lett. 47B (1973) 465.
K. Jeager et al., Phys. Rev. D11 (1975) 2405.
F. LoPinto et al., Phys. Rev. D22 (1980) 573.
A. Sheng et al., Phys. Rev. D11 (1975) 1733.

[31] M. Anikina et al., JINR report E1-84-376 (1984).

[32] R. Szwed, 'Charged particle production in αα, αp, dd, and pp collisions at the CERN ISR', ISR R418 and R419 experiments, talk given at the XVth Symp. on Multiparticle Dynamics, Lund, 1984.

[33] J. Tran Thanh Van, 'New results on the dual parton model', talk given at the XVth Symp. on Multiparticle Dynamics, Lund, 1984.

[34] A. Capella et al., CERN preprint Ref.TH.3700 (1983).

[35] R. Szwed, report CERN-EP/83-50 (1983) and in Proc. Int. Workshop on Gross Properties of Nuclei and Nuclear Excitations XI, Hirschegg, 1983 (GSI, Darmstadt, 1983), p. 119.

[36] W. Busza and A.S. Goldhaber, Phys. Lett. 139B (1984) 235.

[37] H.H. Gutbrod et al., in Proc. Int. Conf. on High Energy Nuclear Physics, Balatonfured, 1983 (KFKI, Budapest, 1983), p. 269.
H. Pugh, 'Equilibration, compression and flow at the Bevalac', talk given at this conference.

[38] See talks given by B. Callen ('Measurement of E_T flow for pp, dd and αα interactions') and M. Tannenbaum ('E_T and π^0 spectra from αα, dd, pp at $\sqrt{s}$ = 31 GeV') at this conference.

[39] A.L.S. Angelis et al., Phys. Lett. 141B (1984) 140.

[40] A. Breakstone et al., CERN ISR R418 and R419 experiments, paper in preparation.

[41] J.W. Cronin et al., Phys. Rev. D11 (1975) 3105.

L. Kluberg et al., Phys. Rev. Lett. 38 (1977) 670.

[42] A. Karabarbounis et al., Phys. Lett. 104B (1981) 75.
W. Bell et al., Phys. Lett. 112B (1982) 271.
A.L.S. Angelis et al., Phys. Lett. 116B (1982) 379.
T. Akesson et al., Nucl. Phys. B209 (1982) 309.

[43] U.P. Sukhatme and G.Wilk, Phys. Rev. D25 (1982) 1978.

[44] M. Staszel and G. Wilk, Z. Phys. C19 (1983) 57.

[45] M. Lev and B. Petersson, Z. Phys. C21 (1983) 155.

[46] H.J. Weber, Phys. Lett. 80B (1978) 232.
G. Horlacher and H. Arenhovel, Nucl. Phys. A300 (1978) 348.

[47] See talks given by T. Ludlam ('Heavy ion collider') and H. Pugh ('New scheme for a heavy ion collider at LBL') at this conference.

[48] M. Gazdzicki et al., JINR report E1-84-444 (1984).

Figure captions

Fig. 1: Compilation of the data on the dependence of the average transverse momenta of negative pions produced in nucleon-nucleon [12,21,22] and nucleus-nucleus [6,8] interactions on the incoming beam momentum. The solid line is plotted to guide the eye.

Fig. 2: The energy dependence of the slope factor E_0, fitted to the proton and pion spectra at $\Theta_{cm} = 90^o$, in Ne-NaF collisions. The cross-sections were parametrized by $\exp(-E_K^*/E_0)$, where E_K^* is kinetic energy in the c.m. frame.

Fig. 3: Ratios of the average number of K_S^o's (top figure) and Λ's (bottom figure) to the average number of negative secondaries produced in the nucleus-nucleus collisions [26-28]. In order to gain statistics, some data having close values of the product of the $A_P \times A_T$ of the mass numbers, were added up and are represented by a single data point.

Fig. 4: The dependence of the ratio of charged kaons to the charged pions measured for the $\alpha\alpha$, αp, and pp interactions by the ISR R807 experiment [29] on the charged multiplicity m; $\langle m \rangle$ denotes mean charged multiplicity.

Fig. 5: Average transverse momentum of Λ's produced in nucleus-nucleus collisions [25,27,28] versus the mass number product of the colliding nuclei.

Fig. 6: Compilation of the data on the dependence of the average transverse momenta of the K_S^o's and Λ's produced in the proton-proton [30] and nucleus-nucleus collisions [25,27,28] on the incoming beam momentum. The solid line corresponds to the pion data presented in the Fig. 1.

Fig. 7: Multiplicity dependence of rapidity density for three different rapidity regions: a) $|y| < 1$; b) $2 < |y| < 3$; c) $4 < |y| < 6$; in $\alpha\alpha$ and pp interactions [12].

Fig. 8: 'Spike' rapidity event registered in $\alpha\alpha$ interactions [32].

Fig. 9: Net proton rapidity distributions for $\alpha\alpha$ collisions for three different multiplicity bins [32].

Fig. 10: Ratio of the negative hadron and π^0 yields as a function of p_T. Compilation of the ISR $\alpha\alpha$ data [42]. The curves represent predictions of the multiple-parton scattering models [43-45].

Fig. 11: Schematic representation of the triggering on central collisions at a fixed-target nucleus-nucleus experiment. For more explanation, see text.

Fig. 12: Schematic representation of the triggering on central collisions in a collider nucleus-nucleus experiment.

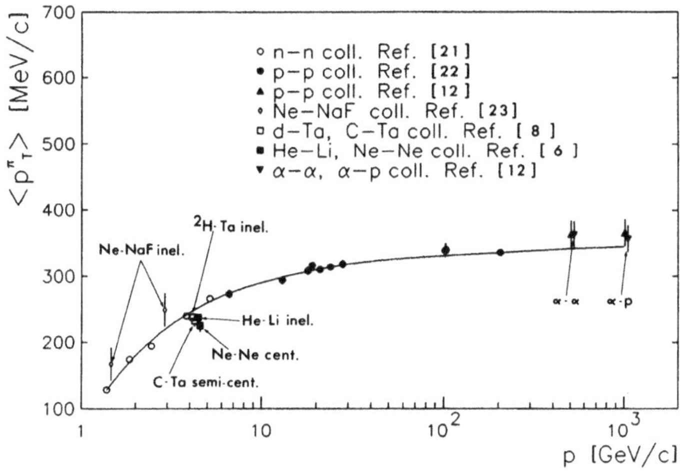

Fig. 1

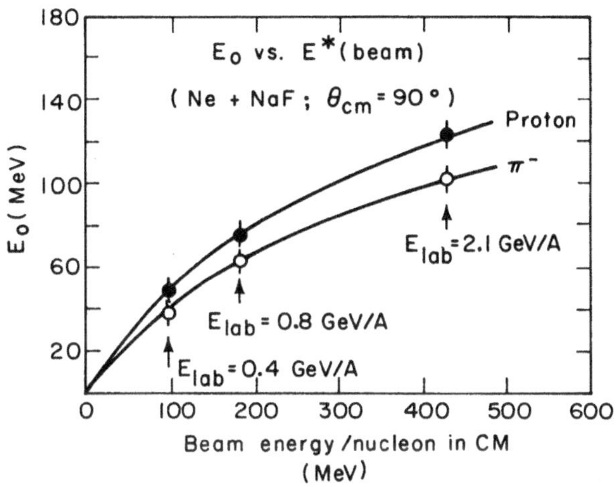

Fig. 2

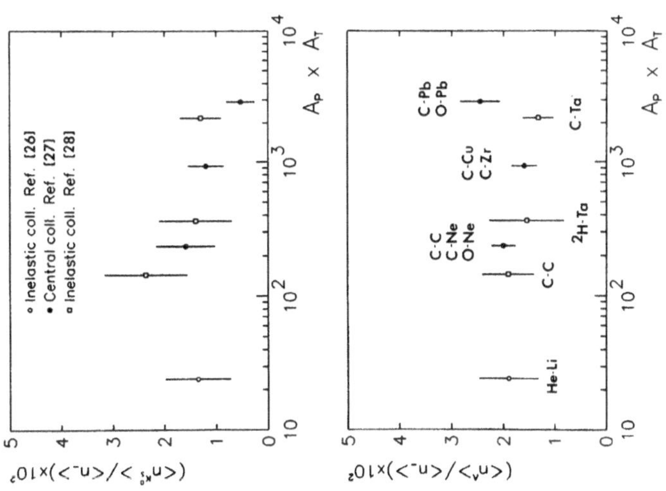

Fig. 3

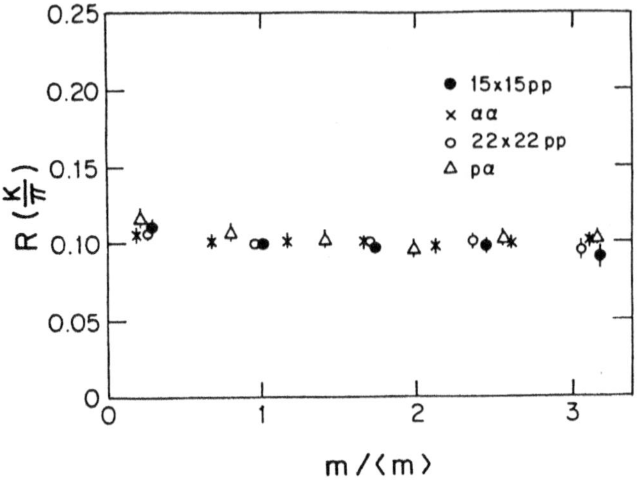

Fig. 4

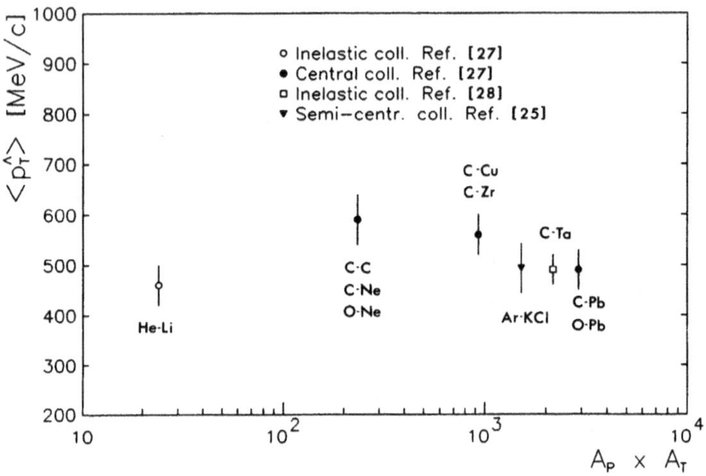

Fig. 5

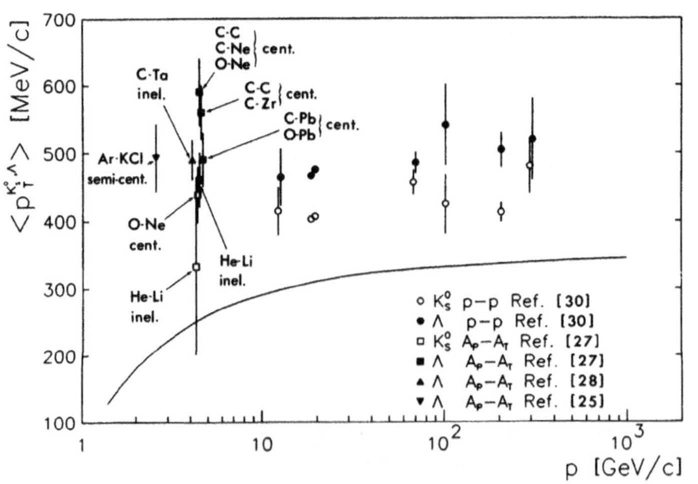

Fig. 6

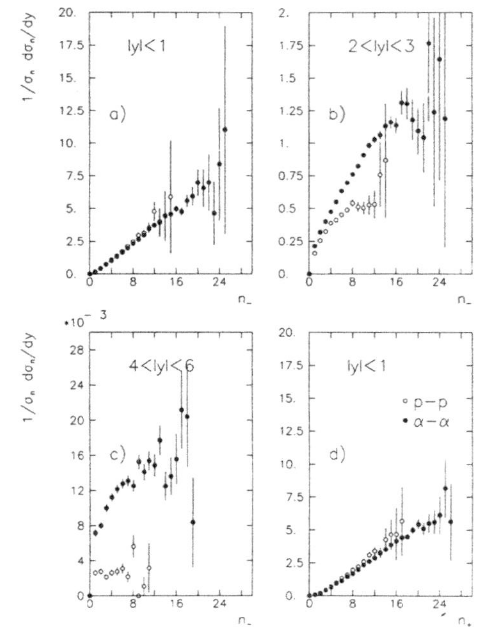

Fig. 7

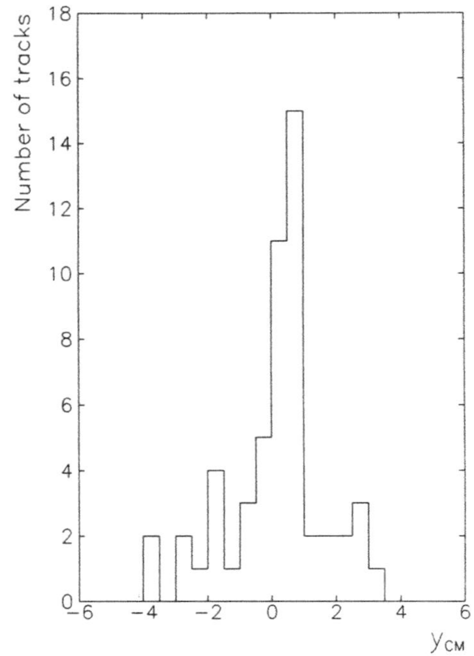

Fig. 8

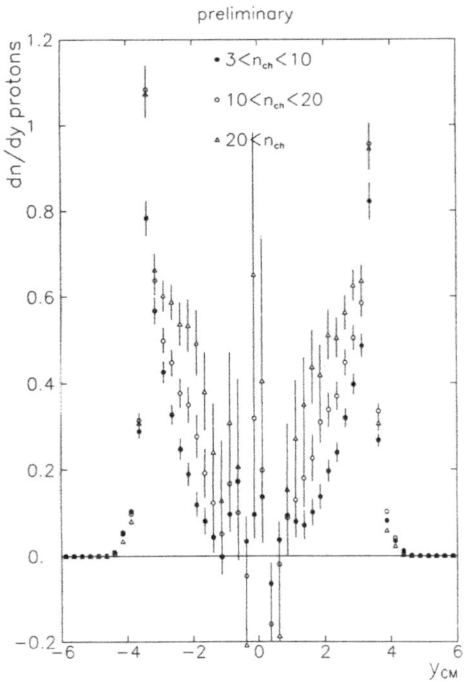

Fig. 9

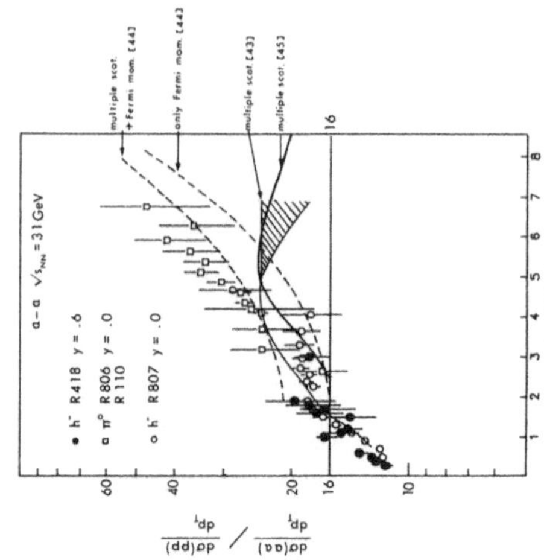

Fig. 10

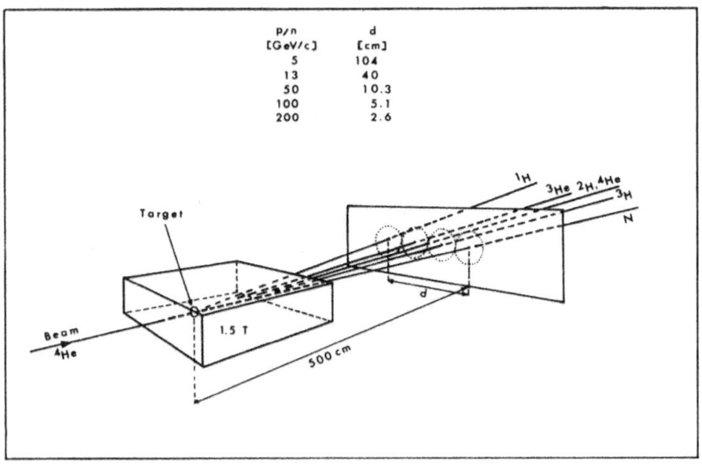

Fig. 11

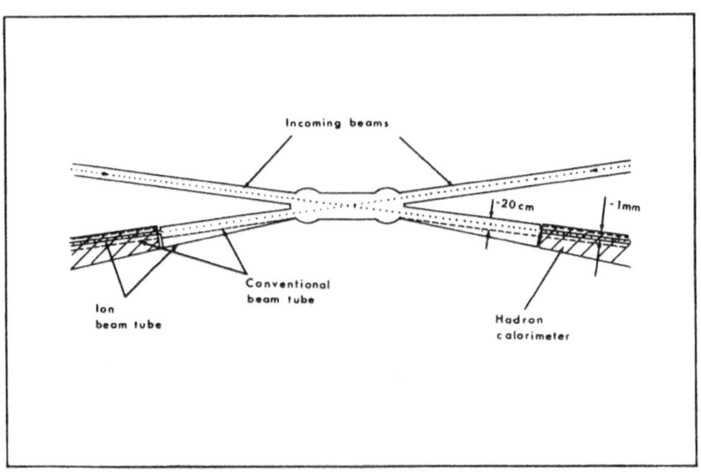

Fig. 12

E_t Spectra for $\sqrt{s} = 31$ A-GeV Light Ion Collisions at the CERN-ISR: Data and Analysis

Bruce Callen

University of Pennsylvania, Philadelphia, PA 19104 USA

Preliminary transverse energy (E_t) cross-sections for p–α, α–α, and d–d collisions measured with the Axial Field Spectrometer hadronic calorimeter are presented. The spectra from these three light ion systems are compared with p–p spectra at the same $\sqrt{s}$, using the Wounded Nucleon Model. We find that the data are well described by this 'participant' model. Although the E_t spectra approach or exceed the p–p kinematic limit for all three systems, no evidence for any anomalous behavior is found. Circularity studies show no signs of significant jet production in any of the spectra, even at the highest values of E_t. We also apply our model to neutral energy data from the BCMOR collaboration at the ISR. Finally, we discuss modifications to the Wounded Nucleon Model.

1. Introduction

Understanding relativistic nucleus-nucleus collisions from first principles has proven so far to be a nearly impossible task. Quantum chromodynamics, which we believe to be a legitimate theory of the strong interaction, is unable to tell us anything definite about low q^2 interactions. Since high-energy nuclear collisions are dominated by such soft processes, we cannot at present understand them at the quark/gluon level. However, several models of nuclear collisions attempt to bypass the problem by explaining the nuclear reactions direct from the observed p–p data. Among these are the wounded nucleon model (WNM)[1,2,3], the dual parton model[4], and the gluon string model[5].

The first ISR light ion run in 1980 offered an excellent chance to compare the predictions of these models with experiment. The Axial Field Spectrometer collaboration took data on multiplicity and charged E_t distributions using the central drift chamber, and found the WNM to give a good fit to both sets of data[2,6]. Following the 1980 run, a full azimuth uranium calorimeter was installed at the AFS, making total energy measurements possible. We present here E_t spectra taken with the calorimeter for the three light ion beams stored during August 1983 (p–α, α–α, and d–d) and compare them with the p–p spectra, all taken at the same $\sqrt{s}$ of 31 GeV per nucleon-nucleon collision. These spectra are not yet corrected for calorimeter resolution. However, preliminary studies indicate that these effects will be small. Fig. 1a shows both the uncorrected and corrected p–p spectra. Based on this, we do not expect large differences in the relative shapes of the various spectra. The work presented in this paper is an extension of our previous efforts on p–α and α–α multiplicity and charged E_t distributions to these new data.

High E_t measurements in nucleus-nucleus collisions are of special interest for two reasons. First, they provide a laboratory in which to study both rapidity and charged/neutral fluctuations. Both experiments described here, AFS and BCMOR, cover a limited rapidity range ($|y| < 1$). Yet both the AFS and BCMOR data extend to very high energies. Because of this, rapidity and charged/neutral fluctuations are needed to account for the large energies seen in such a small rapidity range. Second, reaching high values of E_t offers the hope of observing the hard scattering processes that QCD predicts. Observing jets in a nuclear setting would be interesting by itself. However, detecting such processes at this relatively low $\sqrt{s}$ is known to be difficult even in p–p interactions; jets are much more clearly visible at a $\sqrt{s}$ of 63 GeV[7]. The difficulty of observing jets is compounded by the increased presence of soft particles from spectator interactions, which is enhanced in nucleus-nucleus collisions by at least the ratio of the mean multiplicities.

It has been suggested that the ratio of p-nucleus to p–p E_t cross-sections is 'anomalous', since it is greater than A[8,9]. This question has been studied with the FERMILAB p-A data[9] by Brody *et al.*[3], using two different models of low p_t particle production. Our new data, and the new data of the BCMOR collaboration, extend the testing ground for these models of soft particle production to even higher energies. In addition to the WNM calculations presented here, analysis using the gluon string model and the dual parton model is in progress.

2. Apparatus and Data Reduction

The data were taken with the AFS Hadron Calorimeter, a full azimuth calorimeter made of a U/Cu/Scintillator sandwich. The calorimeter and its trigger system, along with the full AFS, have been described in detail elsewhere[10,11]. The calorimeters are arranged in four walls of uranium/copper, with the corners filled in by four copper arrays. Each wall consists of a 16×12 array of towers, each with a separate electromagnetic and hadronic section. The copper arrays contain 16 towers. The rapidity coverage varies from $|y| \sim 0.6$ to $|y| \sim 1.2$, due to the box-like geometry of the calorimeter and center-of-mass motion arising from the non-zero beam-beam crossing angle. The uran um calorimeter has an energy resolution (σ/E) of 37% $/\sqrt{E}$ for hadronic showers and 16% $/\sqrt{E}$ for electromagnetic showers. The copper calorimeters have a resolution of 48% $/\sqrt{E}$ for hadronic showers.

The low energy data were taken with a 'minimum bias' trigger, which requires either a hit in one of 44 angular sections of a cylindrical scintillator hodoscope surrounding the interaction region, or a hit in each of two sets of 39 'beam-beam' counters surrounding each beam pipe downstream from the interaction region. The rest of the data were taken with an E_t trigger. The signal from each phototube dynode was weighted to correspond to E_t in the center-of-mass frame, and then summed over all calorimeter towers. The trigger also required at least 2 hits in the cylindrical hodoscope. Thus all of the E_t triggered events satisfy the minimum bias trigger as well. All four data sets were taken at a $\sqrt{s}$ of 31 GeV per nucleon-nucleon collision, with integrated luminosities for α–α, p–α, d–d, and p–p respectively, of 0.8, 0.8, 2.7, and 6.0×10^{34} cm^{-2}.

Events were processed through a reconstruction program designed to group adjacent towers into energy clusters, corresponding to individual showers. We then cut on cluster pseudorapidity to give uniform coverage, keeping only those clusters with $|\eta| < 0.7$. To reduce background due to uranium activity, only those clusters with an energy greater than 120 MeV were kept. The E_t for each event was then calculated from the transverse center-of-mass energy of all remaining clusters. A cut was made on the misbalance in p_t in the event in order to eliminate cosmic rays and beam-gas interactions, but did not throw out any significant fraction of real events. Finally, timing information from the calorimeters and the cylindrical hodoscope was used to correct for double events. The resulting spectra, uncorrected for calorimeter resolution effects, are shown in Fig. 1b.

3. Calculating with the Wounded Nucleon Model

We have taken the assumptions contained in the WNM and turned them into an algorithm for using p–p results to predict multiplicity distributions, E_t spectra, etc., for nucleus-nucleus collisions. The major assumptions of our calculations are:

- In p–p collisions, the two nucleons contribute *independent* amounts of transverse energy E_1 and E_2, with probability $f(E_1)f(E_2)\delta(E-E_1-E_2)$, to the total observed energy $E = E_1+E_2$

- **In nucleus-nucleus collisions, every struck nucleon independently contributes energy E_i to the total energy**

$$E = \sum_{\substack{struck \\ nucleons}} E_i.$$

The spectrum of energy contributed by a wounded nucleon is given by the same function $f(E_i)$ as in p–p collisions. Note that in the WNM the relevant parameter is the number of struck nucleons–NOT the number of nucleon-nucleon collisions.

For example, in a p–α collision where all four alpha nuclei are hit by the incident proton, the E_t will be the sum of five independent contributions, one from each wounded nucleon, as opposed to four independent p–p collisions. Using the first assumption, we can determine $f(E_t)$, the E_t contribution from a single wounded nucleon. Since we assume the two nucleons contribute equally and independently, the single nucleon contribution is simply the deconvolution of the p–p spectrum. Fortunately, the p–p data is well fit by the functional form

$$\frac{d\sigma}{dE_t} \sim E_t^x e^{-\alpha E_t}.$$

The deconvolution can then be done analytically. This parameterization of the pp cross-section is shown superimposed on the data in Fig. 1b; it gives $\alpha = 1.16 \pm .04$ and $x = 2.56 \pm .20$. For p–nucleus and nucleus–nucleus collisions, we will need to know the energy spectrum for collisions with 2, 3, etc. wounded nucleons. According to our model, a nucleon contributes only once, regardless of how many individual nucleon-nucleon collisions it participates in. An interaction with n wounded nucleons would have n equal and independent contributions to the E_t spectrum, and thus the total contribution from such events would be an n-fold convolution of $f(E_t)$.

However, just knowing the spectrum from events with n wounded nucleons is not enough. We must also know the probability of a collision involving 2, 3, etc. wounded nucleons. These probabilities, which are labeled a_n, are calculated from a Monte Carlo scattering program. Nucleons are distributed radially according to the spatial density $\rho(r)$ determined from electron scattering data, and randomly in $\cos\theta$ and ϕ. An impact parameter is then chosen, and the nuclei are projected onto each other. Any overlap between nuclei from different projectiles is counted as a hit, and both nucleons are tagged as *wounded.* For a collision between nuclei with nucleon numbers A and B, the normalization requirement is

$$\sum_{n=2}^{A+B} a_n = 1,$$

i.e., that the probability of wounding *any* number of nucleons is unity. The a_ns are obviously dependent on the p–p inelastic cross-section and the nucleon hard-core. Fig. 2 shows the a_n distributions for $\sigma_{pp}^{inel} = 25\,\text{mb}$ and a nucleon hard-core of 0.4 fm.

To find the total cross-section predicted by a WNM calculation, we must sum the contributions from all possible occurences. We can write this in the following form:

$$\frac{d\sigma}{dE_t} \propto \sum_n a_n f^n$$

where f^n denotes the n-fold convolution which is the contribution from collisions involving n wounded nucleons. Thus we sum each n-fold convolution weighted by a_n. We then normalize this sum to the observed total cross-section. The WNM predictions for the three light ion runs describe the data very well over several orders of magnitude, and are shown superimposed on the data in Fig. 1b. Varying the parameters α and x by 1 σ gives an estimate of the error in these calculations. This is shown for the α–α data in Fig. 1b as a band surrounding the WNM prediction.

4. R110 Neutral Energy Data and Jets

We have also applied our analysis to α–α neutral energy data (E_0) taken by the Brookhaven-Cambridge-Michigan State-Oxford-Rockefeller collaboration (Experiment R110) at the ISR[12]. Although they record total neutral energy, as opposed to E_t, the method of analysis is the same. (As a crude reference, the conversion factor from E_t to E_0 is roughly 1.14 for $|y| < 1$.) Fig. 3 shows the R110 p–p and α–α data, along with the WNM predictions. Once again, the main features of the α–α data are predicted by the p–p data and our simple participant model. It may be noticed, however, that the WNM prediction falls somewhat above our α–α E_t data, and below the BCMOR E_0 data. This may be a result of jet production setting in; such processes would be more prominent in BCMOR data, which extends to higher values of E_t than the AFS data.

The WNM is obviously a 'soft' model. The assumption that each struck nucleon radiates independently is of course invalid for hard scattering processes such as jet production. If the any part of the spectrum, especially the high E_t end, had a significant contribution from jets, then we could not expect the WNM to legitimately describe the data. A useful variable for examining the shape of events is circularity, C. For each event, a vector is found such that

$$C = \min_{\vec{C}} \sum \frac{|\vec{p}_i|^2 \sin^2\theta}{|\vec{p}_i|^2}$$

A perfectly isotropic event would have $C = 1$, while a perfect collimated two jet event would have $C = 0$. Fig. 4 shows the mean circularity as a function of E_t for the α–α data. There is no decrease in $\langle C \rangle$ as E_t increases. A more sensitive measure is to look at the circularity distributions as E_t increases. Fig. 5 shows the circularity distributions in E_t bins from 0 to 40 GeV. Even in the highest bin, there is no significant contribution at $C = 0$, and thus no sign of jets. This indicates that soft processes are still dominant, and that we are still legimately testing the WNM at these high values of E_t.

5. Modifications to the WNM

As stated above, our model assumes that each wounded nucleon has the same energy spectrum. It is clear that a more exact model should incorporate energy loss for subsequent scatters by these nucleons. This is supported by the analysis of the FERMILAB p-nucleus data by Brody *et al.*[3]. An empirical fit to the p-A data found that rescaling the energy spectrum for the n th collision by

$$E \longrightarrow E[1 - k(n-1)], \quad k = \frac{1}{12}$$

gave a better description of the data. Fig. 6 shows this modified WNM calculation for the p–α data. This modification improves the high end of the WNM fit.

6. Conclusions

The E_t data presented here extends our previous work on charged E_t and multiplicity distributions to higher energies. Although the data approach or exceed the p–p kinematic limit, we find that the spectra are well described by the Wounded Nucleon Model, with no evidence of any 'anomalous' behavior.

7. Acknowledgements

I would like to thank Dr. M. Tannenbaum for several helpful discussions regarding the BC-MOR data and analysis, and my fellow AFS collaborators for all their help and support.

I would especially like to thank Profs. S. Frankel, W. Zajc, W. Frati, and H. Brody for their support and encouragement.

8. References

1) A. Dar and J. Vary, Phys. Rev. D6 (1972) 2412.
A. Bialas, M. Bleszynski, and W. Czyz, Nucl. Phys. B111 (1976) 461.

2) T. Åkesson *et al.*, Phys. Lett. 119B (1982) 464.

3) H. Brody, *et al.*, Phys. Rev. D 28 (1983) 2334.

4) W.Q. Chao and H.J. Pirner, Z. Phys. C13 (1982) 165.

5) A. Bialas, W. Czyz, and W. Furmanski, Acta Phys. Pol. B8 (1977) 585.
A. Bialas, W. Czyz, and L. Lesniak, Z. Phys. C13 (1982) 147.

6) H. Gordon *et al.*, Phys. Rev. D28 (1983) 2736.

7) T. Åkesson *et al.*, Phys. Lett. 128B (1983) 354.

8) H.E. Miettinen, Nucl. Phys. A418 (1984) 315c.

9) R. Brown, *et al.*, Phys. Rev. Lett. 50 (1983) 11.

10) T. Åkesson *et al.*, Phys. Lett. 118B (1982) 185.
T. Åkesson *et al.*, Phys. Lett. 118B (1982) 193.

11) H. Gordon *et al.*, Nucl. Instrum. Methods 196 (1982) 303.
O. Botner *et al.*, Nucl. Instrum. Methods 196 (1982) 314.
O. Botner *et al.*, Nucl. Instrum. Methods 179 (1981) 45.

12) A.L.S. Angelis, *et al.*, Submitted to Physics Letters B.

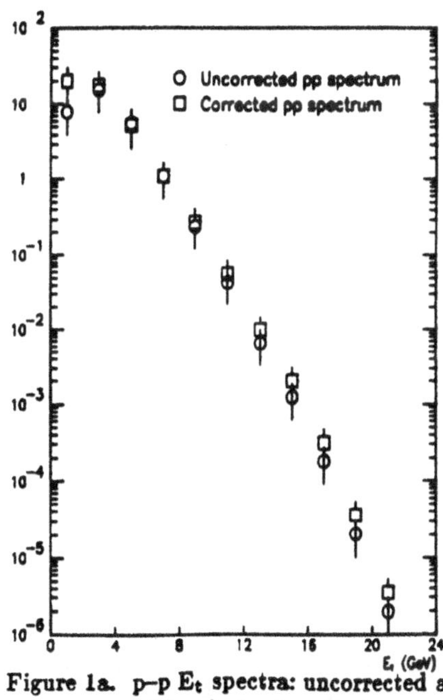

Figure 1a. p–p E_t spectra: uncorrected and with calorimeter effects

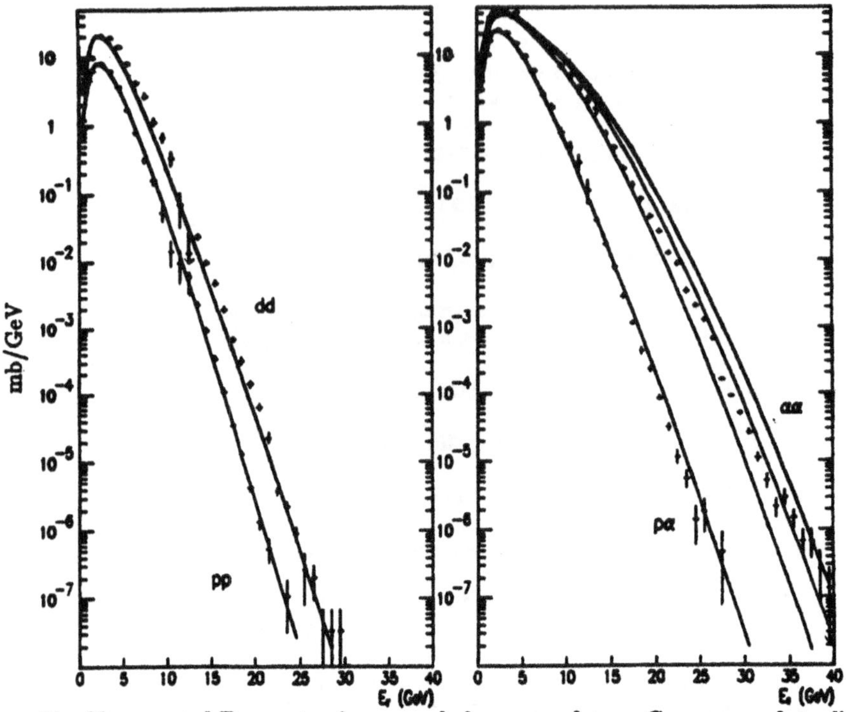

Figure 1b. Uncorrected E_t spectra for p–p , d–d, p–α, and α–α. Crosses are data; lines are WNM predictions. α–α fit is bracketed by 1σ variation in p–p parameters.

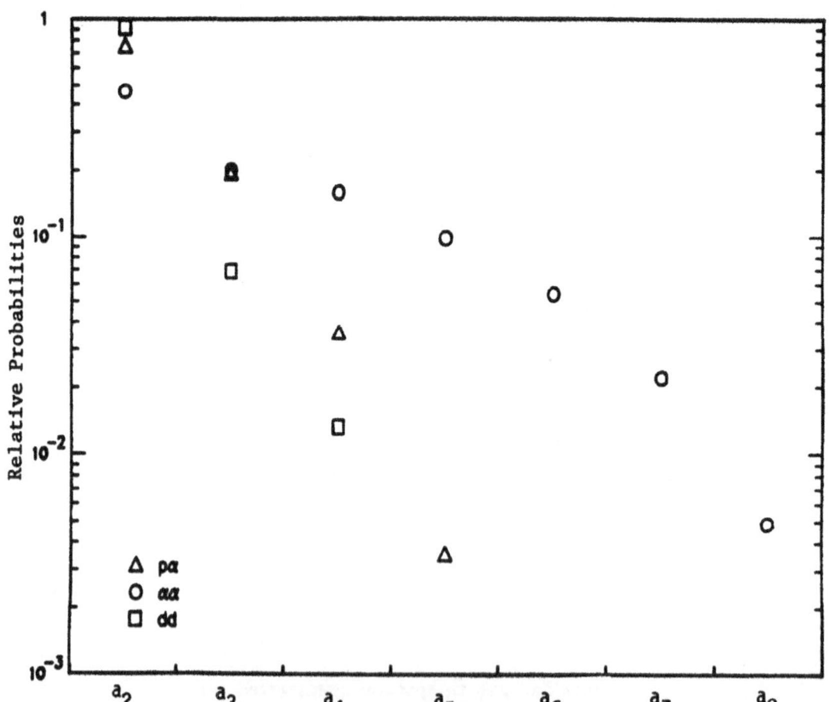

Figure 2. a_n distributions.

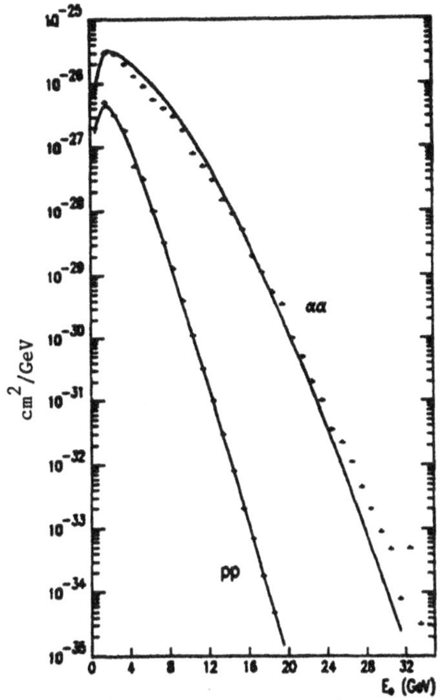

Figure 3. R110 E_0 spectra for p–p and α–α with WNM prediction.

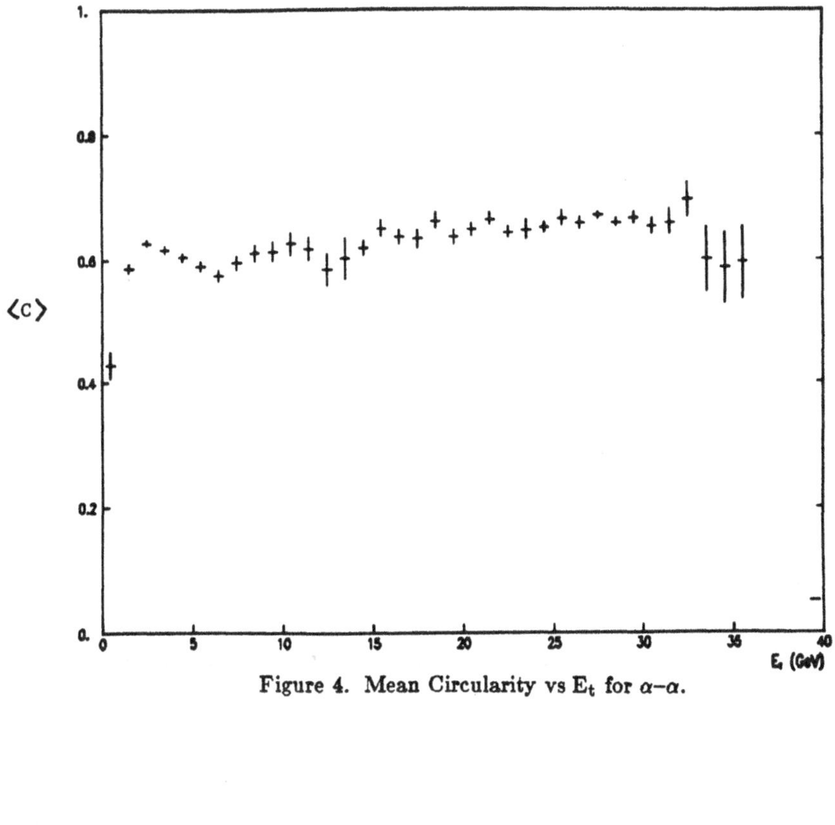

Figure 4. Mean Circularity vs E_t for α–α.

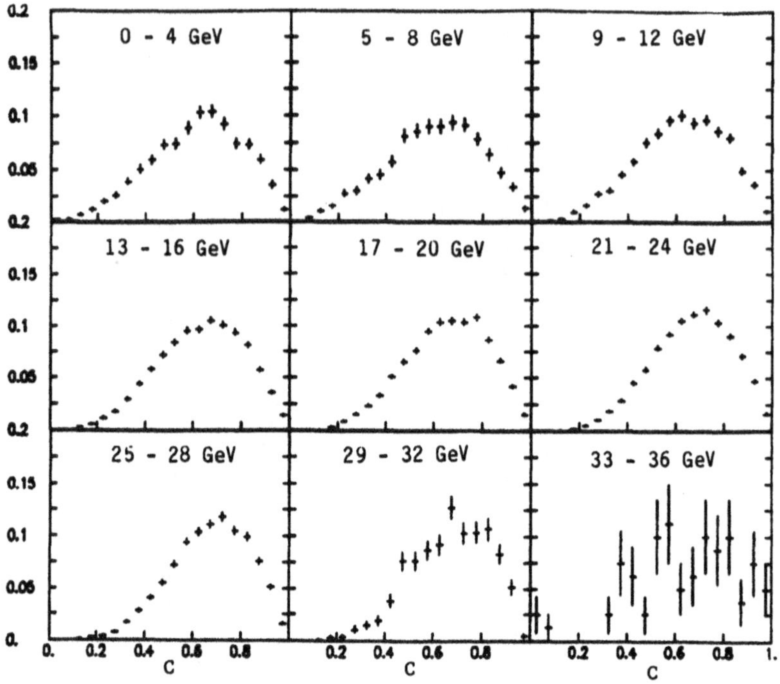

Figure 5. Circularity distribution in E_t bins for α–α.

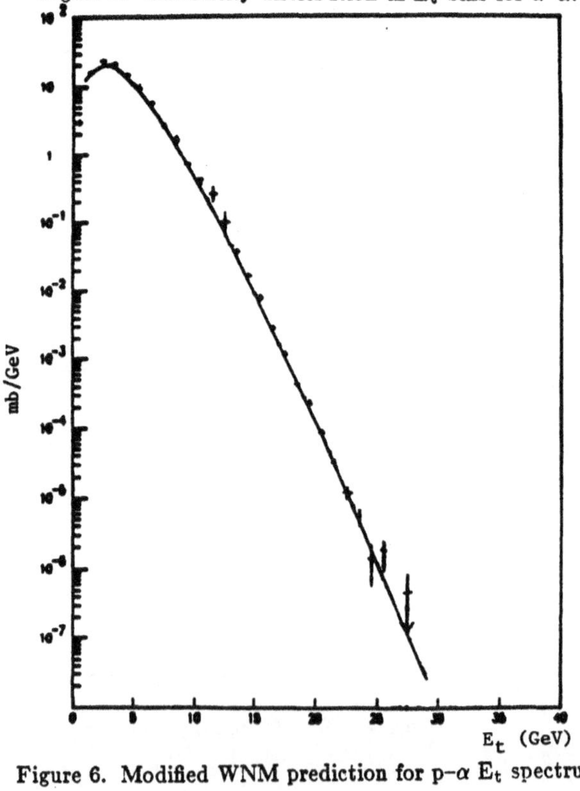

Figure 6. Modified WNM prediction for p–α E_t spectrum.

EQUILIBRATION, COMPRESSION AND FLOW AT THE BEVALAC*

Howel G. Pugh
Nuclear Science Division
Lawrence Berkeley Laboratory
University of California
Berkeley, California 94720

This paper presents data acquired in the past two years by the Streamer Chamber group[1] and the Plastic Ball group.[2] I shall present evidence that at Bevalac energies, head-on nuclear collisions result in stopping; I shall present evidence for collective flow of the interacting nuclei; I shall present evidence that a substantial part of the energy is tied up in compressional energy at the high density phase of the collision; finally, I shall present recent data concerning aspects of the collisions such as interaction volume, particle ratios and temperatures. At the meeting at Bielefeld two years ago, it was not possible[3] to make definitive statements about most of these subjects: the progress in the intervening period has been remarkable.

1. ^{40}Ar + Pb Central Collisions at 0.77 AGeV

Central collisions can be selected by placing a detector of suitable size at 0° and requiring that no projectile fragments are observed in that detector, i.e., the target nucleus completely overlapped the projectile (as seen along the beam direction). For equal mass nuclei this indicates the impact parameter $b = 0$; for a light projectile on a heavy target it indicates $b < r_{target} - r_{projectile}$. The cascade model shows that the participant proton multiplicity M_p can also be used as an approximate measure of impact parameter.

This method has been used to study ^{40}Ar + Pb central collisions at 0.77 AGeV, using the LBL Streamer Chamber facility.[4,5] Figure 1 shows the distribution of participant proton multiplicity M_p, as well as a comparison with the results of an intranuclear cascade calculation.[6] We shall later divide the data into two groups: high M_p and low M_p corresponding to $b < 3$ fm and 3 fm $< b < 5.5$ fm respectively.

To test for stopping, the simplest test is for isotropy in the c.m. system of the participant protons, taken on an event-by-event

*This work was supported by the Director, Office of Energy Research, Division of Nuclear Physics of the Office of High Energy and Nuclear Physics of the U.S. Department of Energy under Contract DE-AC03-76SF00098.

basis. This requires the quantity R, defined by R = $\frac{2}{\pi} \sum_{\nu=1}^{N} |P_\perp|_\nu \Big/ \sum_{\nu=1}^{N} |P_\parallel|_\nu$ to equal unity, where the sums are taken over the number of protons observed in each event. Figure 2 shows a contour diagram of R versus M_p. It is seen that for the lower M_p (larger impact parameters) R is less than unity, indicating incomplete equilibration. However as M_p increases, R crosses unity at about M_p = 40 and even exceeds unity, suggesting some kind of side splash. On the other hand, the cascade model predictions only approach unity so that the nuclei seem to be somewhat more efficient in stopping each other than expected in that model. Figure 3 shows a contour diagram of $\frac{2}{\pi} \langle P_\perp \rangle$ versus $\langle P_\parallel \rangle$ for the high $M_p$cut. The dashed line indicates R = 1.

It is interesting to look beyond this at the event shapes, for which it is convenient first to identify the reaction plane for each event. This is the plane containing the beam direction such that the summed projection of momenta perpendicular to the plane is zero. All events can be rotated into the same reaction plane; no physics is lost by this process, for unpolarized beam and target. Figure 4 shows a contour diagram of proton transverse and longitudinal momenta in the reaction plane for the low M_p cut, 3 fm < b < 5.5 fm (the cut shows the region of detector inefficiency near the target). A systematic event shape is seen, showing an average sideways deflection of the participants. It is remarkable that the events show symmetry about the origin even though there is a different number of participants from the two interacting nuclei.

2. Flow Tensor Analysis and Collective Flow

To characterize the shape in momentum space of each event some further parameterization is necessary. One method is to introduce the flow tensor $T_{ij} = \sum_{\nu=1}^{N} P_{i\nu} P_{j\nu} \omega(\nu)$ where the sum is over all the particles in a given event, i,j, and k are the cartesian coordinates, and ω is a weighting factor. The weighting factor has recently been taken as $\omega_\nu = 1 \Big/ |P_\nu| \sum_{\nu=1}^{N} |P_\nu|$ in the streamer chamber data, so that T represents momentum flow. The plastic ball group has preferred ω_ν = $1/2m_\nu$, yielding energy flow. The results are insensitive to the choice of weighting, which adds confidence that the conclusions are significant.

Such an analysis yields six quantitities characterizing the flow tensor: The lengths of the three axes of an ellipsoid, and three orientation angles. This is a great number of parameters to fit even to the large amount of data obtained, and we restrict ourselves here to the "flow angle," i.e. the value of θ for the longest axis of the ellipsoid, measured relative to the beam direction. Figure 5 shows data obtained by the Plastic Ball group[7] for ^{40}Ca + Ca and ^{93}Nb + ^{93}Nb at 0.4 AGeV. The data for ^{93}Nb show a well-developed peak in the flow angle which increases in angle as the charged particle multiplicity increases, i.e. as the impact parameter decreases. The effect, while present in the ^{40}Ca data, is less clear, demonstrating the value of using heavy nuclei for these studies. Calculations using a cascade code[8] do not show the effect, suggesting that collective phenomena are being observed.[9] Figure 6 shows results similar to those of Figure 4, but taken with the Plastic Ball. Note that the experimental data, which should be symmetric between target and projectile, show detector inefficiencies in the backward hemisphere; these have been taken into account in applying the cascade predictions.

That these results provide evidence for relativistic hydrodynamics is shown by calculations by Buchwald et al.[10] whose results are shown in Figure 7. They are in qualitative agreement with the data. However, it is very difficult to introduce individual particle effects such as detector inefficiencies into hydrodynamic calculations, and much work remains to be done. New data on ^{139}La + ^{139}La and ^{197}Au + ^{197}Au will be available soon.

In addition to the flow angle, the shapes of the momentum (or energy) ellipsoids are of interest. In general events seem to be prolate, with the longest axis closest to the beam direction. However, for $b = 0$ collisions of equal mass nuclei, one might expect an oblate spheroid with its shortest axis along the beam direction, and a flow angle of 90°. This would also give a value greater than unity for the R ratio discussed earlier. The data presented in Figure 5 do not approach $b = 0$ for ^{93}Nb + ^{93}Nb since the charged multiplicity for such events should be about 80. While the situation is not quite analogous, it is therefore necessary to return to the ^{40}Ar + Pb data at 0.77 GeV/n. Here since the radii of ^{40}Ar + ^{208}Pb are roughly 4 fm and 7 fm respectively, the higher multiplicity sample, $b < 3$ fm, corresponds to complete overlap of the two nuclei while the lower multiplicity sample, $3 \text{ fm} < b < 5.5$ fm, is closer to the nuclear peripheries. Figure 8(a) shows the flow angle distributions for the two multiplicity cuts. The low M_p sample shows a flow pattern similar to that seen in Figure 5 for Nb + Nb, which seems to be characteristic of intermediate impact parameters for heavier nuclei. The high

M_p sample is roughly isotropic, while the event shapes are approximately spherical. Thus the anticipated oblate event shape is approached but not definitely observed. Figure 8(b) shows intranuclear cascade calculations[6,9] for the same impact parameter cuts. No collective flow effect comparable to those seen in the data is observed.

3. Collective Effects in the Energy

A detailed analysis of the final state energy shows, for both streamer chamber and plastic ball data, that all the initial energy is accounted for. What would be interesting is a breakdown of the total energy into collective and random components as a function of time during the collision. In the initial state all the energy is collective. An analysis of the plastic ball data on Nb + Nb[7] shows that roughly 10% of the energy in the final state appears in non-isotropic components of the collective flow. Extraction of the radial component is difficult because it can be confused in the data with the thermal component,[11] and its extraction involves a very detailed analysis of the energy spectra.

An estimate of the energy tied up in compression at the maximum-density phase of the collision has been given by Stock, Harris et al.[12,13] They used the streamer chamber to study ^{40}Ar + KCl as a function of energy. At each energy the π^- yield was measured as a function of proton participant number and extrapolated to $b = 0$, as shown in Figure 9. The results shown in Figure 10 as a function of energy indicate as a function of energy a substantial reduction below the predictions either of cascade calculations or of a chemical model, though the theories agree with each other. The interpretation offered is simple: at each beam energy, if the energy available for pion production is reduced by the amount shown by the horizontal arrows in Figure 10 the correct number of pions would be obtained. The deficiency is attributed to energy required to compress the nuclear matter, which is not allowed for in the theoretical results. Since the theories provide the density reached in the collision it is possible to go one step further and derive an equation of state for nuclear matter, as shown in Figure 11. Despite criticism of the procedure in detail, the concept has not been invalidated and it remains as the only means so far of extracting an equation of state from experimental data.

Overall, we thus have a rough idea of how the energy is degraded during the collision, if we are willing to mix results from such disparate experiments as Nb + Nb at 0.4 AGeV and Ar + KCl at 0.5 - 1.8 AGeV. At maximum density, where there can be no flow effects, about 35% of the energy seems to be tied up in compression while 65% must have gone into thermal motion, including pion creation. In the final state, 10% of the energy has been identified in non-radial flow. It would be very valuable to complete this picture.

4. Other New Plastic Ball Data

A high-statistics study of two-proton intensity interferometry has been made for ^{40}Ca + Ca and ^{93}Nb + ^{93}Nb at 0.4 AGeV.[14] Figure 12 shows a fit which includes the effects of Coulomb repulsion, final-state interaction and experimental resolution. Figure 13 shows the systematic behavior of the extracted radius parameter as a function of participant number, defined as N_pA/Z where N_p is the measured number of protons. The fits shown use the formula

$$r = r_o \; (N_pA/Z)^{1/3} / \sqrt{5/2}$$

The factor of $\sqrt{5/2}$ is to convert the radius from a hard-sphere to a gaussian. The proportionality to the cube root of the participant number is presumably an approximation since the emitting volume is hardly likely to be spherical at all impact parameters. Nevertheless the result is extremely suggestive, because the hard sphere radius constant extracted is r_o = 1.9 fm, corresponding to freeze-out at about 25% of nuclear density. The protons appear therefore to be emitted at a very late stage of the reaction.

Some supporting evidence is given by observations of deuteron-proton ratios,[15] shown in Figure 14. The d/p ratio increases with multiplicity, and can be explained if the participant volume increases as $(N_pA/Z)^{1/3}$, as before. The freeze-out here occurs at about 50% of nuclear density.

A possibly related result appears in other data from the same reactions. The proton energy spectra were fitted with Boltzmann distributions and "temperatures" extracted, as a function of charged multiplicity. The results,[16] given in Table 1, show that the temperature is an increasing function of multiplicity. While the numbers have not been explained qualitatively, the authors suggest that the rise may be due to the increasing d/p ratio: when a neutron and proton coalesce to form a deuteron, the number of degrees of freedom diminishes, which increases the temperature (the energy release in the process $n + p \rightarrow d$ is negligible).

Table 1
Temperature as a function of multiplicity[15]
(Plastic Ball--0.4 AGeV)

charged multiplicity		0-10	10-20	20-30	30-40	40-50	50-60
T(MeV)	Ca + Ca	43	48	54	56		
	Nb + Nb	42	46	51	55	60	65

5. Some New Results from the Streamer Chamber on Pion Spectra

An analysis has recently been completed of pion and proton spectra in central ^{40}Ar + KCl collisions at 1.8 GeV/n.[17] The conclusions are as follows:

a) 85% of the pions are emitted isotropically.

b) The remaining anisotropy can be attributed to the "corona effect" observed in equal-mass collisions.

c) The pion spectra can be fitted fairly well with intranuclear cascade calculations, except for a 5% high temperature component.

d) The predominant pion temperature is 58 MeV compared with 118 MeV for the protons and 111 MeV for the 5% pion component.

e) At first sight this is a disaster for any thermodynamic ideas of equilibrium, but once the kinematics of Δ decay are introduced, everything falls into place. The cascade calculations[4] already include the kinematics, and if the source of pions and protons is assumed to be the decay of Δ's in equilibrium with the protons, the thermodynamic model is also reconciled with the data.

f) However, it is not possible on the basis of the present data, and it will always be difficult, to separate the effects of radial flow.[11] Whether the Δ temperature is a true temperature or an effective temperature made by folding a lower true temperature with radial flow makes little difference to the final pion and proton spectra.

6. Conclusions

The amount of data growing out of the Bevalac is qualitatively greater than a few years ago, and we look forward to significant progress in understanding in the next few years also.

Acknowledgments

I am grateful to the members of the Streamer Chamber and Plastic Ball collaborations[1,2] for helping me assemble this paper, and to the Organizers of this Conference for asking me to present it.

This work was supported by the Director, Office of Energy

Research, Division of Nuclear Physics of the Office of High Energy and Nuclear Physics of the U.S. Department of Energy under Contract DE-AC03-76SF00098.

References

1. D. Bangert and W. Rauch, Univ. of Marburg; R. Bock, R. Brockmann, J.W. Harris, M. Maier, A. Sandoval, R. Stock, and H. Strobele, GSI; A.Dacal and M.-E. Ortiz, UNAM, Mexico; G. Odyniec, H.G. Pugh, and L.S. Schroeder, LBL; R.E. Renfordt, D.Schall, and K. Tittel, Univ. of Heidelberg; K.W. Wolf, Texas A M University.

2. H.-A. Gustafsson, M. Doss, A.M. Poskanzer, T. Renner, H. Riedesel, and A. Warwick, LBL; H.H. Gutbrod, B. Kolb, H.-G. Ritter, F. Weik, and H. Wieman, GSI; K.-H. Kampert and H. Lohner, Univ. of Munster; B. Ludewigt, Univ. of Marburg.

3. H.G. Pugh, (Bielefeld Workshop on) Quark Matter Formation and Heavy Ion Collisions, ed. M. Jacob and H. Satz, World Scientific (1982), p. 185

4. H. Stroebele et al., Phys. Rev. C27, 1349 (1983); note that for ^{40}Ar + Pb the c.m. frame of the participant protons has been extracted on an event-by-event basis by using the measured momenta.

5. R.E. Renfordt et al., Phys. Rev. Lett. 53, 763 (1984).

6. J. Cugnon, T. Mitzutani, and J. Vandermeulen, Nucl. Phys. A352 505 (1981);
J. Cugnon, D. Kinet, and J. Vandermeulen, Nucl. Phys. A379, 553 (1982)

7. H.A. Gustafsson et al., Phys. Rev. Lett. 52, 1590 (1984)

8. Y. Yariv and Z. Fraenkel, Phys. Rev. C20, 227 (1979)

9. The paper presented by Cugnon at this conference suggests that the flow effect is present in cascade calculations. However, it is interesting to note that the calculations used for comparison with the streamer chamber data for ^{40}Ar + Pb (presented later in the present paper) used Cugnon's code, and showed no flow.

10. G. Buchwald et al., Phys. Rev. Lett. 52, 1594 (1984)

11. P. Siemens and J. Rasmussen, Phys. Rev. Lett. 42, 880 (1979)

12. R. Stock et al., Phys. Rev. Lett. 49 (1982) 1236

13. J.W. Harris et al., LBL-17404 (1984), to be published.

14. H.-A. Gustafsson et al., Phys. Rev. Lett. 53, 544 (1984).

15. H.H. Gutbrod et al., Phys. Lett. 127B, 317 (1983)

16. H.-A. Gustafsson et al., GSI-84-9 (1984), and to be published.

17. R. Brockmann et al., LBL-17755 (1984), to be published.

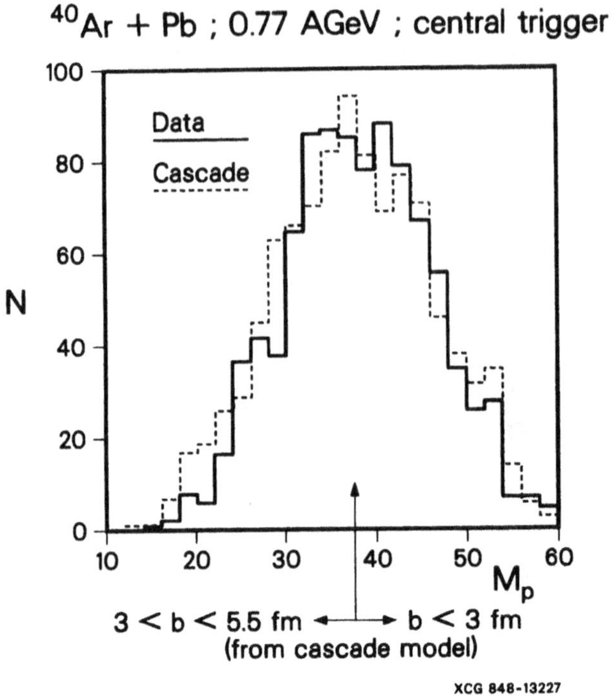

Figure 1

Multiplicity distribution of participant protons and comparison with cascade model predictions.

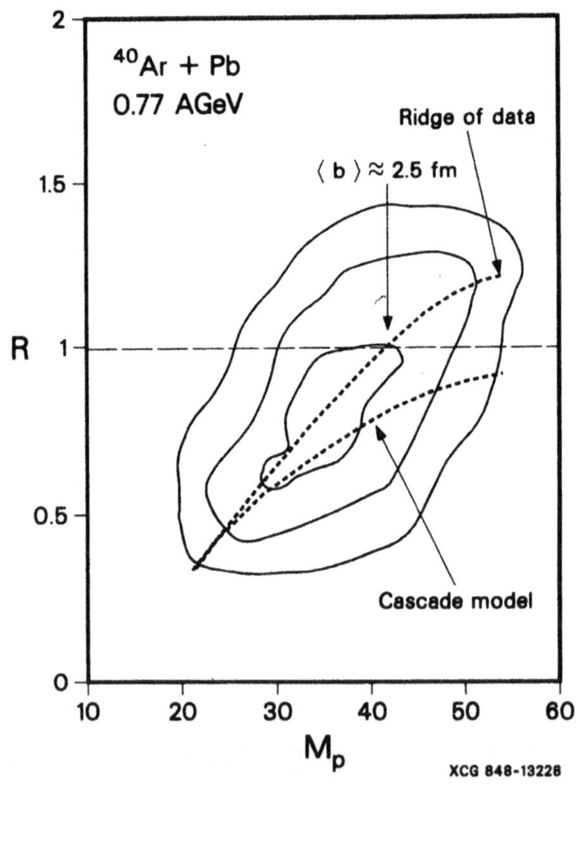

Figure 2

Contour plot of the ratio R versus the participant proton multiplicity. The ratio R should equal unity if equipartition occurs.

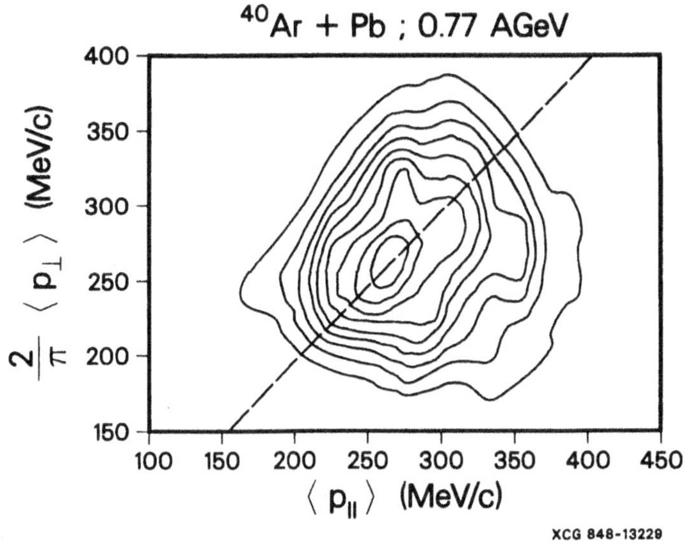

Figure 3

Contour plot of $\frac{2}{\pi}$ $<p_\perp>$ versus $<p_\parallel>$. The line shows the expected locus for complete equipartition.

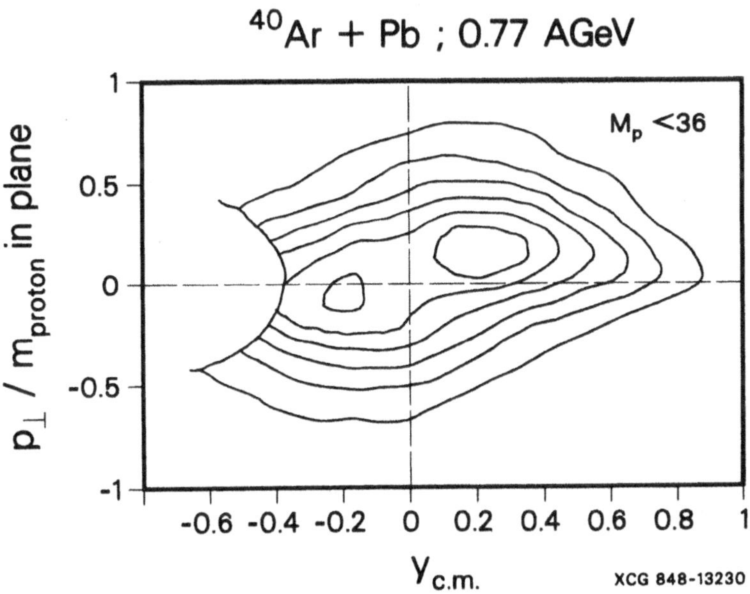

Figure 4

Contour plot of $p_\perp/m_p$ versus y_{cm}, after all events have been rotated so that the reaction planes coincide. Streamer Chamber data.

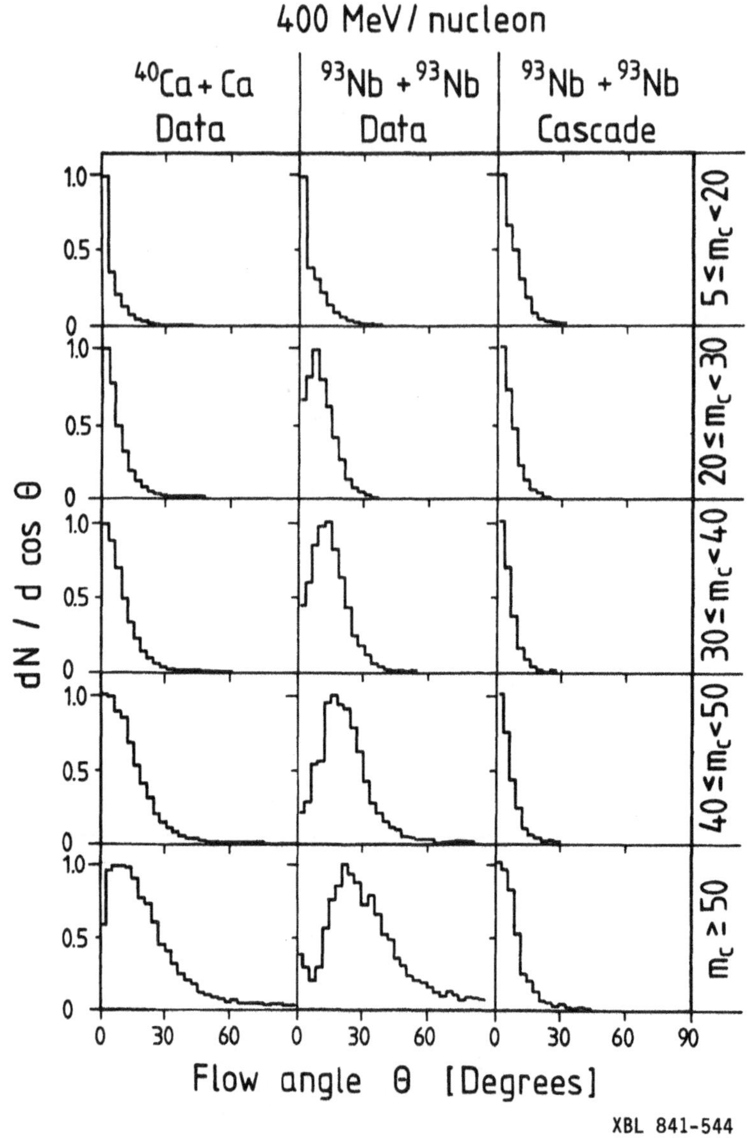

Figure 5

Flow angle distributions for Ca + Ca and Nb + Nb collisions and cascade predictions for Nb + Nb. Plastic Ball data.

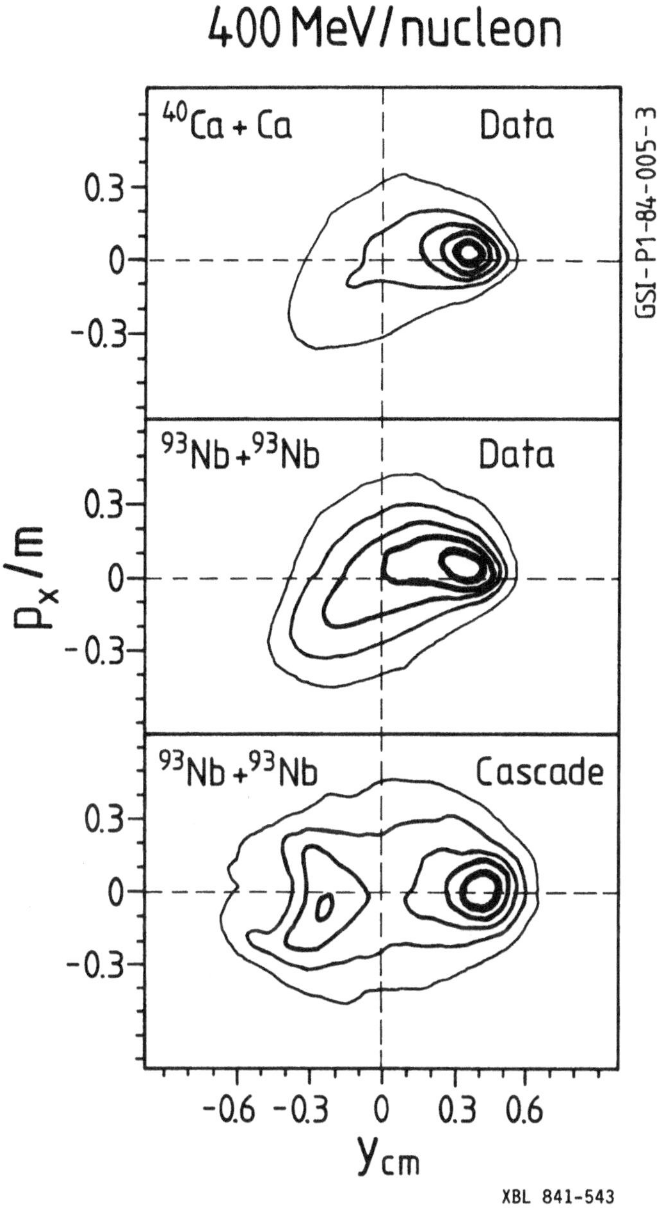

Figure 6

Contour plots of P_x/m versus y_{cm} after all events have been rotated so that their reaction planes coincide. Plastic Ball data.

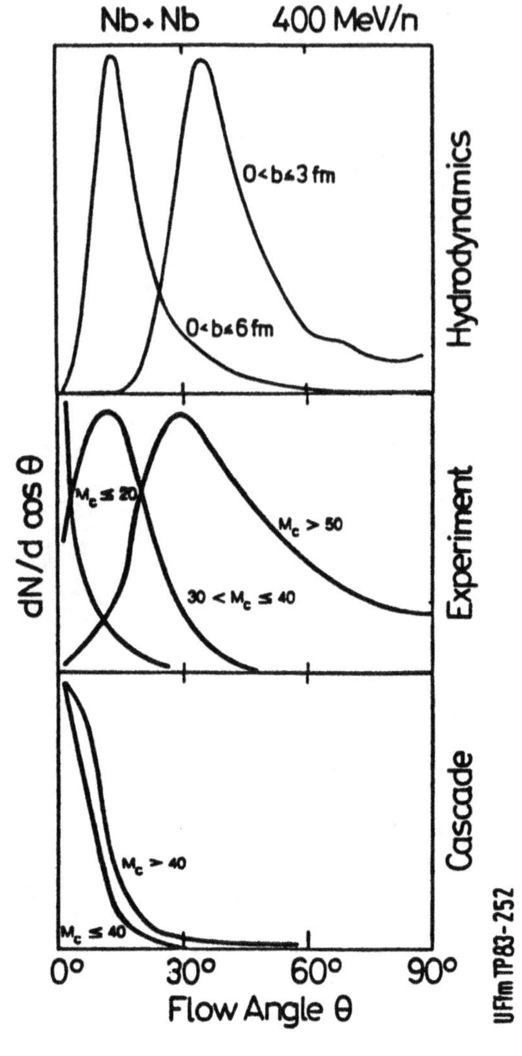

XBL 849-3692

Figure 7

Predictions of flow angle distributions derived from a relativistic hydrodynamic model.

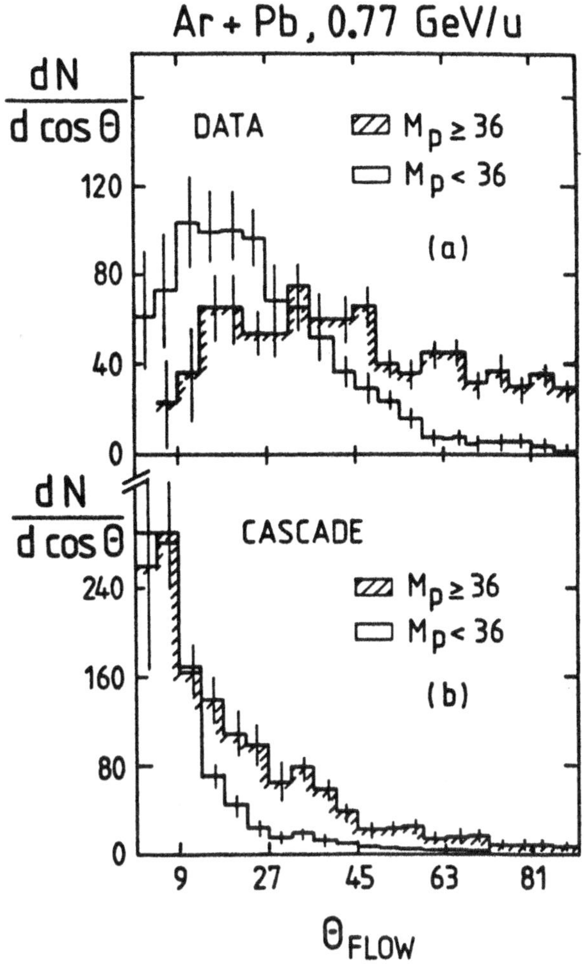

XBL 849-3693

Figure 8

(a) Flow angle distributions for two participant proton multiplicity samples corresponding to b < 3 fm ($M_p \geq 36$) and 3 < b < 5.5 fm ($M_p < 36$). Streamer Chamber data.

(b) Corresponding predictions of the cascade model, corrected for detector acceptance.

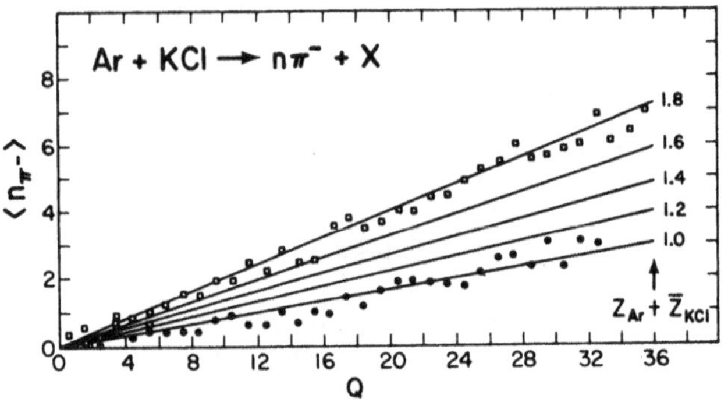

Figure 9

Multiplicity of π^- production versus participant proton multiplicity for various bombarding energies. The value extrapolated to Z_{Ar} + Z_{KCl} corresponds to b = 0.

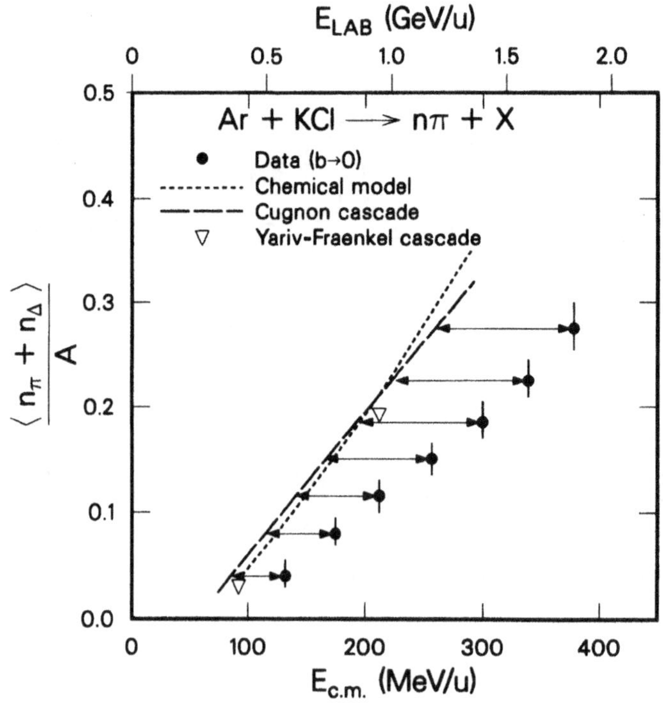

Figure 10

XCG 8312-4933

Pion multiplicities versus E_{cm} for Ar + KCl central collisions. The length of the arrows is assumed to reflect compressional energy.

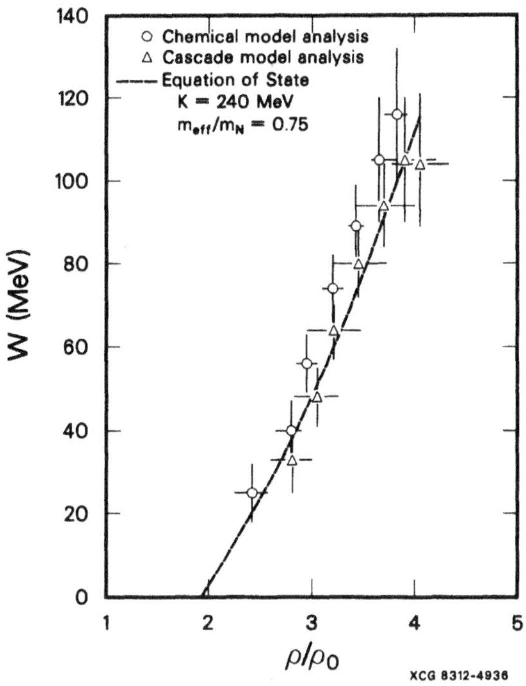

Figure 11

Equation of State derived from Figure 10. W is the compressional energy.

Figure 12

Two proton intensity interferometry data showing a theoretical fit. Plastic Ball data.

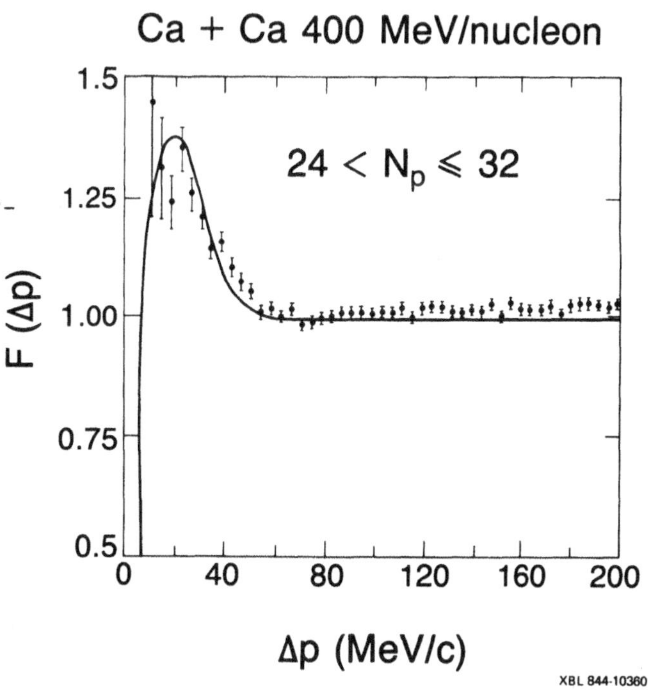

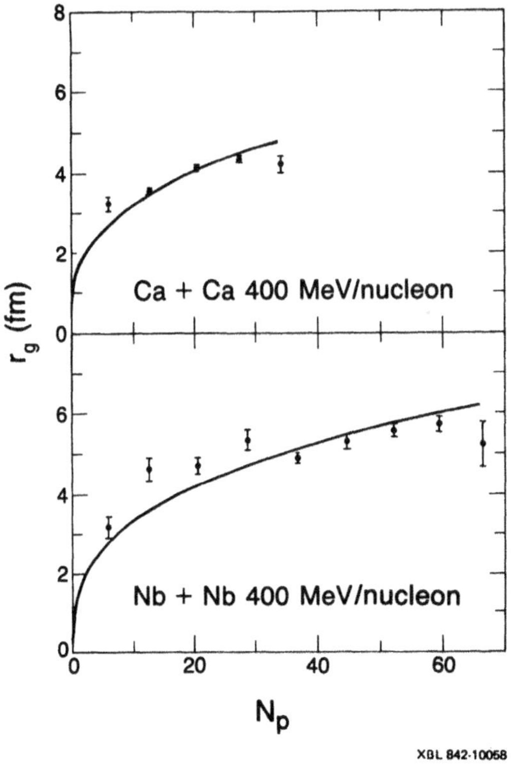

Figure 13

Dependence of radius parameter extracted from two-proton intensity interferometry on the participant proton multiplicity. The curves correspond to $\rho/\rho \sim 0.25$.

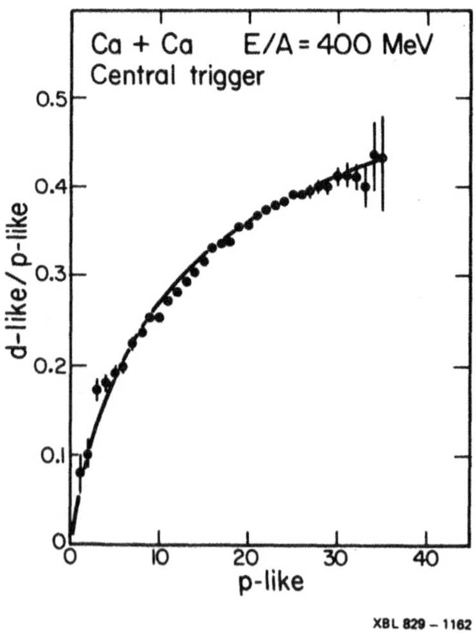

Figure 14

The d (like)/p (like) rates extracted from Plastic Ball data. The curve corresponds to $\rho/\rho_0 \sim 0.5$.

INTERACTIONS OF RELATIVISTIC NUCLEI FROM THE DUBNA SYNCHROPHASOTRON: FRAGMENTATION AND PARTICLE PRODUCTION

J.Bartke*

Joint Institute for Nuclear Research
Dubna, USSR

Interactions of relativistic nuclei have been studied at Dubna since the early seventies. During this period of time the choice of nuclei accelerated in the synchrophasotron was being gradually enlarged and their intensity increased. Putting a new ion source with a 10 Joule CO_2 laser into service led to an increase of the intensity of a beam of carbon ions by a factor of about 10^2 and allowed one to add lithium and magnesium to the last year's list of accelerated nuclei [1]. Table 1 shows nuclear beams available at present (summer 1984). All these are external beams of fully stripped ions. Fast or slow ejection with 400 ms "flat top" are at the user's choice.

Table 1

Type of nuclei	Energy, GeV	Intensity per pulse	Ion source
p	9.0	4×10^{12}	duoplasmatron
d	8.2	1.5×10^{12}	duoplasmatron
d↑	8.2	5×10^{8}	"Polaris" [2]
^3He	17.2	2×10^{10}	duoplasmatron
^4He	16.4	5×10^{10}	duoplasmatron
^6Li	24.6	1.5×10^{8}	Laser [1]
^7Li	23.9	2×10^{9}	Laser
^{12}C	49.2	5×10^{8}	Laser
^{16}O	65.6	5×10^{5}	"Krion" [3]
^{22}Ne	81.0	10^{4}	"Krion"
^{24}Mg	100	10^{5}	Laser

Cryogenic panels [4] are being installed inside the vacuum tank of the synchrophasotron in order to improve the vacuum, what is necessary for the acceleration of heavier nuclei. With all panels in place, what is expected by the year's end, one should be able to go up to $Z = 18$ (argon). For $Z/A = 0.5$ nuclei the maximum kinetic energy is 4.2 GeV per nucleon (momentum $P_{max} \cong 5$ A GeV/c).

* On leave from Institute of Nuclear Physics, Cracow, Poland

In a more distant future there are plans to build the NUCLOTRON, a new synchrotron ring with superconducting magnets, which would be able to accelerate all ions up to about 6 GeV per nucleon [5]. A model superconducting synchrotron for a 1.5 GeV energy for protons, called the SPIN, is now near completion [6].

Physics experiments at the synchrophasotron use various visual track detectors (bubble and streamer chambers, nuclear emulsions) and electronic spectrometers. Results obtained have been published in many papers and were regularly presented at international conferences [7-17]. A recent review was given at the Balaton Conference last year [17].

Having in view the name of this conference: QUARK MATTER'84, I will discuss in this report some experimental results which might be relevant to the quark structure of nuclei and to the possibility of creation of quark-gluon plasma in collisions of relativistic nuclei. This talk will be complemented by Dr. Šumbera's contribution at the experimental parallel session.

General features of the interactions of relativistic nuclei at moderate energies are well described by the so-called "participant-spectator" model [18]. In the terminology of this model, "fragmentation" means the break-up of the spectator parts of interacting nuclei. This process is to a good approximation independent of whatever happens to the "participant part". Experiments with hadrons incident upon a target at rest, and especially those using emulsion technique, furnished much information on nuclear fragmentation [19]. However, this process may be studied in the best way in experiments with relativistic nuclei. A convincing comparison of "at rest" and "in flight" experiments on deuteron fragmentation can be found in ref. [20]. Early studies of fragmentation of ^{12}C and ^{16}O nuclei at Berkeley [21] showed that in the projectile frame the longitudinal momentum distribution of fragmentation products can be described by a Gaussian form with the width depending on the masses of both the fragmentating nucleus (A) and the fragment (F). This latter dependence is of parabolic form

$$\sigma_{p_{\parallel}}^2 = \frac{F(A-F)}{A-1} \langle p_i^2 \rangle ,$$

where $\langle p_i^2 \rangle$ is the mean-square momentum of a nucleon in the fragmentating nucleus [22]. As fragmentation proceeds with a minimal momentum transfer, the fragmentating nucleus will not be highly excited and thus $\langle p_i^2 \rangle$ should be closely related to the Fermi momentum p_F. The relation $\langle p_i^2 \rangle = \frac{3}{5} p_F^2$, quoted in ref. [22], follows from the Fermi gas model of the nucleus which is not a realistic model. The Fermi gas model assumes constant occupation density in momentum space (i.e., $dN/dp \sim p^2$), with a sharp cut-off at $p = p_F$ [23], while the Gaussian shape of the momentum component distributions in nuclear fragmentation indicates a Gaussian momentum distribution of nucleons in a nucleus: $dN/dp \sim p^2 \exp(-p^2/2\sigma^2)$. Here one should not be misled by the use of the Fermi gas model in calculation of the influence of nuclear effects on deep inelastic lepton scattering [24], as for these processes only very high momentum components really contribute, and those are added by the authors of ref. [24] as a "tail" to the basic Fermi-gas momentum distribution.

The transverse momentum distribution of nuclear fragments is slightly wider than that of longitudinal momentum [25] and a certain azimuthal asymmetry in the emission of fragments has been observed [26,27], indicating a finite transfer to the fragmentating nucleus of transverse momentum and possibly also of angular momentum. The longitudinal momentum distribution reflects only the internal motion [25] and is therefore more directly related to the wave function of the nucleus [28].

The fragmentation of relativistic deuterons in a hydrogen bubble chamber (reaction $d + p \rightarrow p + n + p$ at 3.33 GeV/c) has been used at Dubna for studying the deuteron wave function [29]. The best description of the data was obtained with the Gartenhaus-Moravcsik wave function. A similar experiment [30], performed with a high-resolution magnetic spectrometer "Alpha" (reaction $d + C \rightarrow p + \ldots$ at $P = 8.9$ GeV/c) allowed one to reach higher momentum transfers and revealed a discrepancy with the standard nucleon-nucleon interaction form (the "Paris" potential [31]), see Fig. 1. The "Dubna wave function" extracted from these data agrees with that from electroproduction experiments at SLAC [32], Kharkov [33] and Saclay [34]. The observed excess for $k \gtrsim 0.2$ GeV/c (k – relative nucleon momentum in deuteron) is interpreted as a manifestation of the six-quark component in the deuteron wave function: $\psi(d) = (1-\alpha)\psi(pn) + \alpha \cdot \psi(6q)$, with a value of about 10% for the "6q" admixture [35].

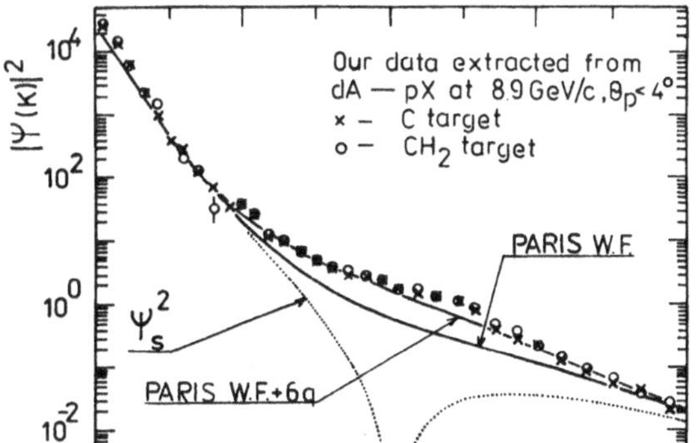

Fig. 1

An investigation of the fragmentation of polarized deuterons [36] is planned at Dubna by the same group.

For heavier nuclei, as pointed out in refs. [37,38], the longitudinal momentum distribution in the fragmentation process with the removal of one nucleon, such as $^4He \rightarrow {}^3He$, or $^{16}O \rightarrow {}^{15}O$, will directly reflect the internal momentum distribution of a nucleon inside the nucleus.

The reaction $^4He \rightarrow {}^3He$ at $P = 8.6$ GeV/c has been studied at Dubna using the 1 m hydrogen bubble chamber [39]. The momentum distribution of the "spectator" ^3He fragments up to about 400 MeV/c can be satisfactorily described using the wave function of the ^4He nucleus given by Bassel and Wilkin [40] and the formalism developed by Kopeliovich and Potashnikova [41]. Alternatively, the charge density distribution as given by Sick [42] may be taken and the spectator momentum distributions obtained by a Monte Carlo procedure. This method allows one to calculate momentum distributions not only for single nucleons but also for nucleon clusters as shown for the reaction $^4He \rightarrow {}^2H$ in ref. [43]. The same group is now studying the reaction $^3He \rightarrow {}^2H$ at $P = 8.0$ and 13.5 GeV/c. Also the "Alpha" spectrometer group has begun to investigate $^3He \rightarrow p$ and $^3He \rightarrow d$ reactions at beam momentum from 6.0 to 13.5 GeV/c with the aim to extract the wave function, especially at higher values of internal momentum. A similar experiment with ^4He is planned. In order to describe the ^3He data, in ref. [44] the hybrid model was appropriately extended. Besides the "6q" admixture, a "9q" contribution was also included at a level of $\sim 10^{-3}$. The reaction $^{16}O \rightarrow {}^{15}O$ at $E/A = 2.1$ GeV was studied at Berkeley [38] with the result that a high-momentum part of the wave function should be modified. A similar reaction $^{16}O \rightarrow {}^{15}N$ at $E/A = 3.5$ GeV is being studied at Dubna using the 1 m hydrogen bubble chamber.

An evidence for the existence of multiquark configurations in nuclei and for their increasing role with increasing A comes also from the analysis of nuclear structure functions. A first indication was provided by experiments on cumulative particle production [45]. We call "cumulative" those processes of particle emission which proceed beyond kinematical limits of a single nucleon-nucleon collision (for a target at rest such processes would require a target heavier than nucleon mass). The emission of pions, kaons and antiprotons at backward angles (i.e., with high values of longitudinal momentum transfer) has been studied at Dubna in proton-nucleus collisions at 8.9 GeV. The results of this experiment are shown in Fig. 2a. It has

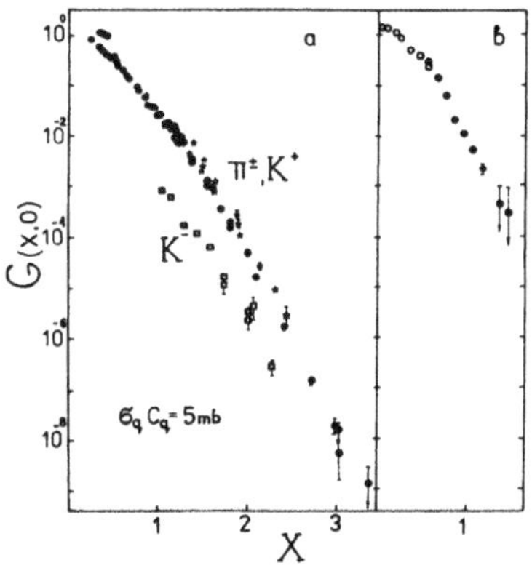

Fig. 2

been found [46] that inclusive cross sections factorize and can be expressed as a product of two universal functions:

$$\frac{E}{A}\frac{d^2\sigma}{p^2dp^2d\Omega} = C\ \varphi(p_\perp^2)\ \ \exp(-X/\langle X\rangle),$$

where $\varphi(p_\perp^2) = 0.9\exp(-2.7\ p_\perp^2) + 0.1$ describes the transverse momentum dependence, and X is a longitudinal scaling variable, defined as

$$X = \frac{(P_I\cdot P_1) - \frac{1}{2}m_1^2}{(P_I\cdot P_{II}) - M_I M_{II} - (P_{II}\cdot P_1)}\ .$$

Here P_I and P_{II} are the four-momenta of colliding nuclei, M_I and M_{II} their masses, and the index 1 corresponds to a secondary particle. For deep inelastic lepton scattering, neglecting masses, X/A_{II} approaches the Bjorken variable $x = Q^2/2M\nu$. The variable X is called the effective cumulative number, and in the rest frame of the fragmentating nucleus it is approximately equal to the minimum target mass $M_{min} = (E - p_{\parallel})/m_N$, where E and $p_{\parallel}$ are the energy and longitudinal momentum of the produced particle, m_N is the nucleon mass. The "cumulative" region corresponds to $X > 1$. Experimental distributions show a universal slope of $\langle X\rangle = 0.14$. Similar results and the same value of the slope have been obtained in the proton-nucleus "cumulative" experiment at 400 GeV at Fermilab [47].

In quark-parton models the function $G(X) = \exp(-X/\langle X\rangle)$ is interpreted as a quark-parton structure function of the nucleus, and the value of $\langle X\rangle$ as an average longitudinal momentum of quarks in the nucleus [48]. At least two arguments can be given in favour of such interpretation of $G(X)$.

Firstly, the same value of $\langle X\rangle$ has been obtained in deep inelastic muon scattering on nuclei. The corresponding prediction was made by Baldin [11,49] and verified by the NA-4 experiment at CERN [50]. These results are shown in Fig.2b, while Fig.3 compiles the values of $\langle X\rangle$ for proton-nucleus "cumulative" experiments and for deep inelastic scattering of 280 GeV muons on nuclei. As deep inelastic lepton scattering is believed to represent interactions with quarks, the universal value of the slope parameter $\langle X\rangle$ for all discussed experiments indicates the quark nature also for cumulative processes.

Secondly, the value of $\langle X\rangle = 1/6$, close to the experimental value of 0.14, follows from quark counting rules. According to refs. [51,52], the expected behaviour is $(1 - X/i)^{6i-3}$ which approaches exp(-6X) already for not very large i .

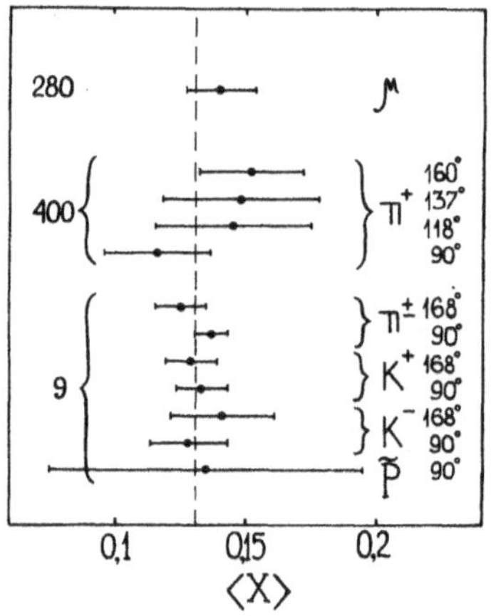

Fig. 3

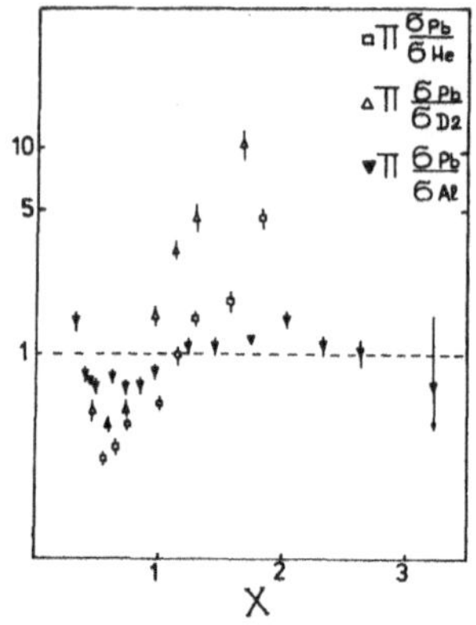

Fig. 4

Ratios of quark-parton structure functions for various nuclei, σ_{Pb}/σ_{D_2}, σ_{Pb}/σ_{He} and σ_{Pb}/σ_{Al}, as obtained from the data on cumulative particle production, are shown in Fig. 4. Their behaviour can be interpreted as a manifestation of multiquark configurations existing in nuclei [48]. The amount of multiquark configurations increases with increasing A and for $A \gtrsim 20$ it seems to stabilize at a certain level.

The above conclusions have been confirmed by experimental results on deep inelastic scattering of muons at CERN [53] and electrons at SLAC [54] on various nuclei. In these experiments it has been found that structure functions of Fe and D nuclei differ strongly, and their ratio cannot be explained by the known nuclear mechanism. This is often called "EMC effect" after ref. [53]. Results of these experiments are, however, restricted to $X<1$.

Of many theoretical papers aiming to explain the EMC effect we shall quote only a few. Faissner and Kim [55] and Kondratyuk and Shmatikov [56] claim that in order to describe the experimental data, the admixture of "12q" clusters in heavy nuclei should be as high as 15-25%. The latter authors stress that "12q" bags are not identical to alpha-particles, the quark distribution in a bag being different from that in a nucleon. Garsevanishvili and Menteshashvili [57], on the other hand, claim to need only "6q" bags and a small admixture of "9q" bags with the same bag radius of $r_q = 0.8$ fm in order to describe the nuclear structure functions both in the $X<1$ and $X>1$ kinematical regions.

Speaking of the fragmentation of relativistic nuclei, one is tempted to say a few words about "anomalons" (nuclear fragments with anomalously high interaction cross sections), which have attracted great attention during the last few years. Now it seems that the odds are against the existence of "anomalons": one can quote here several recent experiments using emulsions [58,59], plastic detectors [60] and multilayer Cherenkov counters [61, 62]. In Fig. 5 the results are shown of the emulsion work by the Alma-Ata etc. Collaboration [58], in which statistics was high enough to allow fragments of all charges from Z = 3 to Z = 10 to be studied separately. No evidence for anomalons can be seen.

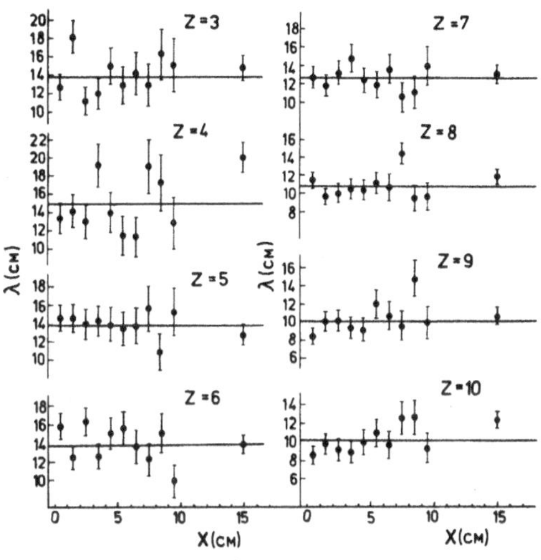

Fig. 5

Concluding the first part of my talk, I would like to state that fragmentation processes of relativistic nuclei are perhaps "ordinary nuclear physics in a fast-moving reference frame" only at very low momentum transfers, while at higher momentum transfers they reflect the quark structure of nuclei. Quark-parton models assuming the existence of multiquark configurations in nuclei seem to be able to explain the behaviour of nuclear form-factors, deep inelastic lepton scattering and particle production in the cumulative region. Substantial differences exist, however, between various formulations of such models.

Coming now to the second topic, which is multiple particle production, I would like to point out that most of the relevant experimental data obtained at Dubna, as well as at Berkeley, are inclusive data (cross sections, multiplicities, single-particle spectra, etc.). Inclusive measurements have recently become the subject of strong criticism as non-discriminative between various theoretical models and thus not useful for our understanding of the underlying physics (see, e.g., [63]). Let us, however, recall that the main result of the large-angle two-proton correlation experiment at Berkeley was a direct observation of quasi-elastic NN scattering processes in collisions of relativistic nuclei [64] (Clean Knock-On, or CKO processes in the terminology of ref. [65]). An evidence has been also obtained that most of the protons suffer more than one collision [66]. A substantial role of NN scattering, both elastic and inelastic, could, however, be expected on the basis of the de Broglie wave length being much shorter than the average internucleon distance [67], and a successful description of experimental data by the cascade model supports this conjecture. So, we have not learnt anything exciting.

Multiple particle production has been studied at Dubna by means of the 2m propane bubble chamber with internal tantalum target plates, exposed to p, d, ^{4}He and ^{12}C beams at P/A = 4.2 GeV/c, and the 2m streamer chamber "SKM-200" with various internal solid targets exposed to ^{4}He, ^{12}C and ^{16}O beams at P/A = 4.5 GeV/c. Small exposures of the bubble chamber to ^{22}Ne and ^{24}Mg beams have been also made for methodical purposes. Main results from the bubble chamber can be found in refs. [68-70], and those from the streamer chamber in refs. [71-73]. Let us look at some new results on interactions of p, d, ^{4}He and ^{12}C with carbon [74]. These reactions were selected from interaction events occurring in propane. Secondary tracks in about 2000 events of each group were measured, and single-particle inclusive distributions were obtained. Table 2 shows the average characteristics of secondary π^- mesons together with the predictions of the Dubna version of the cascade model - DCM [75]. Fig. 6 shows π^- rapidity distributions also compared to DCM. The agreement with the model is good.

Table 2

		pC	dC	HeC	CC
$\langle n_- \rangle$		0.33 ± .02	0.62 ± .03	1.07 ± .05	1.52 ± .07
$\frac{\langle n_- \rangle}{D_-^2}$	expt.	1.14 ± .08	1.23 ± .08	1.11 ± .08	0.85 ± .07
	DCM	1.23	1.15	1.07	0.70
$\langle p_\perp \rangle$	expt.	0.26 ± .01	0.26 ± .01	0.26 ± .01	0.25 ± .01
GeV/c	DCM	0.21	0.23	0.24	0.24
$\langle y_{Lab} \rangle$	expt.	0.85 ± .04	1.00 ± .03	1.04 ± .03	1.10 ± .03
	DCM	0.95	0.98	1.05	1.05

Data on π° production in d-Ta and C-Ta collisions at 4.2 GeV/c per nucleon [76] are certainly worth mentioning. In the JINR propane bubble chamber gamma-quanta can be recorded with an average registration probability of ~8%. On the basis of 248 e^+e^- pairs from d-Ta and 855 e^+e^- pairs from C-Ta interactions it has been found that the average numbers of π° mesons and their multiplicity distributions are the same as the corresponding distributions for π^-. Fig. 7 displays the dependence of the average number of π^o versus the number of π^-. Contrary to NN interactions at the same energy per nucleon where $\langle n_o \rangle$ decreases with increasing n_-, in nucleus-nucleus collisions $\langle n_o \rangle$ increases with n_- up to $n_- = 2\langle n_- \rangle$. This difference in behaviour seems, however, only to reflect the mechanism of multiple NN collisions in nucleus-nucleus interactions.

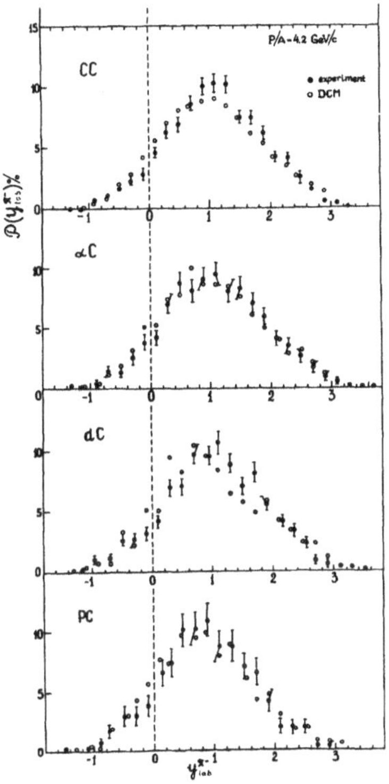

Fig. 6

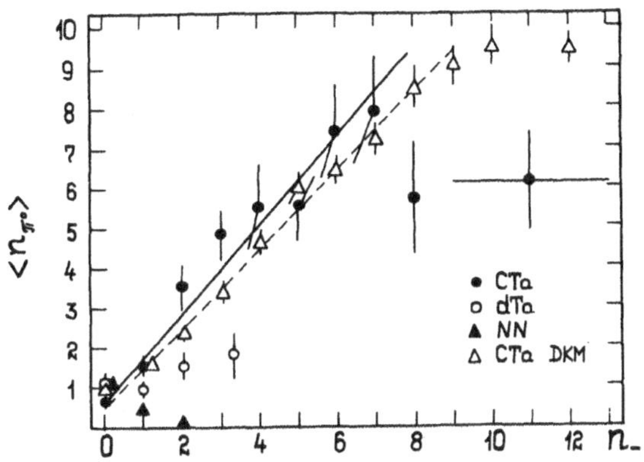

Fig. 7

An interesting question in nucleus-nucleus collisions is how many nucleons participate in the interaction. The number of the interacting nucleons from the projectile nucleus, ν_p, can be determined experimentally by counting non-interacting (spectator) nucleons. In our earlier papers it has been shown that for a given target nucleus the multiplicity of produced pions is proportional to ν_p: $\langle n_- \rangle = a\langle \nu_p \rangle$, which confirms the conjecture about independent interaction in the target of nucleons from the projectile[10]. A direct determination of the corresponding number, ν_t, for the target nucleus is difficult, as low-energy fragments of the target nucleus are not detected. One could, however, use the asymmetry in the emission of secondary particles in order to determine ν_t [77]. The additive quark model with coloured strings of Białas et al. [78] provides a possible theoretical basis for such an approach. According to this model, particles emitted in the central region result from the break-up of coloured strings spanned between interacting quarks. Average velocity of particles created by this mechanism is determined by numbers of quarks (and thus also: nucleons) from the projectile, N_p, and from the target, N_t, "wounded" by the interaction. Assuming that the distribution of particles resulting from the break-up of coloured strings is symmetric, the ratio of the numbers of interacting nucleons from target and projectile N_t/N_p can be found from the formula $N_t/N_p = (\beta_o - \beta_{sc})/(\beta_o + \beta_{sc})$, where β_o is initial nucleon velocity in the NN centre-of-mass system and β_{sc} velocity of the "symmetry system" relative to the NN system. For collisions of various light nuclei (d, ^4He, ^{12}C) with tantalum one obtains the numbers displayed in Table 3 and Fig. 8. The ratio N_t/N_p decreases with increasing A_p as $A_p^{-(0.73 \pm .09)}$. Taking the values of N_p determined in our experiment (see Table 3), one can determine N_t. The obtained values of N_t are given in the lowest raw of Table 3. They show a very weak dependence on A_p.

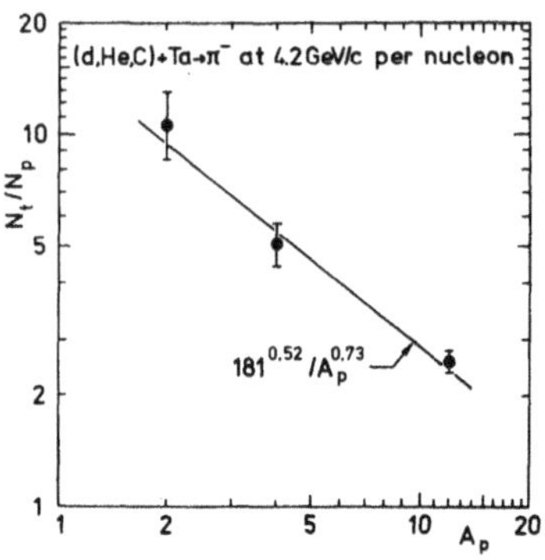

Fig. 8

In C-C interactions at 4.2 GeV/c per nucleon interference correlations have been studied both for $\pi^-\pi^-$ and pp pairs [79]. For $\pi^-\pi^-$ pairs the formalism of Kopylov and Podgoretsky [80] has been used, and for pp pairs - that of Koonin [81] and Lednický and Lyuboshitz [82]. The obtained distributions are shown in Fig. 9a,b. They yield the following values of the r.m.s. radii of the emission volume: $r(\pi\pi) = (2.8 \pm 0.7)$fm for an unbiased sample and $r(\pi\pi) = (3.8 \pm 0.9)$fm for "central" collisions. The corresponding values of r(pp) are about 4.3 and 3.5 fm with an estimated error of about 0.4 fm. Smaller value of r(pp) for "central" collisions (a tendency opposite to that of $r(\pi\pi)$) could be qualitatively explained using the thermodynamical model with two fireballs and taking the production of delta-isobars into account.

Table 3

	d-Ta	He-Ta	C-Ta
N_t/N_p	10.5 ± 1.9	5.0 ± 0.6	2.5 ± 0.2
$\langle N_p \rangle$ $(n_- \geqslant 1)$	$(1.7)^+$	3.3 ± 0.1	7.4 ± 0.1
$\langle N_t \rangle$ $(n_- \geqslant 1)$	$(15.9)^+$	16.5 ± 2.1	18.2 ± 1.1

$^+$ extrapolated values

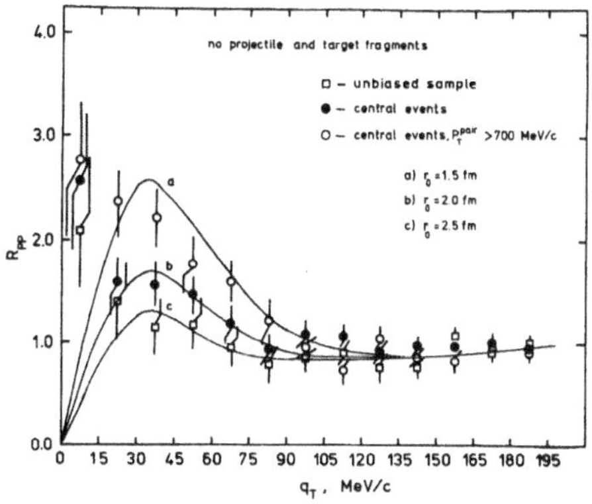

Fig. 9a

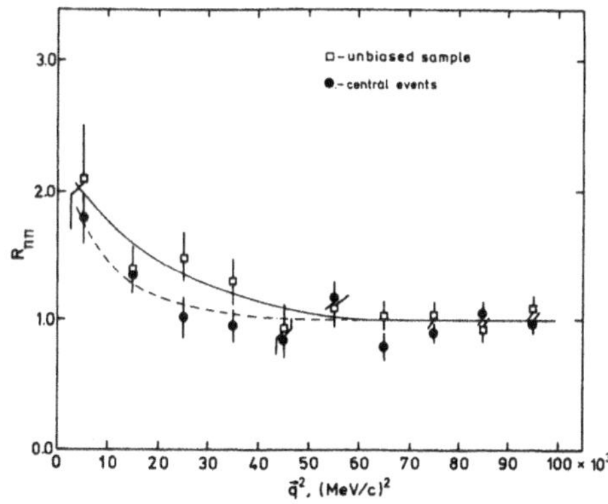

Fig. 9b

In connection with this topic I would like to make some comments on the use of two-particle interferometry [83]. The first point concerns the definition of the radius of the emission volume. In experimental papers various formulae are used for the description of the two-particle correlation function assuming various spatial distributions of the sources (the surface of a sphere, several forms of Gaussian-type distributions), and thus the published values of the parameter called the "radius" of the emission volume cannot be always compared directly with results of other authors. In order to allow such a comparison, we propose to use consistently the <u>root-mean-square radius</u>, as generally accepted for the description of the density distribution inside the nucleon and also for the model description of heavy-ion collisions at lower energies. This would mean the necessity of correcting some of the published results on the radius of the pion emission volume. The conversion factor to the r.m.s. radius is $\sqrt{3/2}$ for the Gaussian-type distribution in the form $\exp(-r^2/R^2)$, $\sqrt{3}$ for the distribution $\exp(-r^2/2R^2)$ and 1.0 when the sources are supposed to be uniformly distributed over the surface of a sphere of radius R. The factor $\sqrt{3/2}$ (or $\sqrt{3}$) should be also applied to the values of the parameter r_o obtained from proton-proton interferometry. It is also essential to compare results obtained in the same reference frame, since the procedure is not invariant under Lorentz transformation.

The second remark concerns the interpretation of the radius of the particle emission volume. The values obtained from both pion and proton interferometry are relatively large as compared to the r.m.s. radii of colliding hadrons (or of the smaller of colliding nuclei) and only weakly depend on the size of colliding objects. The reason is that this method yields the values of the radii corresponding to the emission of particles after they become free of any interactions (decay of resonances, scattering from other hadrons). This would mean that the excited nuclear matter density estimates using pion and proton interferometry results for the radii are only lower limits, not really meaningful for the investigation of the equation of state.

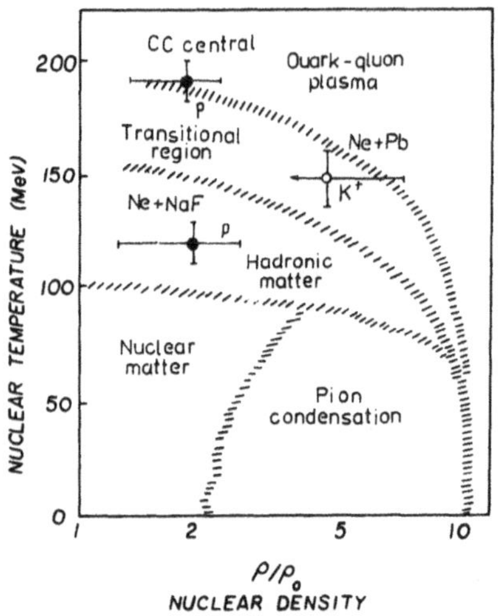

Fig. 10

Nevertheless, having no prompt leptons of photons at our disposal, such attempts are being repeated using hadrons. For C-C central collisions at 4.2 GeV/c per nucleon, and with additional selection $p_{\perp} \geqslant 500$ MeV/c, we have obtained $r^{rms}(pp) = (2.6 \pm 0.4)$fm , which, together with the total number of participating nucleons being about 17, gives estimate of $\rho = (1.8 \pm 0.5)\,\rho_o$ for the density of hot nuclear matter. This value, combined with the "temperature" $T = (190 \pm 10)$MeV estimated from the slope of the secondary proton spectrum, gives a point on the nuclear matter phase diagram as shown in Fig. 10 [84]. Though the above value of T, obtained from the invariant energy spectrum, might be somewhat overestimated, it seems that at moderate energies and for not very heavy nuclei, we may already be not far from a transition of the quark-gluon plasma ....

And this is what one of the little people who supposedly live in Scandinavian woods and rocks (they are called t r o l l e n in Swedish and p e i k o t in Finnish) is thinking about our problems:

REFERENCES

1. Yu.D.Beznogikh et al.: JINR P9-84-246, Dubna, 1984.
2. A.A.Belushkina et al.: JINR E13-80-500, Dubna, 1980.
3. E.D.Donets, A.I.Pikin: Journ.Techn.Phys. 45, 2373(1975).
4. N.I.Balandikov et al.: JINR P8-80-172, Dubna, 1980.
5. A.M.Baldin et al.: Proc.Particle Accelerator Conference, Santa Fe, 1983; IEEE Trans.Nucl.Sci. NS-30, No.4, 1983, p.3247.
6. I.A.Shelayev et al.: Proc.Intern.Conf. on High Energy Accelerators, Fermilab, 1983, p.416.
7. A.M.Baldin: Proc.6th Intern.Conf.on High Energy Physics and Nuclear Structure, Santa Fe and Los Alamos, 1975, p.621; JINR E2-9238, Dubna, 1975.
8. A.M.Baldin: Report to 7th Intern.Conf.on High Energy Physics and Nuclear Structure, Zürich, 1977; JINR E1-11368, Dubna, 1978.
9. A.M.Baldin: Proc. 19th Intern.Conf.on High Energy Physics, Tokyo, 1978, p.455.
10. J.Bartke: Proc. 8th Intern.Conf.on High Energy Physics and Nuclear Structure, Vancouver, 1979, p.481; Nucl.Phys. A335, 481 (1980).
11. A.M.Baldin: Proc.Intern.Conf.on Extreme States in Nuclear Systems, Dresden, 1980, v.2, p.1; JINR E1-80-174, Dubna, 1980.
12. V.S.Stavinsky: ibid, p.126.
13. M.I.Soloviev: ibid, p.165.
14. J.Bartke: Report to Intern.Conf.on Nuclear Physics, Berkeley, 1980; JINR E1-80-754, Dubna, 1980.
15. A.M.Baldin: Proc. 1981 CERN-JINR School of Physics, CERN 82-04,p.1.
16. A.A.Kuznetsov: Proc.Intern.Conf.on Nucleus-Nucleus Collisions, East Lansing, 1982; Nucl.Phys. A400, 493c (1983); JINR E1-82-660, Dubna, 1982.
17. A.A.Kuznetsov: Proc. 6th Intern.Conf.on High Energy Nuclear Physics, Balatonfüred, 1983, p.223; JINR E1-83-334, Dubna, 1983.
18. J.D.Bowman, W.J.Swiatecki, C.F.Tsang: LBL-2908, Berkeley, 1973.
19. P.Zieliński : GSI-82-13, Darmstadt, 1982.
20. B.S.Aladashvili et al.: Nucl.Instr.Meth. 129, 109 (1975).
21. D.E.Greiner et al.: Phys.Rev.Lett. 35, 152 (1975).
22. A.S.Goldhaber: Phys.Lett. 53B, 306 (1974).
23. G.F.Bertsch: Phys.Rev. C15, 713 (1977).
24. A.Bodek, J.L.Ritchie: Phys.Rev. D23, 1070 (1981); Phys.Rev. D24, 1400 (1981); UR-772, Rochester, 1981.
25. S.Nagamiya, M.Gyulassy: Adv.Nucl.Phys. 13, 201 (1983).
26. G.M.Chernov et al.: Nucl.Phys. A412, 534 (1984).

27. L.E.Bengus et al.: Pisma Zh.Eksp.Teor.Fiz. 38, 353 (1983).
28. L.Anderson et al.: Phys.Rev. C28, 1224 (1983).
29. B.S.Aladashvili et al.: Nucl.Phys. B86, 461 (1975).
30. V.G.Ableev et al.: Nucl.Phys. A393, 491 (1983); A411, 541(E) (1983); Pisma Zh.Eksp.Teor.Fiz. 37, 196 (1983) /English translation: JETP Pisma 37, 233 (1983) /.
31. M.Lacombe et al.: Phys.Lett. 101B, 139 (1981).
32. P.Bosted et al.: Phys.Rev.Lett. 49, 1380 (1982).
33. I.L.Agranovitch et al.: Yad.Fiz. 25, 1123 (1977) /English translation: Sov.J.Nucl.Phys. 25, 1117 (1977)/.
34. M.Bernheim et al.: Nucl.Phys. A365, 349 (1981).
35. A.P.Kobushkin: Yad.Fiz. 28, 495 (1978) /English translation: Sov.J.Nucl.Phys. 28, 252 (1978)/.
36. A.P.Kobushkin, L.Vizireva: J.Phys.G: Nucl.Phys. 8, 893 (1982).
37. T.Fujita, J.Hüfner: Nucl.Phys. A343, 493 (1980).
38. J.Hüfner, M.C.Nemes: Phys.Rev. C23, 2538 (1981).
39. V.V.Glagolev et al.: Phys.Rev. C18, 1382 (1978).
40. R.H.Bassel, C.Wilkin: Phys.Rev. 174, 1179 (1968).
41. B.Z.Kopeliovich, I.K.Potashnikova: Yad.Fiz. 13, 1032 (1971) /English translation: Sov.J.Nucl.Phys. 13, 592 (1971) /.
42. I.Sick: Lepton Scattering, presented at Conference on Few Body Systems and Nuclear Forces, Graz, 1978.
43. B.S.Aladashvili et al.: Acta Phys.Slov. 31, 29 (1981).
44. V.V.Burov, V.K.Lukyanov, A.I.Titov: JINR P2-83-749, Dubna, 1983; Subm. to Z.Phys.A.
45. V.S.Stavinsky: ECHAYA 10, 949 (1979).
46. A.M.Baldin et al.: JINR E1-82-472, Dubna, 1982.
47. N.A.Nikiforov et al.: Phys.Rev. C22, 700 (1980).
48. A.M.Baldin: JINR E2-83-415, Dubna, 1983.
49. A.M.Baldin et al.: JINR 1-82-28, Dubna, 1982.
50. I.A.Savin: Proc. 6th Intern.Seminar on High Energy Physics Problems, JINR D1, 2-81-728, Dubna, 1981, p.223.
51. S.J.Brodsky, R.Blankenbecler: Phys.Rev. D10, 2973 (1974).
52. S.A.Gerzon: Yad.Fiz. 38, 1313 (1983).
53. J.J.Aubert et al.: Phys.Lett. 123B, 275 (1983).
54. A.Bodek et al.: Phys.Rev.Lett. 51, 534 (1983); 50, 1431 (1983).
55. H.Faissner, B.R.Kim: Phys.Lett. 130B, 321 (1983).
56. L.Kondratyuk, M.Shmatikov: ITEP-13, Moscow, 1984.
57. V.R.Garsevanishvili, Z.R.Menteshashvili: JINR E2-84-314,Dubna,1984.
58. Alma-Ata, Bucharest, Dubna, Dushanbe, Erevan, Gatchina, Košice, Kraków, Leningrad, Moscow, Tashkent, Tbilisi, Ulan-Bator Collaboration: Proc. 18th Intern.Cosmic Ray Conf., Bangalore, India,

1983, Late paper volume, paper HE 2.2-30.
B.P.Bannik et al.: Pisma Zh.Eksp.Teor.Fiz. 39, 184 (1984).
59. A.Z.M.Izmail et al.: Phys.Rev.Lett. 52, 1281 (1984).
60. W.Heinrich et al.: Phys.Rev.Lett. 52, 1401 (1984).
61. J.D.Stevenson, J.A.Musser, S.W.Barwick: Phys.Rev.Lett. 52, 515 (1984).
62. T.J.M.Symons et al.: Phys.Rev.Lett. 52, 982 (1984).
63. M.Gyulassy: LBL-16800, Berkeley, 1983, Proc.BNL Workshop 1983, Nucl.Phys.A, in press.
64. S.Nagamiya et al.: Phys.Lett. 81B, 147 (1979).
65. R.L.Hatch, S.E.Koonin: Phys.Lett. 81B, 1 (1978).
S.E.Koonin: Phys.Rev.Lett. 39, 680 (1977).
66. I.Tanihata et al.: Phys.Lett. 97B, 363 (1980).
67. S.Nagamiya: Proc.Workshop on Quark Matter Formation and Heavy--Ion Collisions, Bielefeld, 1982, p.281.
68. E.O.Abdrahmanov et al.: Z.Phys. C5, 1 (1980).
69. H.N.Agakishiev et al.: Z.Phys. C12, 283 (1982).
70. H.N.Agakishiev et al.: Z.Phys. C16, 307 (1983).
71. V.D.Aksinenko et al.: Nucl.Phys. 324, 266 (1979).
72. V.D.Aksinenko et al. Nucl.Phys. A348, 518 (1980).
73. A.U.Abdurakhimov et al.: Nucl.Phys. A362, 376 (1981).
74. H.N.Agakishiev et al. JINR P1-84-35, Dubna, 1984; subm. to Yad.Fiz.
75. K.K.Gudima, V.D.Toneev: Nucl.Phys. A400, 173 (1983).
76. G.P.Gulkanian et al.: JINR P1-83-805, Dubna, 1983; subm. to Yad.Fiz.
77. J.Bartke et al.: JINR E1-84-408, Dubna, 1984.
78. A.Białas, W.Czyż, L.Leśniak: Phys.Rev. D25, 2328 (1982).
79. N.Akhababian et al.: JINR E1-83-670, Dubna, 1983; Z.Phys.C, in press.
80. G.I.Kopylov, M.I.Podgoretsky: Yad.Fiz. 15, 392 (1972).
81. S.E.Koonin: Phys Lett. 70B, 43 (1977).
F.B.Yano, S.E.Koonin: Phys.Lett. 78B, 556 (1978).
82. R.Lednicky, V.L.Lyuboshitz: JINR E2-81-453, Dubna, 1981; Yad.Fiz. 35, 1316 (1981).
83. J.Bartke, M.Kowalski: Phys.Rev.C, in press.
84. L.A.Didenko et al.: JINR E1-84-354, Dubna, 1984.

E_T° AND π° SPECTRA FROM $\alpha\alpha$, dd, AND pp INTERACTIONS AT $\sqrt{s}_{NN}$ = 31 GeV AT THE CERN ISR

BNL-CERN-Michigan State-Oxford-Rockefeller (BCMOR) Collaboration[1]

Presented by

Michael J. Tannenbaum
Brookhaven National Laboratory
Upton, NY 11973, USA

Inclusive Single Particle Production

Inclusive single particle production at large transverse momentum, P_T, from heavy nuclear targets was first studied by Cronin and collaborators[2] at Fermilab in the period 1973-1977, basically because of the unavailability of liquid hydrogen targets. They measured the invariant cross section $Ed^3\sigma/dp^3$, for inclusive $\pi^\pm$ $K^\pm$ $p,\bar{p}$ production near 90° from the interaction of 200 to 400 GeV/c protons with an assortment of nuclear targets and found that the atomic weight dependence at fixed P_T could be represented by a power law:[3]

$$\frac{\sigma_{pA}(P_T)}{\sigma_{pp}(P_T)} = A^{\alpha(P_T)} \quad .$$

The exponent $\alpha(P_T)$ varied with P_T, the particle produced and the nucleon-nucleon center-of-mass energy, $\sqrt{s}_{NN}$, and reached values in excess of $\alpha(P_T) = 1$. This was dubbed the "Anomalous Nuclear Enhancement." For $\pi^\pm$ production, $\alpha(P_T)$ increased with P_T, reaching a maximum value at $P_T \sim 4.6$ GeV/c which varied between $\alpha = 1.18$ at $\sqrt{s}_{NN} = 19.4$ GeV and $\alpha = 1.13$ at $\sqrt{s}_{NN} = 27.4$ GeV.

The next step in the study of the collisions of nucleii at high energy occurred at the CERN ISR in 1980, with runs on $\alpha\alpha$ and $p\alpha$ interactions at nucleon-nucleon center-of-mass energies of $\sqrt{s}_{NN} = 31$ GeV for $\alpha\alpha$ and $\sqrt{s}_{NN} = 44$ GeV for $p\alpha$. Several experiments took data on inclusive single particle production.[4] The conclusion was that the anomalous enhancement died away for $p\alpha$ collisions, $\alpha \lesssim 1$ at $\sqrt{s}_{NN} = 44$ GeV, but came on even stronger for $\alpha\alpha$ collisions at $\sqrt{s}_{NN} = 31$ GeV. The ratio of the cross section for inclusive production of a π° at a given P_T in $\alpha\alpha$ and pp collisions varied from a value of ~ 18 in the range of $3 \leq P_T \leq 4$ GeV/c to ~ 24 to 30 in the range $4 \leq P_T \leq 5$ GeV/c to ~ 30 to 40 in the range $5 \leq P_T \leq 7$ GeV/c. The best quoted error[5] on this ratio was $\sim 20\%$

at $P_T \sim 6$ GeV/c. It is expected that the error will be reduced by a factor of 3 when results from the 1983 ISR run become available. If the ratio of the cross sections is assumed to obey a power law $(A_1A_2)^{\alpha(P_T)}$, then values of α as high as 1.32 ± 0.05 have been observed for inclusive pion production at $P_T \sim 6$ GeV/c in $\alpha\alpha$ collisions. The present situation is more thoroughly described in Martin Faessler's review article.[4] One conclusion that I draw from these results is that pA collisions are different from AA collisions. For a somewhat different point of view, see Sherman Frankel's presentation to this conference.[6]

Other Results Previous to 1984

In the period 1980 to 1983, some additional data became available. Among the most noteworthy were the results from the Multiparticle Spectrometer Group at Fermilab[7] who presented the first measurement of transverse energy, E_T, emission in the central region of pA collisions. In this type measurement, the transverse energy, $E_T = E_i \sin\theta_i$, is summed over all the particles emitted on an event into a fixed but large solid angle, typically $\Delta\phi = 2\pi$, $\Delta y = \pm 1$, where y is the rapidity measured in the c.m. system. A frequency distribution, or spectrum of E_T over all events is then computed. The E_T spectrum covered the range $1 \leq E_T \leq 21$ GeV for proton collisions on several nucleii from hydrogen to lead, at $\sqrt{s}_{NN} = 27.4$ GeV, and showed that the parameterization A^α was still reasonable. However, the exponent α had risen to values as high a $\alpha = 2.0$. The A.F.S. Group at CERN also presented a charged E_T spectrum,[8] reconstructed from charged particles observed on minimum bias triggers from the 1980 $\alpha\alpha$ and $p\alpha$ runs at the ISR. The spectrum falls 4 orders of magnitude over the charged E_T range from 0 to 12 GeV.

E_T° Results from the 1983 CERN ISR Run

This brings us to August 1983, when the second (and final) light ion run took place at the CERN ISR. This run was a considerable improvement over the 1980 run, with $\alpha\alpha$, $p\alpha$, dd and pp collisions all studied at $\sqrt{s}_{NN} = 31$ GeV. The $\alpha\alpha$ luminosity achieved at the low-β intersection I-1 was $1.0 \times 10^{30}\,cm^{-2}s^{-1}$. The integrated luminosity was a factor of 16 higher than the 1980 run (Table I).

Integrated Luminosity (I-1)

	1983	1980
αα	$1.0 \times 10^{35} cm^{-2}$	$6.0 \times 10^{33} cm^{-2}$
dd	$4.2 \times 10^{35} cm^{-2}$	--
pp	$5.3 \times 10^{35} cm^{-2}$	$1.9 \times 10^{36} cm^{-2}$ (June 1979)

Table I.

Preliminary results from this run were already presented at the 1983 Quark Matter Conference.[9,10] I shall now present the final results[11] for the total neutral energy spectrum, E°_{TOT}, for pp, dd and αα collisions at $\sqrt{s}_{NN}$ = 31 GeV as measured by the BCMOR collaboration,[1] ISR experiment R110.

In this experiment, the total neutral energy emitted in the central region was measured using an electromagnetic shower counter which detected, but did not separately resolve, the photons from the decays of π° and η° particles ($\pi^{\circ}\to\gamma\gamma$, $\eta^{\circ}\to\gamma\gamma$). The apparatus, Figure 1, consisted of a superconducting solenoid with a field of 1.4 Tesla inside of which were placed a barrel of 32 scintillation counters (A), a system of cylindrical drift chambers with 3 dimensional coordinate readout (DCM1-5) and four modules of lead-scintillator shower counters. The detection of electromaganetic showers was completed by two lead-glass modules located outside the magnet in the azimuthal region not covered by the shower

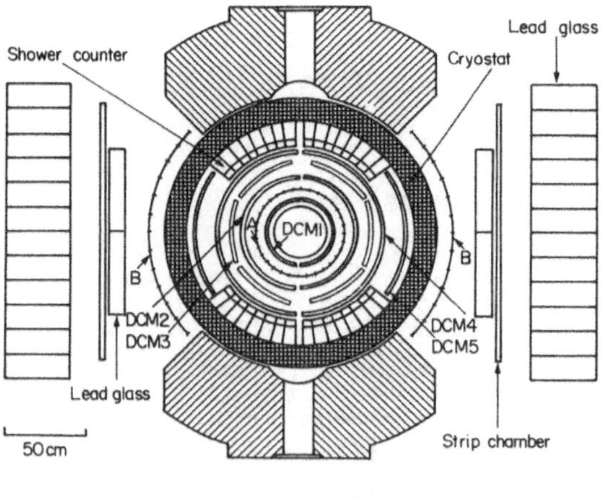

Figure 1

counters. The total thickness of the coil and cryostat was 1.0 radiation length, mostly aluminum, so that γ rays could penetrate without having their energies significantly degraded. The center-of-mass acceptance in which the neutral energy was detected covered 90% of 2π in azimuth with an average rapidity acceptance inside this region of $\Delta y = \pm 0.9$.

Each shower counter module was 1.5 m long, subtended 50° in azimuth and ±1.1 units of rapidity centered on $y = 0$, and was segmented azimuthally into eight counters. Each counter was a total of 14 radiation lengths thick which was subdivided in depth into front and back compartments (1:3) which were independently viewed by photomultipliers at both ends. The lead-glass modules each consisted of a front array of 34 blocks, 4 radiation lengths thick, and a back array of 168 blocks, 15 x 15 cm by 17 radiation lengths thick, arranged in a stack 12 blocks high by 14 wide. The angular acceptance of each lead glass module was 57° in azimuth and ±0.7 units of rapidity. The r.m.s. energy resolution, $\Delta E/E$, of the lead glass was $(4.3/\sqrt{E} + 2)$% and that of the shower counters was $(16/\sqrt{E})$%, with E in GeV. Charged particles were measured over the full azimuth in the c.m. rapidity interval $|y|<1.2$ using the drift chambers. The spatial resolution was about 350 μm, yielding a momentum resolution of $\Delta P_T/P_T = (7\%)P_T$, with P_T in GeV/c. Further details of the apparatus can be found elsewhere.[11,12]

The apparatus was triggered by summing all the energy in the lead glass and shower counters, with the total required to be above a threshold. In order to cover the wide dynamic range, nominal thresholds of 1, 5, 10, 15, 20 and 25 GeV were used. The threshold was applied again in the off-line analysis, using more accurate calibration information.

The following selection criteria were applied to the events:

(i) If more than one distinct time cluster were found during the apparatus recording time, the event was rejected as a multiple event. (ii) Events were required to have at least two charged tracks with a vertex within the interaction region. (iii) There must be some energy sharing between the front and back compartments of the lead glass and shower counters. Criteria (ii) and (iii) removed a small contamination due to cosmic ray and beam-gas interactions.

The longitudinal (along the beam axis) position of the energy deposition in the shower counters was calculated from the time difference or energy ratio of the photomultiplier signals at each end. The energy (ε) and transverse energy ($\varepsilon_T = \varepsilon \sin\theta$) in the N-N CM system were calculated for each lead-glass block and shower counter, assuming the energy was deposited by massless neutral particles originating at the vertex reconstructed by charged tracks. For each event these energies were summed to give the total neutral energy $E^{\circ}_{TOT} = \sum_i \varepsilon_i$ or transverse neutral energy $E_T^\circ = \sum_i \epsilon_T^i$. The energy deposited by charged tracks in the shower counters was estimated and subtracted from E_{TOT}° and E_T° for each event. Then a frequency distribution of the number of events per GeV of E_{TOT}° and E_T° was made and divided by the integrated luminosity L to obtain the spectra $(1/L)(dN/dE_{TOT}^\circ)$ and $(1/L)(dN/dE_T^\circ)$, as shown in Fig. 2. The data cover 10 orders of magnitude!

Only the statistical errors are shown in Figure 2. Due to the charged track energy correction there is also a systematic error in the $\alpha\alpha$ data for $E_{TOT}^\circ > 6$ GeV and the pp data for $E_{TOT}^\circ > 3$ GeV by which the $\alpha\alpha$ and pp data can be displaced together in E_{TOT}° by ± 0.45 GeV and ± 0.25 GeV, respectively. The systematic error in the slope, or logarithmic de-ative, of the $\alpha\alpha$, dd and pp data due to this correction is less than 1%. Other systematic errors in the data of Fig. 2 include an overall uncertainty on the E_{TOT}° scale of ± 5%. The data have not been corrected for resolution smearing or finite bin width, which would be equivalent to an overall energy scale shift of less than 1.8%. In addition, owing to a non-linearity introduced in the shower counter response by multiple hits within the same shower counter, the quoted E_{TOT}° values are high by 3% at $E_{TOT}^\circ = 30$ GeV and 1.5% at $E_{TOT}^\circ = 15$ GeV. This effect is three times larger in the E_T° spectra because multiple hits cause an error in angle as well as energy. Thus, the E_{TOT}° spectra are used for further analysis, since the systematics are better, and in this detector $\langle E_{TOT}^\circ \rangle \lessapprox 1.15 \langle E_T^\circ \rangle$.

The pp and $\alpha\alpha$ spectra shown in Fig. 2 both become exponentially decreasing for values of $E_{TOT}^\circ > 5$ GeV and 12 GeV, respectively. The logarithmic derivatives are -1.26 ± 0.01 GeV^{-1} for pp and -0.83 ± 0.02 GeV^{-1} for $\alpha\alpha$, including all the systematic corrections mentioned above except for the scale error which introduces a common ± 5% uncertainty in the values of the slopes. The ratio of the $\alpha\alpha$ to the pp sepctra rises monotonically from a factor of 7 at $E_{TOT}^\circ = 1.5$ GeV to 10^5 at $E_{TOT}^\circ \gtrsim 19$ GeV. Note that the $\alpha\alpha$ data extend beyond the nucleon-nucleon kinematic limit. This indicates the inadequacy of the parameterization $R(\alpha\alpha/pp) = (A_1A_2)^\alpha$ for the E_{TOT}° spectrum. The exponent α just increases linearly with E_{TOT}° until 19 GeV, when the pp data run out, and would be undefined beyond the kinematic limit.

It is not really suprising that the parameterization which worked for inclusive single particle production should be inadequate for the description of the total energy spectrum in nucleus-nucleus collisions. Single particle spectra are thought to be the result of a single hard collision of nucleons, and thus are sensitive mainly to successive collision effects as an incident nucleon passes through a nucleus before making a hard collision. By contrast, E_{TOT}° is the sum over many particles in the detector, so it is directly sensitive to multiple simulta-

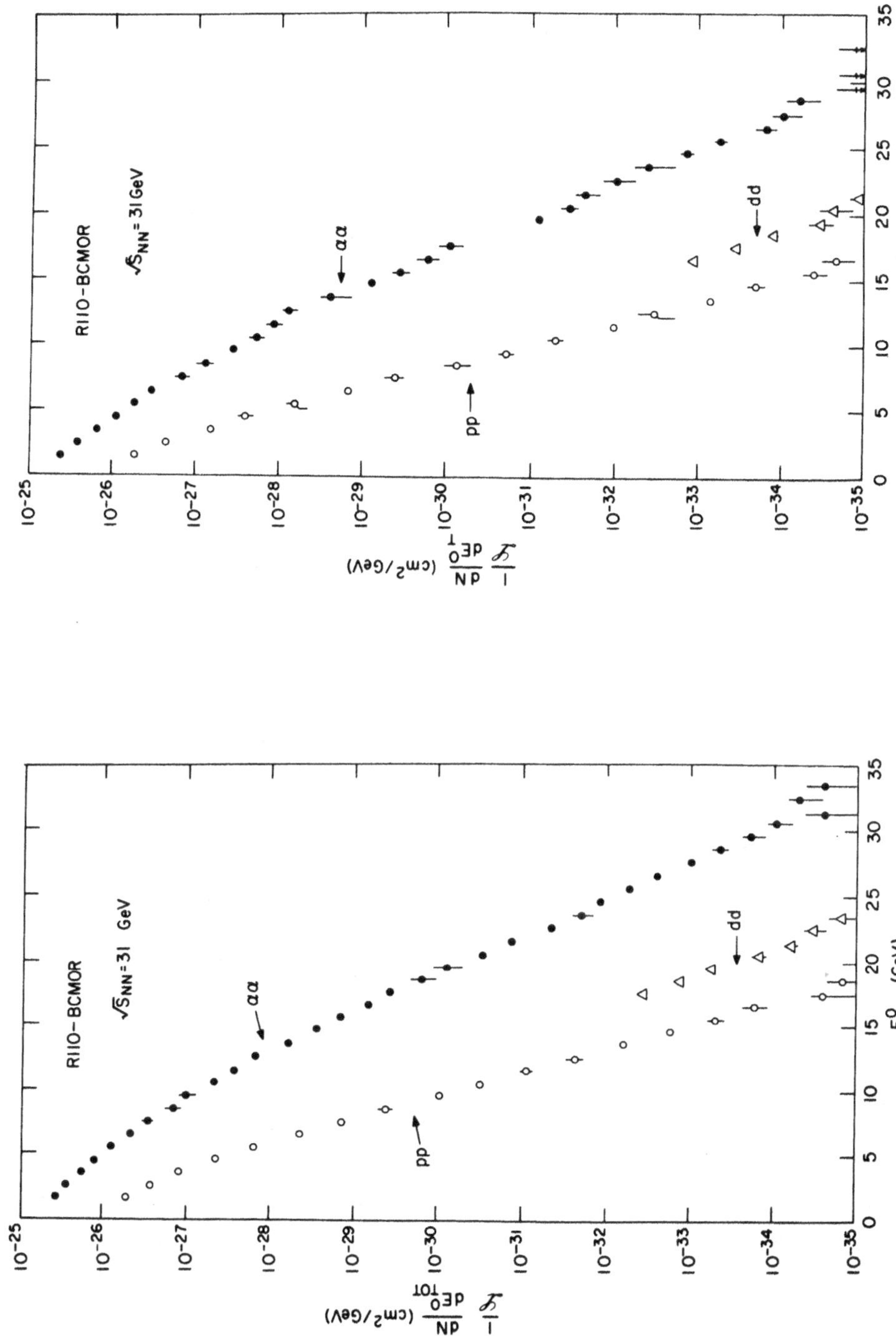
R110-BCMOR
√s_NN = 31 GeV
αα
pp
dd
(1/𝓛) dN/dE⁰_T (cm²/GeV)
E⁰_T (GeV)
R110-BCMOR
√s_NN = 31 GeV
αα
pp
dd
(1/𝓛) dN/dE⁰_TOT (cm²/GeV)
E⁰_TOT (GeV)

Figure 2

neous (or parallel) nucleon-nucleon (N-N) collisions in the interacting nucleii.

Therefore, we attempted to analyze the E°_{TOT} spectra of $\alpha\alpha$ and dd interactions as the result multiple parallel N-N collisions, with each N-N collision producing the observed E°_{TOT} spectrum of pp interactions. To do this the pp spectrum was treated as the probability function for the collision of two nucleons:

$$f_1(E) = \frac{1}{\sigma^{pp}_{in}} \left(\frac{d\sigma}{dE}\right)_{pp}$$

where $\sigma^{pp}_{in} = \int_0^{\sqrt{s}} (d\sigma/dE)_{pp} dE$ and the $f_1(E)$ is the differential probability for the emission of energy E in dE in our detector for a single nucleon-nucleon collision. Then, $f_n(E)dE$, the probability of observing E in dE for n such collisions overlapped, is simply a convolution integral:

$$f_n(E_0) = \int_0^{E_0} dE\, f_1(E)\, f_{n-1}(E_0-E)$$

where $0 \leq E_0 \leq n\sqrt{s}_{NN}$. Here, E_0 represents the energy emitted on n collisions. The first term inside the integral is the probability of emitting energy E on one collision and the second term is the probability of emitting a total of E_0-E on n-1 collisions.

It should be noted that all the n-collision spectra are normalized to unity; only the shape is determined. They can be renormalized to the probability or cross section for n simultaneous parallel nucleon-nucleon collisions, if this is known from a model. Alternatively, the data can be used to fit for the n-collision cross sections σ_n, n = 1,2,...m as parameters:

$$\frac{1}{L}\frac{dN^{\alpha\alpha}}{dE^{\circ}_{TOT}} = \sum_{n=1}^{m} \sigma_n\, f_n(E^{\circ}_{TOT}) \ . \qquad (1)$$

However, it is instructive to plot the individual $f_n(E)$ normalized to the observed pp interaction cross section in our detector:

$$\frac{1}{L}\frac{dN_n}{dE} = \sigma^{pp}_{in}\, f_n(E) \ .$$

This is shown in Figure 3 for n = 1,2,4&5 collisions.

With this "equal probability" normalization, the E°_{TOT} spectrum from two simultaneous pp collisions is an excellent description of the avail-

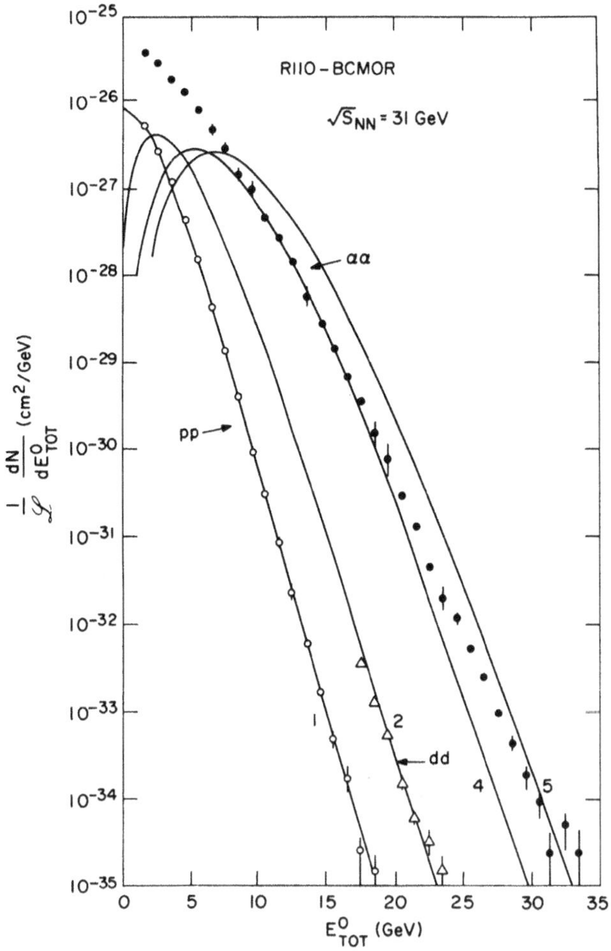

Figure 3

able dd data and similarly the E°_{TOT} spectrum from four simultaneous pp collisions fits the αα data over the range $8 \leq E^{\circ}_{TOT} \leq 17$ GeV. At higher values of E°_{TOT} the slope of the αα spectrum is too flat to be explained by this extreme case in which 4 individual nucleon-pairs are simultaneously interacting, or even by 5 simultaneous nucleon-pair collisions. Above 25 GeV, the slope of the αα data, -0.83 ± 0.02 GeV^{-1}, is much flatter than the -1.00 ± 0.01 GeV^{-1} slope of the 5-collision curve in this E°_{TOT} range.

This analysis can be made more sophisticated by fitting the full E°_{TOT} spectrum in αα collisions to a sum of n-collision convolutions of the observed pp spectrum, as indicated in Equation (1). This is greatly facilitated by using a gamma distribution to represent the N-N probability distribution $f_1(E)$. The pp spectrum in Figure 2 is fit to the function

$$f_1(E) = \frac{\alpha}{\Gamma(p)} (\alpha E)^{p-1} e^{-\alpha E} \tag{2}$$

for the parameters p and α. It should be noted that p>0, α>0, $f_1(E)$ is normalized to unity over the range $0 \leq E \leq \infty$ and that Γ(p) is the gamma function of p, = (p-1)! if p is an integer. The function fits the pp data very well with the result:

$\alpha = 1.41 \pm 0.01\ GeV^{-1}$ $\qquad p = 2.50 \pm 0.06$

$\sigma^{pp}_{in} = 13.1 \pm 0.3$ mb $\qquad \chi^2 = 24.6/15$ d.o.f.

The energy spectrum produced by n simultaneous independent N-N collisions, for n = 1,2,3...., is then simply given by the function:

$$f_n(E) = \frac{\alpha}{\Gamma(np)} (\alpha E)^{np-1} e^{-\alpha E} \tag{3}$$

with the values of α and p determined above. All the $f_n(E)$ remain normalized to unity.

The AFS group[13] has successfully used the "Wounded Nucleon Model" to relate the multiplicity distributions of pp, αp and αα collisions at $\sqrt{s}_{NN}$ = 31 GeV. They predicted the relative cross sections for n nucleon-nucleon collisions per αα interaction:

$$r_n \equiv \frac{\sigma_n}{\sigma^{\alpha\alpha}_{in}} \quad .$$

The main distinctive feature[13] of the "Wounded Nucleon Model" is that it only counts the number of struck nucleons, and that a nucleon contributes only once to the production of particles no matter how many times (≥1) it is successively struck. These predictions can be applied directily in Equation (1), with Equations (2) and (3) used for $f_n(E)$, if n is allowed to take on half-integer as well as integer values. For instance, n = 2 corresponds to 2 N-N collisions, or 4 wounded nucleons, while n = 2½ corresponds to 5 wounded nucleons.

The results of the fits are shown in Figure 4.

The solid line on the pp data is the fit to Equation (2) given above. The solid line on the αα data is the best fit using the n-collision probabilities derived by the AFS collaboration[13] from their charged multiplicity data. This fits the E°_{TOT} spectrum out to 10 GeV,

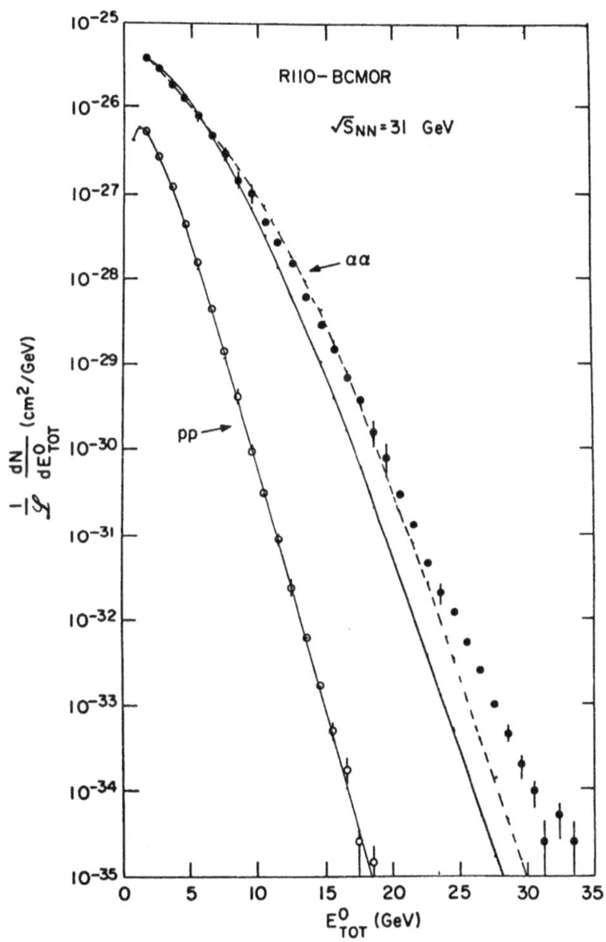

Figure 4

or about 1½ orders of magnitude down in cross section. The broken line on the $\alpha\alpha$ data allows the σ_n, n=1,1½,2....4, to take on the values which give the best fit to the data. This curve fits the E°_{TOT} spectrum out to 20 GeV, or over 4½ orders of magnitude. It is instructive to note that the r_n parameters of the best fit are just the original AFS[13] r_n parameters, plus the 4-collision curve normalized to σ^{pp}_{in} as shown in Figure 3. The more sophisticated analysis specifies the shape of the $\alpha\alpha$ spectrum and thus allows the total cross section for neutral energy (plus 2 tracks) in our detector to be determined, $\sigma^{\alpha\alpha}_{in} = 131 \pm 3$ mb. The best fit value for the 4-collision cross section is 13 mb which equals σ^{pp}_{in} or $\sim 10\%\ \sigma^{\alpha\alpha}_{in}$. The tail above 20 GeV in E°_{TOT} is still not explained with this model.

The following conclusions can be drawn from the data and fits shown in Figure 4.

The wounded nucleon fit with conventional parameters[13] explains the data out to E°_{TOT} = 10 GeV, about 90% of the $\alpha\alpha$ interaction cross section in our detector. This is the domain of the average collision, which may be of interest to cosmologists or to cosmic ray experiments with limited statistics.

From 10 GeV $\leq E^{\circ}_{TOT} \leq$ 20 GeV, the data can be described by the simultaneous parallel collision of all the nucleons in the interacting nucleii, 4 N-N collisions for the $\alpha\alpha$ case, with a cross section approximately equal to the pp interaction cross section. This corresponds to 10% of the $\alpha\alpha$ interaction cross section in our detector. This is the region of "Central Collisions." However it appears that the probability for all the nucleons to interact in an $\alpha\alpha$ collision is much larger than previous estimates.[13]

Above 20 GeV, there is an unexplained exponential tail at a level of 10^{-5} of the total interaction cross section. The exponential slope of this tail is too flat to be explained even by the extreme case in which all 4 individual nucleon-pairs simultaneously interact in $\alpha\alpha$ collisions. The energy density in this tail rises to values greater than 15 GeV per unit of rapidity,

$$\frac{dE^{\circ}_{TOT}}{dy} \geq 15 \text{ GeV} .$$

Using a formula given by McLerran,[14] this corresponds to a spatial energy density of over 1 GeV/fm^3, which is beginning to approach the values required for the formation of the quark-gluon plasma.[14]

Another necessary condition for plasma formation is that the energy be distributed as lots of soft particles, corresponding to long-distance phenomena, rather than as high P_T jets which are short-distance effects. Preliminary analysis of the αα and pp data of Figure 2 for the average sphericity, s, of the events as a function of E_T° is shown in Figure 5.[12] The pp data begin to show a dropoff in mean sphericity above 12 GeV, indicating the beginning of some "jet" contribution to the events. The αα data, with a barely perceptible drop of sphericity out to 26 GeV in E_T°, show no indication of "jettiness." The scale of "jettiness" has been determined for this detector[12] from pp collisions at $\sqrt{s}$ = 63 GeV, where jets are observed for E_T° > 20 GeV (pp collisions!). A uniform azimuthal and rapidity distribution of energy deposition corresponds to an average sphericity, <s> $\sim$0.53; while 20% jet contribution lowers <s> to $\sim$0.42, and 60% jets corresponds to <s> $\sim$0.22. Thus the αα spectrum at the highest values of E_{TOT}° may have a jet contribution of at most a few percent. This supports the conclusion that over the entire spectrum the neutral energy observed in the central region of αα interactions at $\sqrt{s}_{NN}$ = 31 GeV is caused by the emission of many uniformly distributed and presumably soft particles, with jet emission playing a negligible role. It should also be noted that the existence or non-existence of jets has no bearing on our analysis of the αα spectrum in terms of multiple independent pp collisions, since only simultaneous, or parallel, pp collisions were considered.

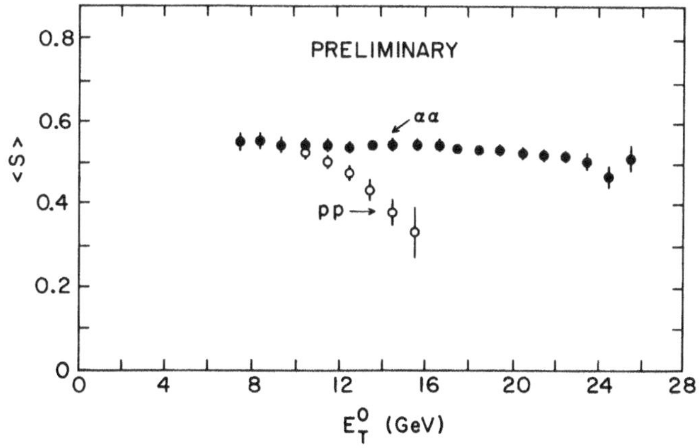

Figure 5

REFERENCES

1. The BNL-CERN-Michigan State-Oxford-Rockefeller Collaboration members are: A.L.S. Angelis, G. Basini, H.-J. Besch, R.E. Breedon, L. Camilleri, T.J. Chapin, C. Chasman, R.L. Cool, P.T. Cox, Ch. von Gagern, C. Grosso -Pilcher, D.S. Hanna, P.E. Haustein, B.M. Humphries, J.T. Linnemann, C.B. Newman-Holmes, R.B. Nickerson, J.W. Olness, N. Phinney, B.G. Pope, S.H. Pordes, K.J. Powell, R.W. Rusack, C.W. Salgado, A.M. Segar, S.R. Stampke, M. Tananka, M.J. Tannenbaum, P. Thieberger, J.M. Yelton.
2. D. Antreasyan, et al., Phys Rev D19 (1979) 764, and references quoted therein.
3. Note that I use the generic term $\sigma(P_T)$ to represent the invariant cross section.
4. See the review article by M.A. Faessler, "Experiments with Alpha Particles at the CERN Intersecting Storage Rings," CERN-EP/84-64, to appear in Physics Reports.
5. A.L.S. Angelis, et al., Phys Lett 116B (1982) 379.
6. S. Frankel, "High P_T Nuclear Cross Sections," these Proceedings.
7. B. Brown, et al., Phys Rev Letters 50 (1982) 11.
8. H. Gordon, et al., Phys Rev D28 (1983) 2736.
9. M. Tanaka, A.L.S. Angelis et al., Nucl Phys A418 (1984) 321c.
10. W. Frati, Nucl. Phys A418 (1984) 191c.
11. A.L.S. Angelis, et al., Phys Lett 141B (1984) 140.
12. A.L.S. Angelis, et al., CERN-EP/84-46, to be published in Nucl Phys B.
13. T. Åkesson, et al., Phys Lett 119B (1982) 464.
14. L. McLerran, these Proceedings.

Characteristics of JACEE Heavy Ion Events at Energies Above TeV/Nucleon

JACEE collaboration #

T.H.Burnett[i], S.Dake[b], J.H.Derrickson[g], W.Fountain[g], M.Fuki[d], J.C.Gregory[h], T.Hayashi[g], T.Hayashi[b], R.Holynski[j], J.Iwai[i], W.V.Jones[f], A.Jurak[j], J.J.Lord[i], C.A.Meegan[g], O.Miyamura[c], H.Oda[b], T.Ogata[a], T.A.Parnell[g], E.Roberts[h], T.Saito[a], S.Strausz[i], T.Tabuki[a], Y.Takahashi[g], T.Tominaga[c], J.W.Watts[g], B.Wilczynska[j], R.J.Wilkes[i], W.Wolter[j] and B.Wosiek[j]

a.Institute for Cosmic Ray Research,University of Tokyo b.Kobe University c.Osaka University d.Okayama University of Science e.Waseda University f.Louisiana State University g.Marshall Space Flight Center,NASA h.University of Alabama in Huntsville i.University of Washington j.Institute for Nuclear Physics,Krakow

1. Introduction

Five balloon flights have been performed up to now in order to get samples of ultra-relativistic heavy ion collisions by cosmic ray. Four of them have been used for detection of the events at energies above TeV/N. Total exposure of them have reached 200 m^2.str.hr and number of observed non-peripheral collisionswith ΣE_γ >5TeV exceeds a hundred. Main purpose of this report is to present some characteristics observed in this sample of events. Besides, topics in preliminary analysis of JACEE-3 experiment in which iron interactions at energies 20 to 60 GeV/N have been detected by emulsion-counter hybrid system, are briefly presented.

2. Some features of iron induced collisions at energies 20 to 60 GeV/N in preliminary analysis

Apparatus used in JACEE-3 experiment is hybrid of emulsion and counter system. Details of instruments are presented in a talk by T.Parnell in this conference[1]. This system enable us to select events in energy region 20 to 60 GeV/N and iron induced collisions in this region were sampled in the first stage. Status of the data sample is summerized in Table 1. In observed 74 events with charge Z $\geq$ 22, number of collisions in CNO targets(Aclyl and CR39) are 52 and this corresponds to cross-section of 2 barn. Number of collisions in emulsion and lead are 9 and 13, respectively. In the table,multiplicity and dispersion from raw data(experimental bias is not corrected) are also presented. In the preliminary analysis for these events, following features are to be noted.

(a)Existence of high pseudo-rapidity density events and an event azimuthal structure

Fig.2 shows two high pseudo-rapidity density events,

Ti(26GeV/N)+Em→255ch with $dN/d\eta_{Max}$=90 and Ti(41GeV/N)+Pb→268ch with $dN/d\eta_{Max}$=120. These pseudo-rapidity density are roughly the same level of high multiplicity events observed in TeV/N region.[2] In the figure curves are calculations of Multi-Chain Model(MCM)[3] including pion production threshold. Both events have more tracks in the central region in comparison with the calculation. An event having clear azimuthal structure is shown in Fig.3. Such structure may be relevant to collectiveness of interaction in this energy region.

(b)Fe-CNO collisions with less iron spactators

In the iron interaction with light target (Aclyl,CR39), number of wounded nucleons of iron is estimated in the following way. According to Goldhaber s formula, average value of pseudo rapidity for iron spectators is calculated. On the other hand, we separate very forward tracks and calculate average pseudo-rapidity. We take more tracks till both average value approximately coincide. Then these very forward tracks are identified as spectators of iron. From the spectators, the wounded nucleon number of iron is given by subtraction from initial mass number with correction of A/Z ratio. Fig.4 shows the distribution of wounded nucleon number of iron. Curve is a calculation of Glauber model.[4] Although comparison of absolute value is not taken seriously, we see a lump of events(12 events) in the region $25 < W_{Fe} \leq 56$. Among this twelve events, six events have more tracks in central region in comparison with the MCM assuming head on collision. This feature may be interesting in connection with the stopping power problem.[5]

3.1 High multiplicity and pseudo-rapidity density events in ultra-relativistic region

Seven events having charge multiplicity larger than 400 have been observed up to now as shown in Table 2. Two of them have been already published. In Fig.5 pseudo-rapidity distribution of an event having highest multiplicity, Fe(1 ∿ 2TeV/N)+Pb→1050^{+100}_{-30} ch , is presented. As shown in the figure, the distribution is single bump and maximum of $dN/d\eta_{Max}$ is 240. Solid curve in the figure is a calculation of MCM assuming impact parameter b=3fm. As already discussed by several authors,[6] multiplicity and pseudo-rapidity distribution are consistent with a class of model based on superposition of nucleon-nucleon or quark-quark collision. However, as discussed in later, these high rapidity density suggest formation of dense product in which superposition may not be justified.

3.2 Transverse momentum and energy density of central product in the events in ultra-relativistic region

Transverse momentum of photon is determined by cascade shower analysis. Except for very forward tracks in high multiplicity events, each cascade generated by a photon is well separated. Due to detection threshold of shower(30 GeV), photons in target side are difficult to be observed. Transverse momentum distributions are measured in central and forward region. Although shape of the distributions are different from event to event, we find a group of events which have greater average transverse momentum in comparison with that of hadron-hadron or hadron-nucleus collision at corresponding energies. Especially a part of high multiplicity events listed in Table 2 have larger average transverse momentum. In Fig.6 transverse momentum distribution of photons in the event Ca(100TeV/N)+CNO→760ch+(300+α)γ is presented as an example. Dashed curve shows normal exponential distribution corresponding to $\bar{p}_T(\pi^0)$=400 MeV/c(π^0 is assumed to be main source of photons). The event has much flatter distribution corresponding to $\bar{p}_T$ =700 MeV/c. A question is whether increase of average p_T can be understandable by superposition picture. Number of jets(jet multiplicity) may be greately enhanced in case of nucleus-nucleus collision by factor of $A^{4/3}$ and this might increase average p_T. However ,we already know that dN/dη exceed a hundred times of that of proton-proton collision in some high multiplicity events and soft component is also greately enhanced. Thus explanations by enhancement of jet component seems to be difficult. Another trivial effect in superposition picture is additional transverse momentum from non-colinearity of nucleon-nucleon or quark-quark collision in the frame of nucleus-nucleus collision. This is due to Fermi-motion or multiple scattering of nucleon or quark. In Fig.6 dotted curve shows a calculation of MCM in which transverse momentum distribution of leading cluster is incorpolated in a form $\exp(-p_T^{LC}/\nu Q_0)$. Where ν is average number of collisions for a leading cluster in a nucleus of counter side. For secondary particles in the central region, this effect is small if Q_0 is 1 GeV/c or so. The reason is that for the central particles additional transverse momentum from transverse boost due to non-colinearity of leading cluster collision is small.

Beside with the existence of events having greater average tranverse momentum, a correlation seems to exist between transverse momentum and energy density of central product. The energy density of central product has been discussed with some variety.[7,8] For our purpose, the simplest one by Bjorken ;

$$\varepsilon = \frac{3}{2}\sqrt{p_T^2 + m_\pi^2}\,(dN/d\eta)/c\pi A_{Min}^{2/3}$$

is chosen. Differences between other formulae are within factor of three. Such systematic multiplication does not change feature of correlation. In Fig.7 plot in average transverse momentum and energy density is presented. The transverse momentum of photon is measured in central and central to forward region for some events. As shown in the figure, rising of average transverse momentum beyond energy density few GeV/fm^3 is appreciable. Although further examinations should be required, this seems to reflect change of state of collision product.[9]

A comment is added for the high energy density events shown in Fig.7. Energy and mass number distribution of the high energy density events is shown in Fig.8. As shown in the figure , several light ion events at energy around 10 TeV/N are included. Dashed line represent cut of and numbers attached are the number of oberved events. This point should be clarified by further sampling of events.

3.3 Power spectrum analysis on azimuthal distribution

Fluctuation in particle distribution has been discussed as a candidate for signature of the phase transition.[10] Although there remains a lot of problem such as order of the transition,final state hadronic interaction and etc., attempt to extract clue of transition from fluctuation seems to be helpful to understand nature of heavy ion interaction. Recently, an attempt has been made by Takagi by using power spectram analysis to pseudo-rapidity distributions of two high multiplicity events.[11] It has been shown that at several frequencies power is strong in comparison with noise level. In this report, results of power spectrum analysis for azimuthal distributions of two high multiplicity events are presented.

Power spectrum is given by strength of fourier components of the distribution. From discretised fourier analysis for the azimuthal distribution with m bins, they are defined as ,for even m,

$$A_k = |a_k| \;, (k=0,\cdots,m/2) \quad \text{and} \quad B_k = |b_k| \;, (k=1,\cdots,m/2)$$

with

$$a_k = (2/m)\Sigma n_j \cos(2\pi kj/m) \text{ and } b_k = (2/m)\,\Sigma n_j \sin(2\pi kj/m)$$

where n_j is number of track bin j (j=0,$\cdots$ m-1). In order for comparison, Monte Carlo events having the same multiplicity are generated. Here we assume flat azimuthal distribution and total multiplicity is adjusted to that of the real events. Probability distributions of A_k and B_k in the Monte Carlo events are almost Gaussian with same witdth except for that of k=m/2 and look like white noise.

Results of power spectrum analysis for two high multiplicity events, Fe(1$\sim$2TeV/N)+Pb$\rightarrow$1050ch and Ca(0.5TeV/N)+Pb$\rightarrow$680ch,are presented

in Fig.s9a,b . To avoid experimental bias from uncertainty of determination of collision axis and from inclination of the axis, tracks in very forward and backwrad region are eliminated(θ cut). 690 and 396 tracks are multiplicities by this θ cut for Fe and Ca events. For the Fe event, power deviates strongly from the noise level(average of Monte Carlo power) at four frequencies. For comparison a example of power which shows maximum deviation in ten Monte Carlo events, is given in Fig.9b. Probability to find a event having the observed large deviation is estimated as follows. In the Monte Carlo generations number of events of which strongest fourier component is comparable or greater than that the real event and next strongest component is comparable or greater than that of the real event and so on for the third and fourth components, is counted. From this procedure, we find in average 50 such events in every five thousand events. Therefore, a hundred Monte Carlo generations give one event having similar deviation from average noise level.[12]
As we select one of seven high multiplicity events arbitrarily, this comparison indicates that the event include nontrivial structure. Similar analysis for the Ca event shows that fifty Monte Carlo generations gives one event having similar power level. Within present analysis, the results look like encouraging and further improved analysis seems to be required. Inclusion of final state interaction might be next step and after that characteristic frequencies could be discussed.

4.Summary

Characteristic features of the sample of cosmic ray heavy ion collision at energies above TeV/N are reported. A group of high multiplicity and high pseudo-rapidity density event is found. Besides, some events have larger average transverse momentum in comparison with that of hadron-hadron or -nucleus collision and a correlation between energy density of central product and transverse momentum seems to exist. As an trial of fluctuation problem in heavy ion collision, power spectrum analysis for azimuthal distribution of some high multiplicity events are also presented. Resultant power spectrum seems to have non-trivial structure in comparison with Monte Carlo events with flat azimuthal distribution.

References

1) JACEE , "Nucleus-Nucleus Interaction Studies Above 20 GeV/AMU Using a Hybrid Counter-Emulsion Chamber(JACEE-3)" presented in Quark Matter 84 by T.Parnell.

2) T.H.Burnett et al. JACEE collaboration, Phys.Rev.Letters 50 (1983) 2062.
3) K.Kinoshita, A.Minaka and H.Sumiyoshi, Z.Phys. C8 (1981) 205.
4) A.Bialas, M.Bleszynski and W.Czyz, Nucl.Phys. B111 (1976) 461.
5) W.Busza and A.S.Goldhaber, Proceedings of the Quark Matter '83 Nucl.Phys. A418 (1984) 635c.
6) H. Sumiyoshi, Phys.Lett. B131 (1983) 241,
A. Capella,C.Pajares and A.V.Ramallo, preprint TH.3700-CERN (1983)

7) J.D.Bjorken, Phys.Rev. D27 (1983) 140.
8) M.Gyulassy and T.Matsui, Phys.Rev. D 29 (1984)419.
9) M.Gyulassy, Proceedings of the Quark Matter '83 Nucl.Phys. A418 (1984) 59c.
10) L.Van Hove, preprint TH.3592-CERN (1983),
M.Gyulassy,K.Kajantie,H.Kurki-Suonio and L.Mclerran, Nucl.Phys. B237 (1984) 477.
11) F.Takagi, Phys.Rev.Lett. 53 (1984) 427.
12) There is an error in original estination of the probability.

Table 1

Target	No of events	<N>	D
Aclyl,CR39	52	40	40
Emulsion	9	86	60
Pb	13	79	65

* total number of events:74

** Bias is not corrected

Table 2

Projectile	Target	E/N(TeV)	Nch	N_γ	dN/dη(Max)
Si	Em	4	1030 ±30	>170	180
Ca	CNO	100	760 ±30	>300	100
Fe	Pb	1 ∿ 2	1050^{+100}_{-50}		240
Si	Pb	4	>700		
Ca	Pb	.5	680		140
Fe	Pb	1.5	500		
Ar	Pb	1	416	>76γ +16π^0	120

Address for correspondence; Institute for Cosmic Ray Research, University of Tokyo, 188 Tanashi, Tokyo, Japan

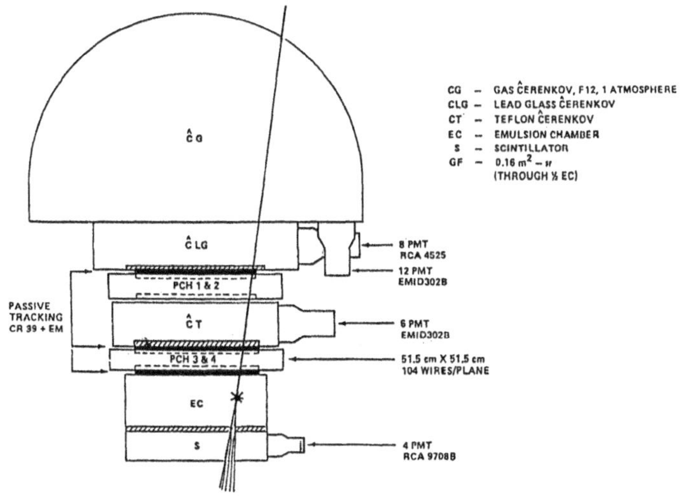

Fig.1 Apparatus of JACEE-3

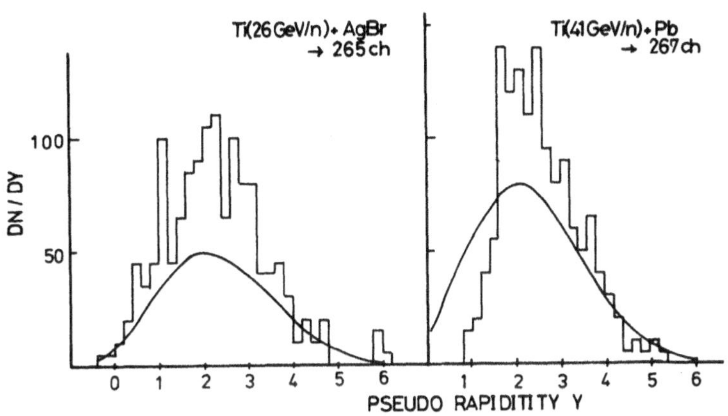

Fig.2 Two high pseudo-rapidity density events

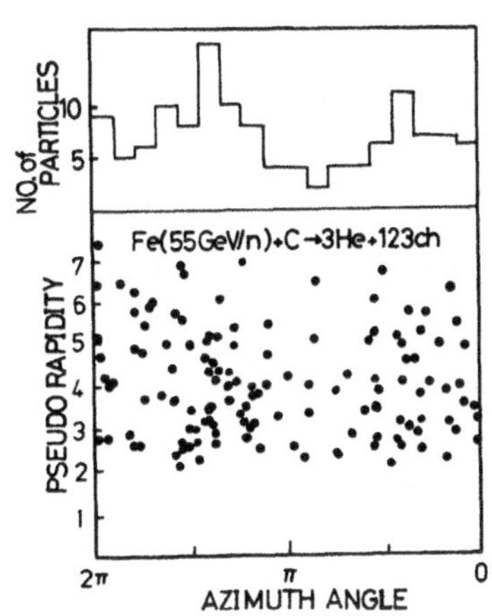

Fig.3

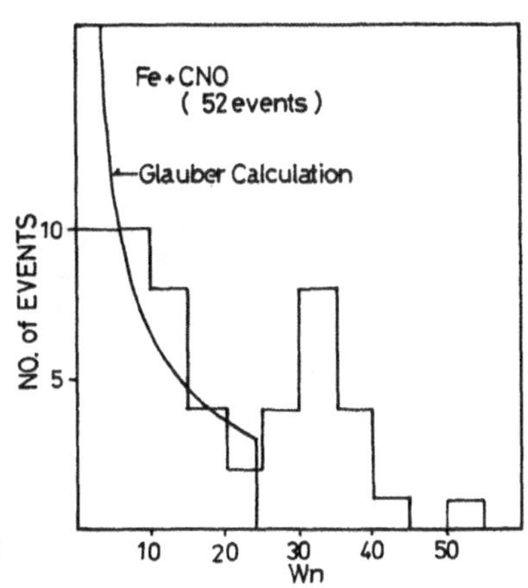

Fig. 4

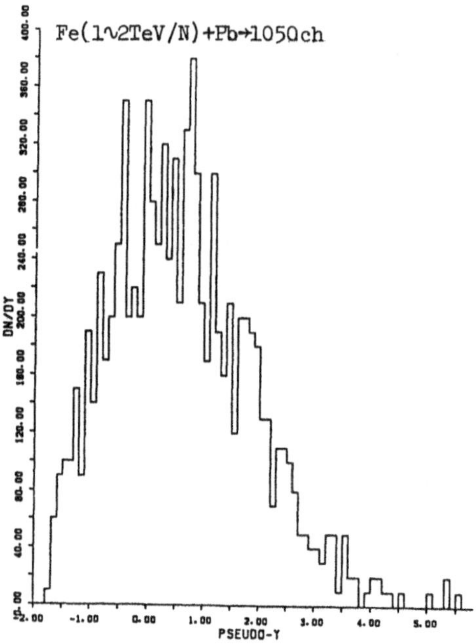

Fig. 5

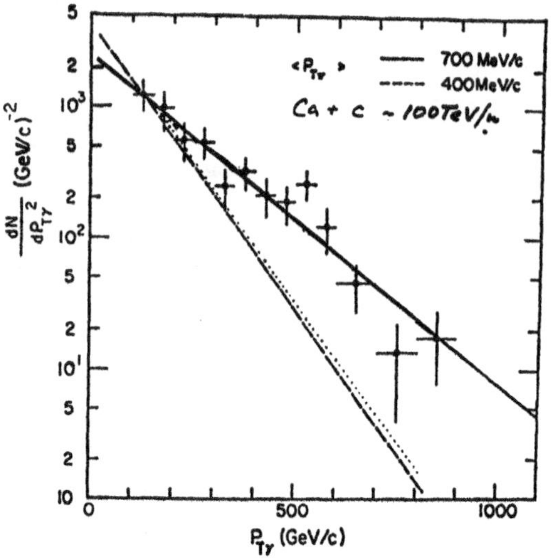

Fig. 6

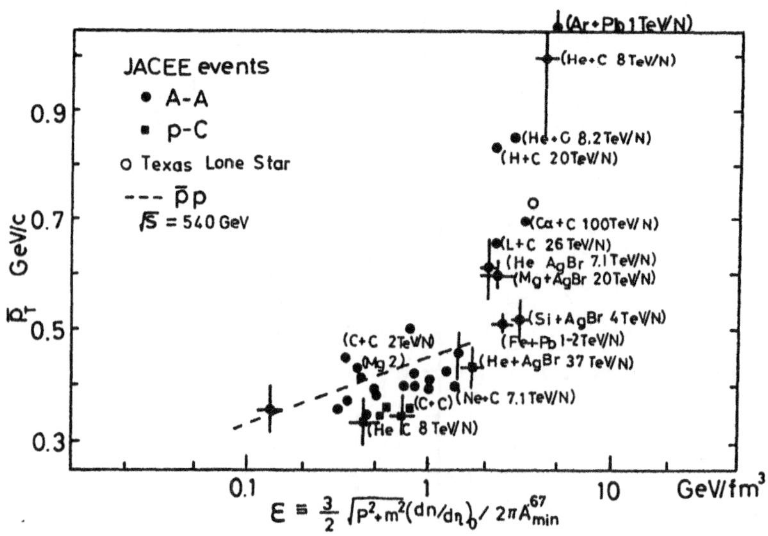

Fig. 7

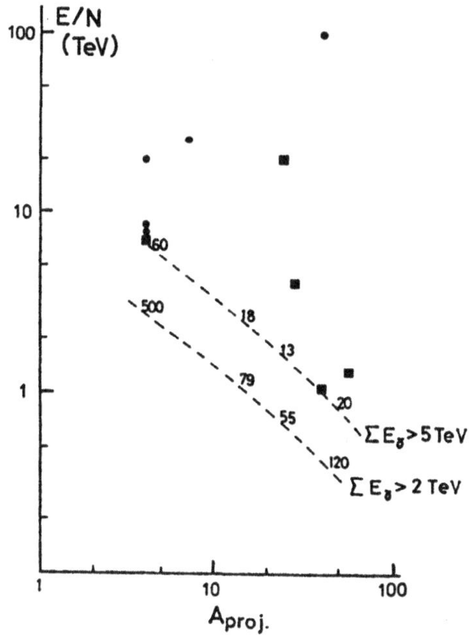

Fig. 8 Events having $\epsilon > 1$ GeV/fm^3

Black circls represent events with light target.

Black squares show events in emulsion and lead.

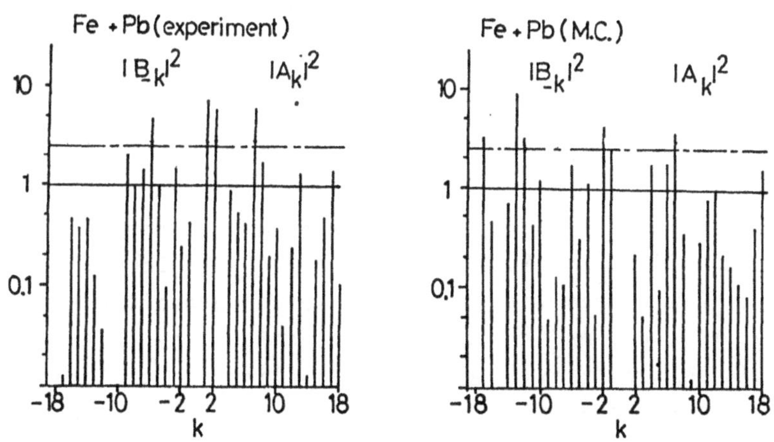

(a) (b)

Fig. 9

DETECTORS FOR HIGH ENERGY NUCLEUS-NUCLEUS COLLISIONS*

L. S. Schroeder
Nuclear Science Division
Lawrence Berkeley Laboratory
University of California
Berkeley, California 94720

1. Introduction

In this talk I want to review the recent Workshop on Detectors for Relativistic Nuclear Collisions, held last March 26-30, 1984, at the Lawrence Berkeley Laboratory. For a complete summary of the meeting consult the Proceedings of the Workshop.[1] Besides providing you with an overall impression/flavor of the Worskhop, I also hope to instill in you a sense of the considerable strides that have been made in coming to grips experimentally with the very complex collisions that we will encounter at higher energies. But at the same time, it must be realized that we still have a long way to go to fully exploit the potential of this new domain of high energy and baryon density.

2. Reasons and Goals for the Workshop

There were several reasons for having the dedicated Detector Workshop. As we proceed into the mid-1980's and beyond, the prospects for higher energy nucleus-nucleus collisions are quite promising. Specifically, at Brookhaven a transfer line between the Tandem Van de Graaff and the AGS will provide light ions ($A \leq 32$) at energies up to 15 GeV/nucleon in 1986.[2] At the same time, an upgrade of the PS injector by the GSI/LBL collaboration will provide ^{16}O ions in the CERN SPS at energies of 15-225 GeV/nucleon beginning in 1986.[3] In late 1983, the Nuclear Science Advisory Committee of the Department of Energy and the National Science Foundation recommended[4] that the next major nuclear physics facility in the United States should be a relativistic nuclear collider to explore the domain of high energy densities and hopefully the production of deconfined hadronic matter--the quark-gluon plasma. Finally, several national laboratories in the United States have active programs underway to look at collider options, including RHIC at Brookhaven,[5] the Mini-Collider at Berkeley,[6,7] and a 10 GeV/nucleon on 10 GeV/nucleon

*This work was supported by the Director, Office of Energy Research, Division of Nuclear Physics of the Office of High Energy and Nuclear Physics of the U.S. Department of Energy under Contract DE-AC03-76SF00098.

collider at Oak Ridge.[8]

Since new fixed target heavy-ion facilities will be available by 1986, only two years away, we felt that it was both timely and necessary to start looking at the detector requirements imposed by these much higher energies well in advance of their ability to deliver beam on target. With the somewhat more distant future for colliders, an early start on detectors is also called for. Specifically, the goals of the Workshop were:

- Build on the substantial base of information generated for detectors at the previous Quark Matter meetings,[9,10] fixed target detectors discussed at Bielefeld and at an SPS fixed target workshop[11] in December 1982, and collider detectors at Brookhaven (QM'83).
- Consider detectors for both geometries, i.e.,
  fixed target: spanning energies from 15-225 GeV/nucleon
  collider: covering the range from 2 + 2 to 100 + 100 GeV/nucleon
- Want to identify any detector elements thought to be on the critical path for investigation of the quark-gluon plasma which require an R&D effort.

3. Structure of the Working Groups

In order to provide a central focus for the Workshop it was decided to concentrate on the detection of the quark-gluon plasma as our theme. However, it should be remembered that the conclusions arrived at for detectors are really in the context of high energy and baryon density achievable in central high energy nucleus-nucleus collisions and apply to a potentially broader and richer range of phenomena than just the quark-gluon plasma per se.

The individual working groups (five in all) were roughly set up along lines of observables of the quark-gluon plasma.[12,13] Figure 1 is the now familiar representation of the evolution of a central nucleus-nucleus collision in the temperature (T) and relative baryon density (ρ/ρ_0) plane. Two imagined trajectories, one emphasizing the baryon-rich plasma, the other the meson-rich (so-called "transparency") regime are shown. In order for a detection scheme to be sensitive to the presence of the quark-gluon phase, measurements of direct radiation from the plasma itself (region 1) or from the rehadronized phase (region 2) will be required. Since the plasma is not thought to be stable, radiation from both regions could be present in each event. A device measuring hadrons in the final state would

Phase Diagram With Nuclear Collision Trajectories

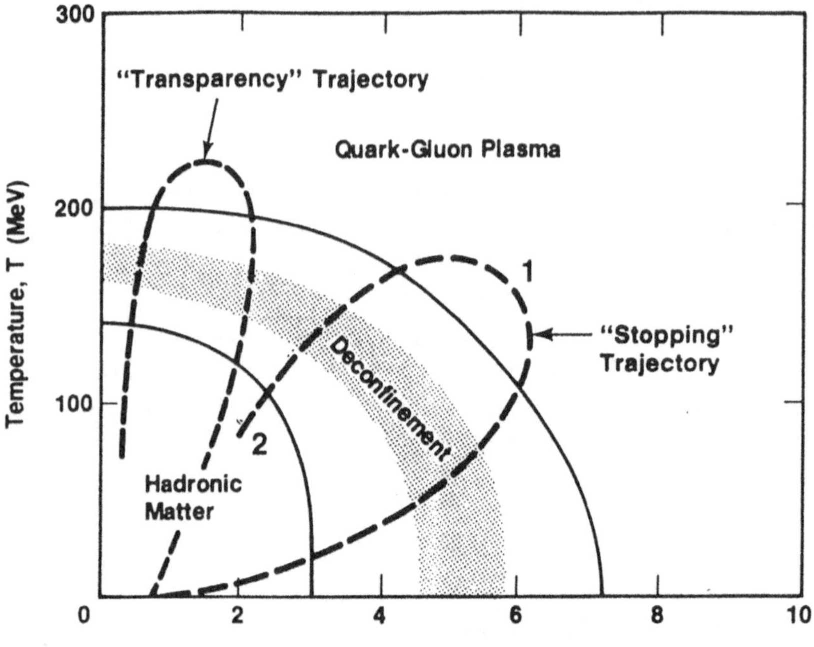

Experimental Signatures

Region 1:	**Direct radiation from plasma (photons, leptons, lepton pairs)**
Region 2:	**Recondensed hadronic matter (π, K, p, $\bar{p}$,...) Possible free quarks, diquarks**

Figure 1: Phase diagram for nuclear matter showing trajectories corresponding to the "transparency" (baryon-free) and "stopping" (baryon-rich) zones of the nuclear collision process.

detect hadrons arising from ordinary hadronization processes found, for example, in nucleon-nucleon collisions, and hadrons produced by quark and gluon recombination processes in the plasma phase. Theoretical guidance will certainly be required to differentiate between these sources.

For the Workshop we identified three areas of detector measurement or technique which were used as the focus for individual groups. These were:

- Hadrons--generally thought to include detection of strange particles and antibaryons, i.e., few particle measurements, often with a reduced solid angle detector system.
- Event parameters--experiments with large ($\sim 4\pi$) solid angle coverage, measuring many of the final state particles in order to determine such experimental observables as the average charged particle multiplicity <M>, the pseudo-rapidity ($dn/d\eta$), or rapidity (dn/dy) distributions as function of multiplicity, fluctuations in particle density, etc.
- Penetrating probes--experiments involved in measuring photons, leptons, lepton pairs, etc., which are thought to be useful in probing the hot, compressed stage of the collision process.

For each of these categories, there was a separate working group for fixed target and collider geometries. Each working group had conveners who organized and guided the work of the group. The conveners were:

	Hadrons	Event Parameters	Penetrating Probes
Fixed Target:	C. Gruhn (LBL) I. Otterlund (Lund)	K. Foley (BNL) A. Poskanzer (LBL)	J. Carroll (UCLA) M. Goldberg (Syracuse)
Collider:	W. Carithers (LBL) T.J.M. Symons (LBL) G. Young (ORNL)		T. Ludlam (BNL) L. Madansky (Johns Hopkins)

These are the people who did the work and are the ones who made the Workshop a success.

4. Background Information--A Starting Point

Before proceeding to the results, I first want to illustrate some of the experimental problems and discuss related instrumentation questions that we know will have to be dealt with to unravel the dynamics of central nucleus-nucleus collisions.

The first and most obvious fact is that central nucleus-nucleus collisions (always assume the heaviest nuclei for maximizing production of the plasma) will yield very high multiplicities of final state particles. Already at Bevalac energies the GSI/LBL Plastic Ball experiment observes ≈100-200 charged particles being produced. This is shown in Figure 2 for the case of Au + Au collisions at a laboratory kinetic energy of 1.05 GeV/nucleon (0.234 GeV/nucleon in the c.m.). In this energy range the final states are still dominated by nucleons. At about 10 GeV/nucleon in the laboratory, roughly the energy available at the AGS, we expect several hundred mesons (mostly pions) will be created, comparable to the number of nucleons. At the energies of the now famous Si + Ag JACEE event,[14] ≈4-5 TeV/nucleon in the laboratory, we expect a thousand or more mesons to be created and are in a "meson-dominated" regime. The multiplicities described above <u>have</u> <u>been</u> <u>observed</u>! Recent event simulations are also finding similar numbers.[1]

The large multiplicities that will be present naturally lead us to ask instrumentally oriented questions. These include:

- To what extent will tracking of individual particles still be possible (useful) in such a high multiplicity environment?
- As the multiplicity increases, do we eventually just rely on calorimetry for our event information?
- How to receive and process the large amount of information associated with these events?
- How to trigger on central collisions?
- Are there existing detector designs that can do (almost do) the job?
- Does experience with big high energy detectors such as UA1 or UA2 provide some of the answers to our questions?
- Finally, what are the signatures of the quark-gluon plasma?

Having formulated these questions, we hoped to provide some but not all of the answers.

5. General Comments on Detectors in Fixed Target and Collider Geometries

As pointed out previously, the individual working groups divided into separate categories (hadrons, event parameters, penetrating probes) and concentrated on detectors in either a fixed target or collider geometry. Before discussing the highlights of the working groups, I want to discuss some general features of detectors which

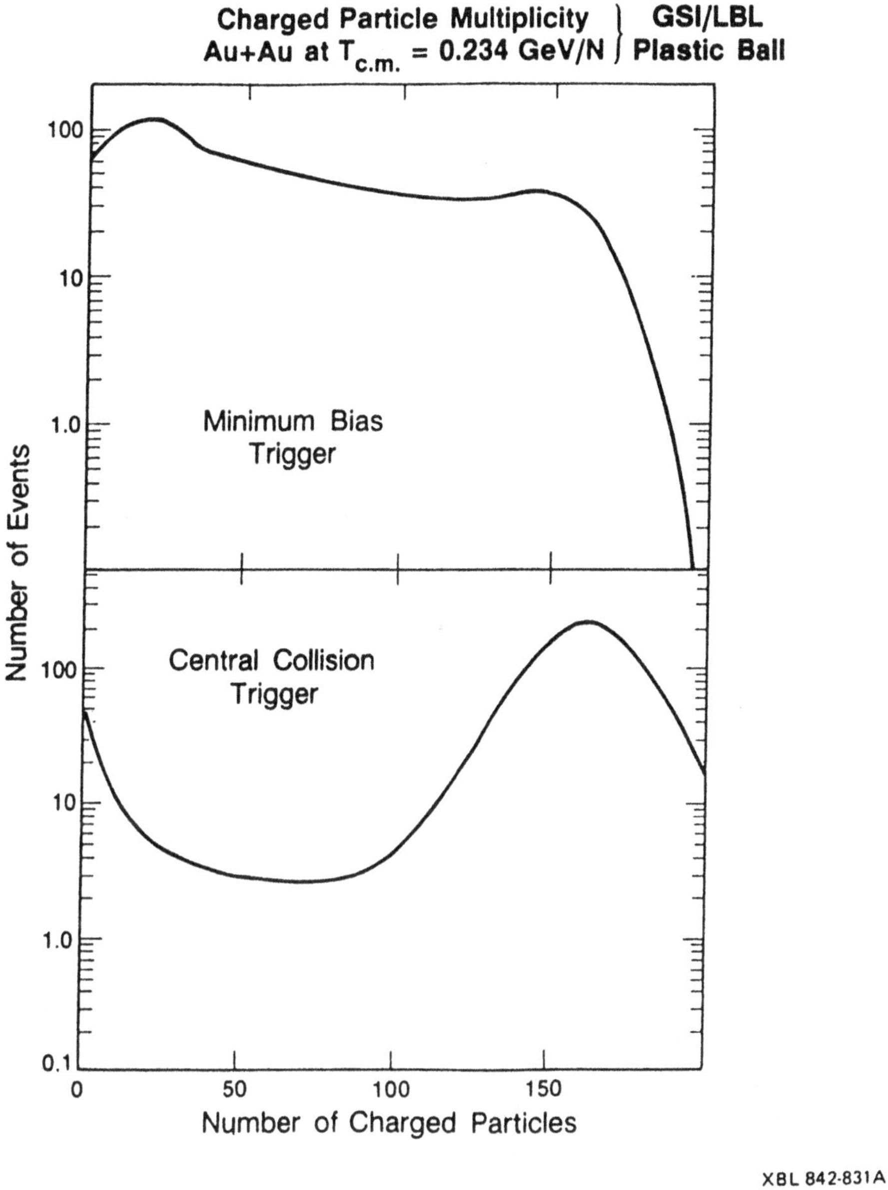

Figure 2: Charged particle multiplicity distribution for Au + Au collisions studied by the GSI/LBL Plastic Ball collaboration at the Bevalac.

are intrinsic to: i) the geometry of the experiment (fixed target or collider) and ii) the energy range associated with these geometries.

The fixed target groups considered experiments spanning the energy range of E_T = 15-225 GeV/nucleon. This corresponds to a rapidity interval between target and projectile of $\Delta y \approx 3.5$ at 15 GeV/nucleon, and $\Delta y \approx 6.2$ at 225 GeV/nucleon. Figure 3 indicates the present concept of the collision process, where the regions are divided into spectator and participant zones. For $E_T \leq 225$ GeV per nucleon no clear baryon-free central region will be developed. An additional feature of event detection in the fixed target mode is the strong kinematic focussing of particles in the forward direction. In particular, the spectator nucleons associated with the projectile will be emitted into a narrow cone of $\Delta\Theta < 1^{o}$ about the beam axis at 15 GeV/nucleon, and $\Delta\Theta < 0.1^{o}$ at 225 GeV/nucleon. This means that one will be able to measure particles emitted in the target and central rapidity regions, but that the projectile associated nucleons will be much harder. However, this focussing does have one major advantage: a lack of fast forward-going projectile fragments within the interval $\Delta\Theta$ can be used as part of a central collision trigger (e.g., use in veto mode--if projectile spectator fragment observed, veto event--not central). The strong center of mass boost experienced with fixed target geometry is also advantageous to short-lived particles such as K^0_s, Λ^0--greatly improves their detection efficiency. Finally, in the fixed target mode the "effective luminossity" (event rate = $\sigma\mathcal{L}$) can be made large, just by using a thicker target; this enhances the possibility of observing processes with very low cross-sections. The thickness must of course be chosen to accommodate other considerations such as keeping the conversion of γ-rays (mostly from $\pi^0, \eta \rightarrow \gamma\gamma$) under control.

In collider geometry the particles are emitted over the complete 4π solid angle. This greatly reduces the number of particles per unit solid angle, but does place an added burden of requiring a much larger detector system (therefore more expensive) if one is interested in obtaining information on the complete event. For colliding beam experiments the interaction region can be up to a meter in length. With such an extended source and assuming high interaction rates (i.e., high luminosity) some tracking of the particles will be required just to guarantee that they come from a single vertex. Since one can not place a counter directly in the beam, other methods of selecting central collisions than the one discussed above for the fixed target mode will have to be developed (e.g., large multiplicity, high transverse energy or momentum, etc.).

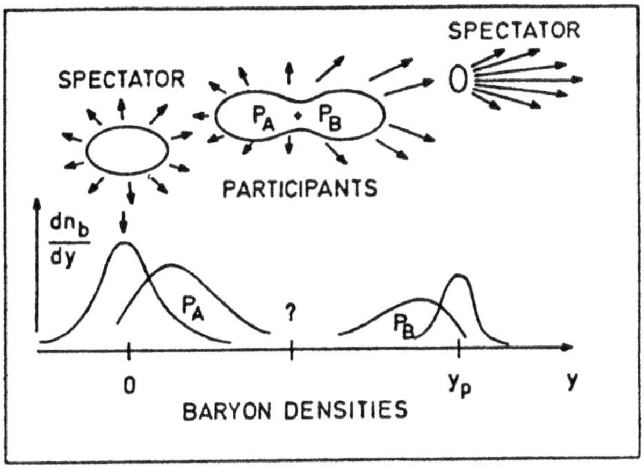

Figure 3: Characterization of the collision process showing spectator and participant regions.

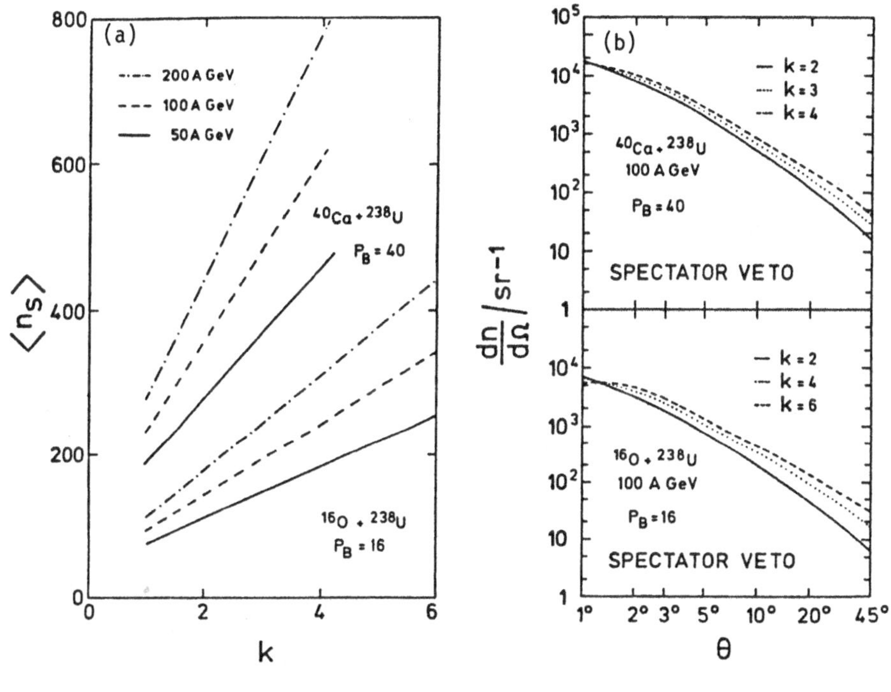

Figure 4: (a) Charged particle multiplicities for ^{16}O + A and ^{40}Ca + A collisions obtained by extrapolation of pA data, where k = number of participant nucleons/number spectator nucleons.
(b) Expected number of charged particles per steradian.

Now we turn to a review of the results from the individual working groups. I will try to cover the highlights of each group and again refer you to the Workshop Proceedings for a much more complete version.[1]

6. Hadrons-Fixed Target

This group considered those fixed target experiments involving measurements of the single-particle or few-particle inclusive type. Experiments of this genre would generally be restricted to reduced solid angle coverage (i.e., not 4π) with associated detectors (e.g., multiplicity array, calorimeter, etc.). Examples of the experiments considered by the group were: i) measurement of strange particles and antibaryons for flavor tasting of the plasma, ii) Hanbury-Brown/Twiss (HBT) like-particle interferometry to extract information on the space-time evolution of the system, and iii) production of heavy flavors such as $\phi(1020) = \phi(s\bar{s})$, $\psi(3100) = \psi(c\bar{c})$ as a probe of the quark content of the plasma.

Considerable attention was paid to the question of how to trigger on central collisions. One possibility is the so-called "spectator veto" where one vetos events which produce near beam-rapidity particles in the projectile fragmentation region. In order to understand the influence of such a central collision tag, extrapolations based on proton-emulsion data[15] have been performed for nucleus-nucleus reactions considered by the hadron group. Figure 4 shows the results of these extrapolations for charged particle multiplicities (4(a)) and the number of particles expected per steradian (4(b)) as a function of the laboratory emission angle, Θ. In each case, the distributions were calculated for the spectator veto.

Figure 5 shows a relatively large solid angle spectrometer being designed by Gruhn et al.[16] for the detection of strange baryons and antibaryons. Particular attention is being given to Ω production since being an (sss) system it could be a very sensitive probe of the plasma phase of the collision. The main elements of the detector are a superconducting magnet (3.5-7 T) to sweep away low momentum particles from the forward cone, a multiplicity array for centrality tagging, and a downstream projectile calorimeter in which low energy deposition would be used to signal a central event. The heart of the apparatus is a micro-TPC (μTPC) which will be used to recognize and identify both the initial strange particle decay (e.g., $\Xi^- \rightarrow \Lambda^\circ \pi^-$) and the subsequent decay ($\Lambda^\circ \rightarrow p\pi^-$). Three planes of drift chambers are used in front of the μTPC to increase momentum resolution.

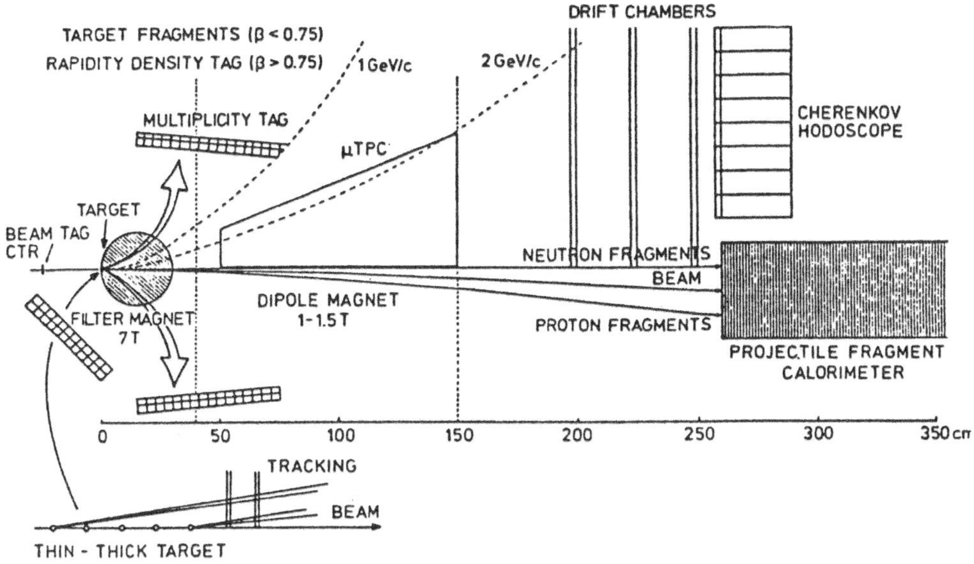

Figure 5: Proposed layout of an experiment to study strange baryon and antibaryon production with ^{16}O beams at the CERN SPS (Ref. 16).

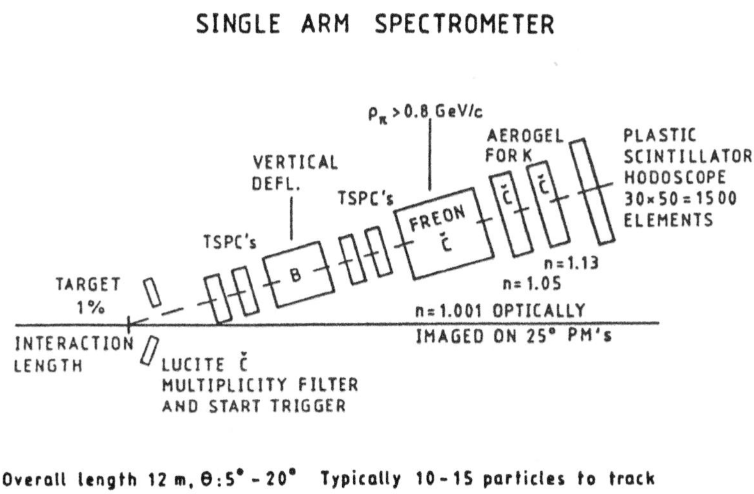

Figure 6: Single arm spectrometer being considered for possible use at the AGS (15 GeV/nucleon) with light ion beams (A ≤ 32).

Small solid angle devices were also considered. They offer several advantages, including: i) excellent particle identification, ii) good momentum resolution, $\Delta p/p \lesssim 1\%$, and iii) are capable of taking high beam intensity. Figure 6 shows an example of an inclusive small solid angle ($\Delta\Omega \approx 10$ msr) spectrometer being considered for studying charged particle spectra ($\pi^{\pm}, K^{\pm}$, p, $\bar{p}$, Λ decays) in the central rapidity region ($5^{\circ} \lesssim \Theta_{Lab} \lesssim 20^{\circ}$ for ^{32}S + Au collisions at 15 GeV/nucleon). For 10^7 particles on target per sec up to 10 particles per event could be tracked and identified in the spectrometer. The main elements of the detector are the track segment projection chamber (TSPC) for tracking, a large aperture vertical deflection magnet, followed by several arrays of threshold Čerenkov and aerogel counters. A lucite multisegmented Čerenkov multiplicity filter would be used to tag central events. The device would be employed to measure inclusive spectra of strange and non-strange particles. Comparison of these spectra at several beam energies would provide information on the "nuclear temperature" of the participant zone and the strange particle yields have of course been suggested as a probe of the plasma phase.[12,13,17] This spectrometer could also be used for interferometry measurements (HBT) of π's and K's. In this context, W. Zajc led a small working group which considered interferometry in detail.[18] The good news appears to be that rates for HBT studies will be large (one gains as the square of the multiplicity). But the high multiplicities present as yet unsolved tracking problems. This group estimated the time needed to reach 10^5 events as summarized below:

	2π pairs	2K pairs
15 GeV/nucleon	10 hr.	1000 hr.
200 GeV/nuclon	1 hr.	100 hr.

These estimates are for ^{16}O collisions on a heavy (Pb or U) target.

The hadron group also considered streamer chambers,[1,3] solid state nuclear track detectors[1] and emulsion chambers[1] as suitable detectors for fixed target nuclear collision studies. The group raised the following detector issues which need further study: i) at high multiplicities secondary interactions in detector material can cause a problem for pattern recognition, ii) reduced coverage ($\Delta\Omega$) at forward angles may be necessary to accommodate large particle fluxes--impacts detector segmentation, and iii) groups using streamer chambers need to look further into CCD and holographic read-outs (~40 μm resolution needed). Finally, they identified the following areas as needing R&D: i) particle identification (RICH's, etc.) and ii) Č segmentation required due to density of final state particles (how best to do it? how to focus Č light?).

7. Event Parameters--Fixed Target

This group concentrated on detectors for fixed target work which have nearly 4π solid angle (i.e., complete azimuthal coverage). The four types of detectors and the range of physics that each would cover is indicated in Table 1. Beams of both 15 GeV/nucleon ^{32}S at the AGS and 225 GeV/nucleon ^{16}O at the CERN SPS were discussed.

Table 1

Event Parameter		Idealized Detector	MPS III	Non-Magnetic	Streamer Chamber
Multiplicity:					
	Neg	X	X		X
	Total	X	X	X	X
$dN/d\eta$		X	X	X	X
dN/dy:	Neg	X	X		X
	Others	X			
$\pi^{\pm}$ energy		X			
π° energy		X		X	
E_T		X		X	
P_T		X	X		X
Shape:	equilibration	X	X	X	X
	flow	X	X	X	X
	jets	X	X	X	X
Particle corr:	π^-	X	X		X
	K, $\bar{p}$	X	X		
	Others	X			

I will only review their results for the so-called "idealized" detector. The MPS III is being described elsewhere in these proceedings,[2] and the non-magnetic detector and streamer chamber are closely related to the detectors being designed by the GSI/LBL PS 190 collaboration and are described elsewhere in these proceedings.[3]

The idealized dipole detector conceived by this group is shown in Figure 7. It consists of a target surrounded by a cube of CCD detecting planes which act as vertex detector and multiplicity counter inside the uniform magnetic field region. Three-dimensional tracking chambers are spread throughout the volume of the magnet, with electromagnetic and hadronic calorimeters (located inside return yoke of dipole coils) and time-of-flight counters providing angular coverage (down to $\Theta_{Lab} \approx 20°$). Downstream of the magnet are additional tracking chambers, possible RICH detectors for particle identification and more time-of-flight and calorimetry.

Figure 7: Idealized detector.

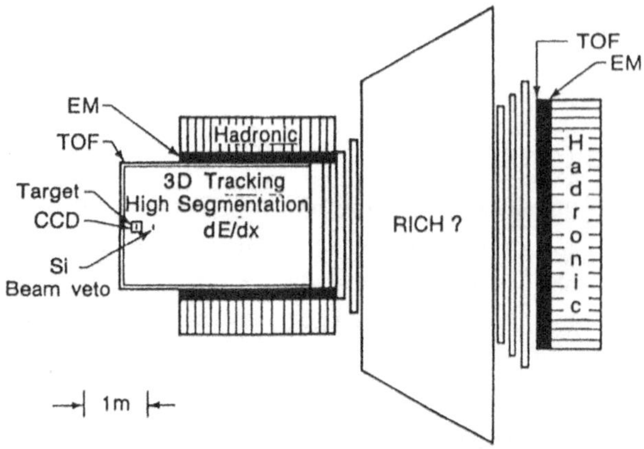

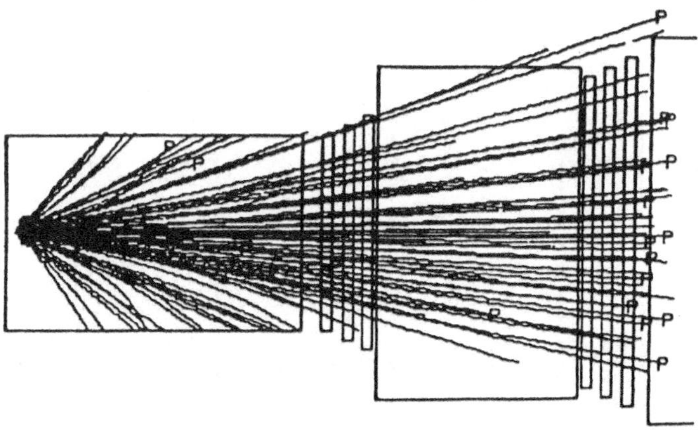

Figure 8: (a) 15 GeV/nucleon ^{16}O + Au central collisions tracked in the non-bend plane of the idealized detector.

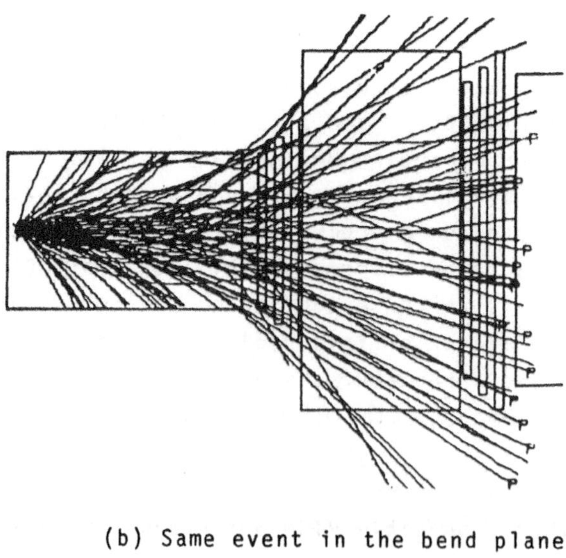

(b) Same event in the bend plane with a field of 0.5T.

As suggested earlier, the numbers of particles will be large. For example, the results of a HIJET calculation[19] for ^{16}O + Au at 15 GeV/nucleon for impact parameters b $\leq$ 4 fm, suggests several hundred particles. To see what this means for a detector like the idealized system of Figure 7, the group has simulated what a typical central event at 15 GeV/nucleon would look like. Figure 8(a) shows the event in the non-bend plane where pattern reconstruction is easier (tracks are straight lines). Figure 8(b) shows the same event in the bend plane (assume 0.5 T field). Clearly at 225 GeV/nucleon, due to the larger number of particles and their stronger forward focus, it will be even harder. The group suggests that a system like the European Hybrid Spectrometer which uses a series of dispersing dipoles spaced along the beam line would possibly be a better solution. They further concluded that heavy beam (Au + Au) fixed target experiments look very difficult at 15 GeV/nucleon and only calorimetry at 225 GeV/nucleon looks feasible, even for ^{16}O.

They identified the following areas as requiring substantial R&D: i) CCD vertex dectectors are completely untested for heavy ions, ii) RICH detectors have been built but not proven in high multiplicity environment, iii) three-dimensional tracking chambers, iv) new development in calorimetry of soft hadrons needed, v) thin μTPC's, Si-beam anti's and time-of-flight barrel arrays, and vi) pattern recognition problems require much more software development.

8. Pentrating Probes-Fixed Target

This group considered systems designed to detect directly produced electromagnetic probes (photons, leptons, lepton pairs) in light ion collisions from 15-225 GeV/nucleon. In looking at the physics addressed by such detectors, they followed a rough set of guidelines provided by L. McLerran[1,13]:

Transverse Mass ($m_T^2 = m_{pair}^2 + p_T^2$)	Possible Source
1) $m_T \lesssim 50$ MeV	coherent emission from local charge fluctuations
2) $50 \lesssim m_T \lesssim 500$ MeV	hadronic decays; some coherent effects
3) $500 \lesssim m_T \lesssim 3$ GeV	direct emission from plasma
4) $3 \lesssim m_T \lesssim 10$ GeV	approach to equilibrium; structure function of quarks and gluons (perturbative QCD)
5) $m_T \lesssim 10$ GeV	Drell-Yan

Although several different detectors were identified by the group, I will only mention plans to measure high mass muon pairs using a combination (NA34) of a spectrometer (first discussed at the Bielefeld meeting[9] by W. Willis and collaborators) designed for proton-nucleus experiments and the existing NA3 spectrometer to be used as the downstream muon spectrometer. Its elements are shown in Figure 9. Their approach employs strong filtering (≈10 interaction lengths) after the target to remove hadrons and limit decays of π's and K's into muons. In this way they hope to get around problems arising from high multiplicities, lepton identification and combinatorics. Multiple scattering in the filter will of course impose limits on the mass resoution and p_T of the dimuon pair. The front-end of the detector consists of a vertex detector and hadron calorimeter. In addition, they will also incorporate electron identification in the front-end so that the low-mass spectrum (m_{ee} < 1000 MeV) can be studied. However, at the time of the Workshop, studies indicated that due to the large multiplicities expected (they quote ~300 charged particles within a 5° core for $^{16}O + {}^{195}Pt$ at 225 GeV/nucleon), individual electron tracking and identification may not be possible. Furthermore, the 150 π°s expected/central event will yield between 1 and 2 e-pairs via Dalitz decay alone--complicating the process. A possible future development is the addition of a pair spectrometer (placed in an opening in the calorimeter section) to measure soft photons and electrons.

The group developed Table 2, to assess whether experiments are do-able; and if so in what kinematic domains; and if proposed.

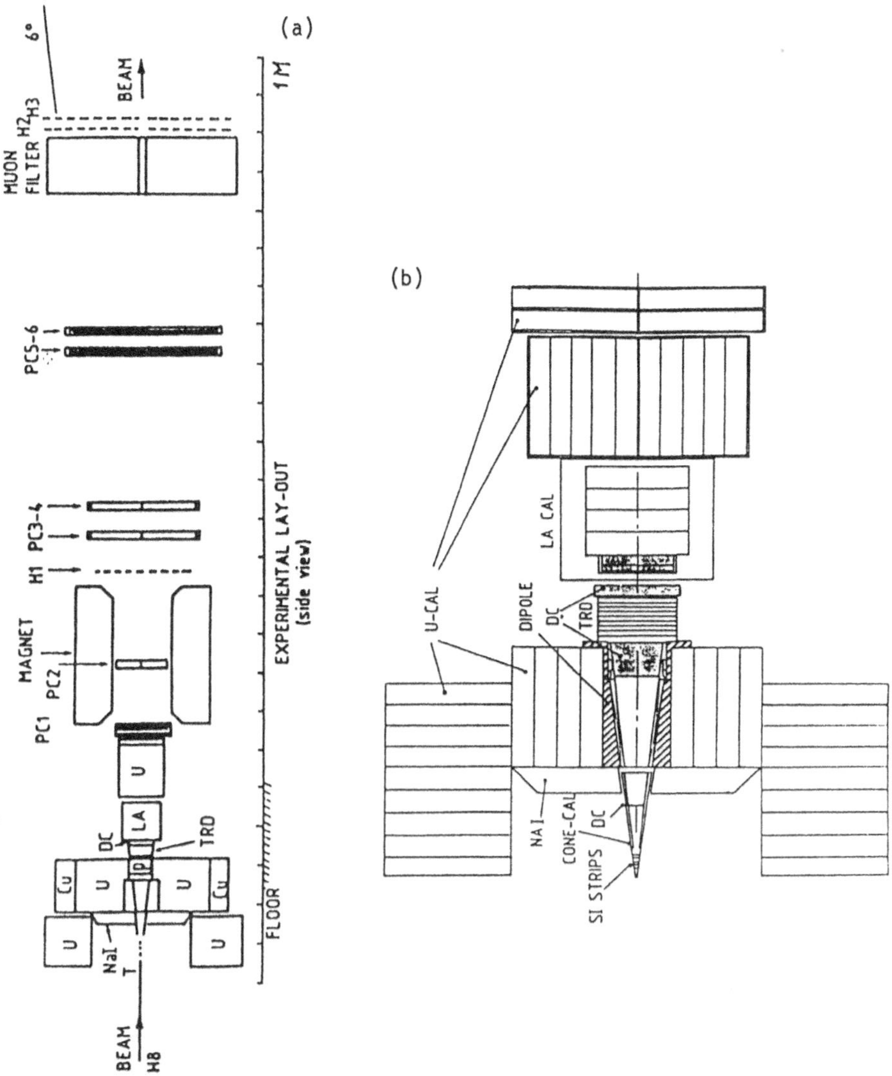

Figure 9: NA34 detector being proposed for light ion experiments at CERN. (a) Full detector, (b) expanded view of the near target detector.

Table 2

CODE: DO ≡ DO-ABLE
PROP ≡ existing PROPOSAL
DET ≡ existing DETECTOR
Y = YES N = NO R = RUMORED

Accelerator	AGS			SPS		
Probe	DO	PROP	DET	DO	PROP	DET
Hard Gamma (same info as D-Y pairs but higher) yield)	NO (π° reject too hard)			MONEY (low yield requires dedicated mach. then OK for $p_T>2.5$ GeV/c)		
Medium Backward Gamma	Y	R	N	Y	R	UA1
Soft Gamma, $X_f\sim 0$	Y	N	N	Y	N	N
Lepton Pairs						
Large Mass	Y ($M_{e^+e^-}>1$ GeV)	N	N	Y R	Y	NA3($X_f\sim 0$) NA10($X_f\sim 1$)
ϕ,ψ	(conclusions same as for large mass pairs directly above)					
Low Mass, 0.4-1 GeV	Y (if $N_{ee}/N_{\pi}>10^{-5}$)	R	N	MONEY Y	N N	N($X_f\sim 0$) N($X_f\sim 1$)

9. Penetrating Probes--Collider

This group employed the same criteria for the physics and kinematic regimes of interest as described in section 8, the difference being that they were concerned with experiments done in the colliding beam mode, where the particles of interest spread out over 4π. Figure 10 shows us a picture of the geometry and kinematics in the collider frame of reference (positions of detector elements are shown). The average momenta associated with different rapidity values are not large (<1-2 GeV/c) for values of y near midrapidity. Note that the extended source present in the collider will require some close-in tracking capability in order to differentiate between multiple event vertices. For purposes of their work, the group considered collider detectors at 3 + 3 GeV/nucleon (Δy = 2.8) and at 100 + 100 GeV/nucleon (Δy = 10.6), with luminosities rising from $\mathcal{L} \sim 10^{24}$ cm^{-2}sec^{-1} at the lower end to $\mathcal{L} \sim 10^{27}$ cm^{-2}sec^{-1} at the upper end. Fig. 11 shows results of a HIJET calculation[19] of the number of particles per unit rapidity (as function of y) at three different collider energies for central Au + Au collisions.

The group did not undertake the design of a specific apparatus;

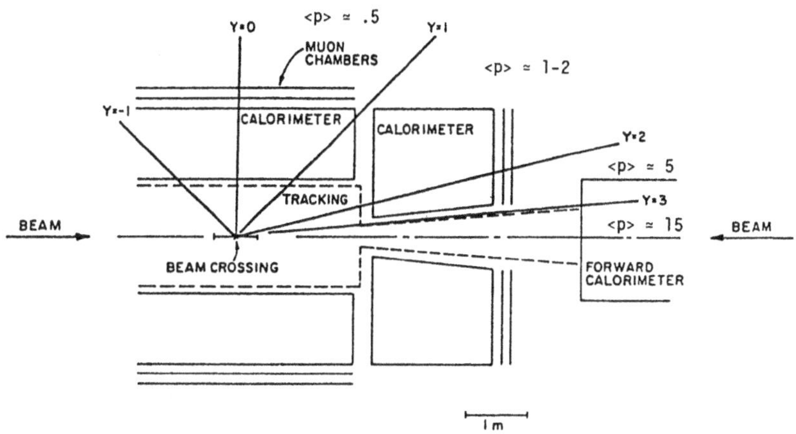

Figure 10: Schematic layout of the forward half of a collider detector showing the angular correspondence with pseudorapidity ($y = -\ell n\ (\tan\ \Theta/2)$); this corresponds to the rapidity of particles whose $m \ll p_T$). The approximate average momentum, <p>, is given for each rapidity interval.

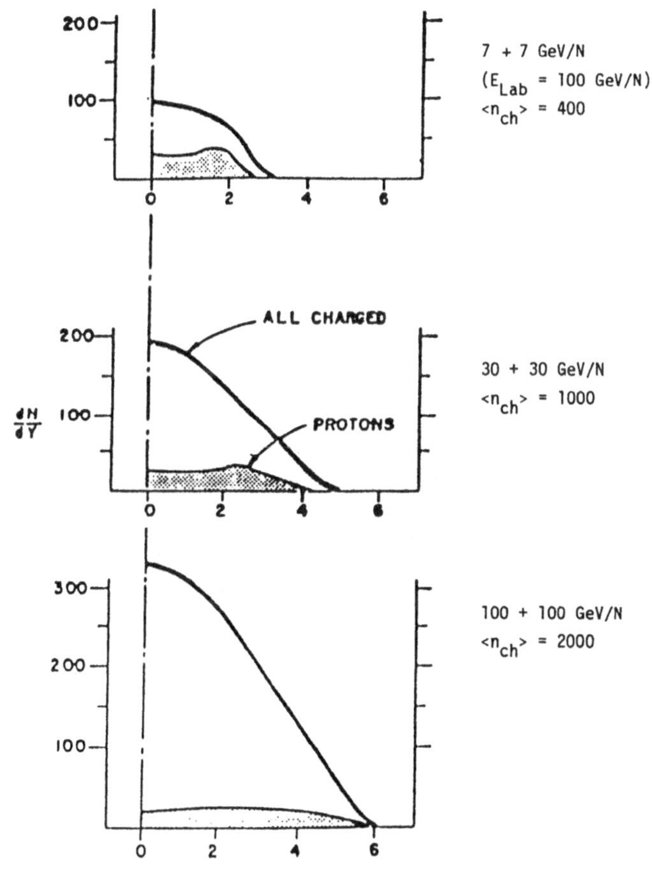

Figure 11: HIJET predictions for central Au + Au collisions at three different CM energies.

however, they did conclude that a device like that shown in Figure 12 (a symmetric version of the Bielefeld detector[9]) had many of the features required to make it a useful dimuon detector. That is, it should be compact (to filter out hadrons) and be able to detect muons with $p_\mu > 5$ GeV/c (momentum needed for effective hadron filtering). Also if one is interested in studying the small pair mass and transverse momentum chraacteristics of the plasma, one is restricted to $|y| > 2$, where the geometry and kinematics are similar to fixed target experiments, consistent with the design in Figure 12. The practical limit for pair mass muons in the central region in a collider would be $M_{\mu\mu} > 3$ GeV--not the region most strongly populated by a thermalized quark-gluon plasma.

Low-mass electron pairs for probing the plasma will be hard due to the high multiplicities expected. The group suggested that any electron detector should exploit the reduced numbers of particles expected at midrapidity. Difficulties arise due to tracking in this environment, and an electron/hadron rejection of $\gtrsim 10^5$ is needed to be able to identify direct electrons. A specialized detector with relatively small aperture at $\Theta_{c.m.} = 90^\circ$ appears to be the best choice. They noted that electron identification techniques are not easily incorporated in a large acceptance system, such as the one shown in Figure 12.

High p_T photons ($p_T \gtrsim 8$ GeV/c) are rare processes in hadron-hadron collisions and require high luminosities to be seen. However, they may be enhanced in central nucleus-nucleus collisions and should be looked for (again, luminosity will be the dominant issue). Shower counters, capable of resolving individual photons, will be required, all in a high multiplicity environment. They conclude that much work remains in this area before experiments are undertaken.

One final point worth noting: they discussed the possibility that jet-production[20] may be an interesting new signature for the plasma. Collider energies in excess of 50-75 GeV/nucleon will be required before such studies can even be started.

In the area of R&D the group identified: i) charged particle tracking and high resolution wire chambers capable of resolving individual particles in high track density and multiplicity environment, ii) vertex detectors (e.g., silicon strip detectors), iii) techniques of particle identification in the background of many particles, and iv) more information needed on effects of radiation damage in nuclear beam experiments (including effects of slow neutrons and heavy fragments).

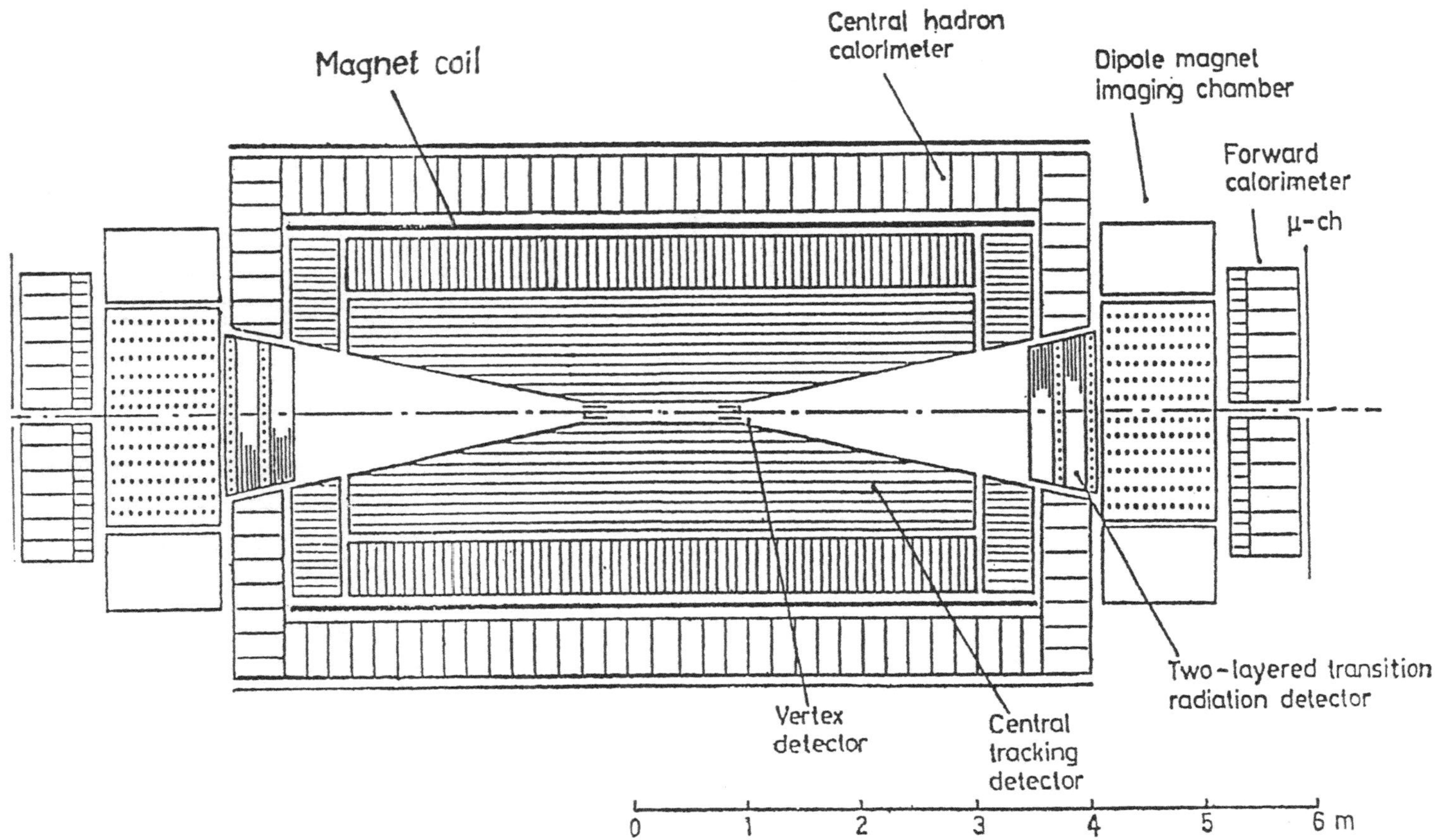

Figure 12: A 4π detector for high energy colliding beam operations derived from a workshop study for a fixed target device at the CERN SPS (Ref. 10). To make this detector, two of the SPS fixed target designs were placed back-to-back.

10. Hadrons and Event Parameters--Collider

This group considered detectors that could address the broad range of needs of both hadron and event parameter experiments. For hadron studies they considered detection of baryons, mesons and their antiparticles with the idea of obtaining information on: i) strange/nonstrange and baryon/antibaryon production ratios, ii) relative yields of particles with $|s| = 0, 1, 2, 3$, iii) $\pi\pi$, KK, pp intensity interferometry, and iv) studies of multipion and multikaon systems. Event parameter measurements include: i) multiplicity and energy and momentum flow as functions of rapidity and azimuthal emission angle, ii) correlations between dN/dy and $\langle p_T \rangle$, and iii) fluctuations in these observables. Also considered were the similarities and differences for collider detectors operating between 2 + 2 GeV/nucleon and 100 + 100 GeV/nucleon. One immediate comment is that if the plasma is characterized by a temperature of, say T = 200 MeV, the average energies of particles emitted from such a thermal source will be relatively low (e.g., π's ≃ 500 MeV, K's ≃ 400 MeV). This means that the detector (particularly at midrapidity) will generally be sampling soft hadrons. This puts an added burden on calorimetry for these particles and on the survival probability for those particles which are unstable (e.g., ≈88% of the π^+'s produced from a 200 MeV plasma will survive a 3 meter flight path, and ≈52% of the kaons).

Their general aim was to define a relatively simple collider detector, one which employs tracking over a limited solid angle while providing calorimetry over the remaining nearly 4π. In this scheme, the target/projectile fragmentation regions were not considered, since at collider energies most of the fragmentation products remain inside the beam pipe. Figure 13 shows the general layout of such a detector. It consists of the following elements: i) a multiplicity barrel surrounding the interaction region capable of measuring $dN/d\eta$ with good accuracy, ii) electromagnetic and hadronic calorimetry over at least the central two units of rapidity ($40° \lesssim \Theta_{c.m.} \lesssim 140°$), and iii) an open region into which one could place either a visual detector like a streamer chamber or a charged particle spectrometer with the ability to track up to 20 particles. Thus, the general philosophy was to use the calorimetry section for triggering (high p_T or E_T) and the streamer chamber or spectrometer arm for measuring single and few particle correlations. The capabilities of a streamer chamber are well known for relatively high multiplicities and for observing decays of unstable particles like $\Lambda° \rightarrow p\pi^-$. The magnetic spectrometer arm would have to possess good tracking and

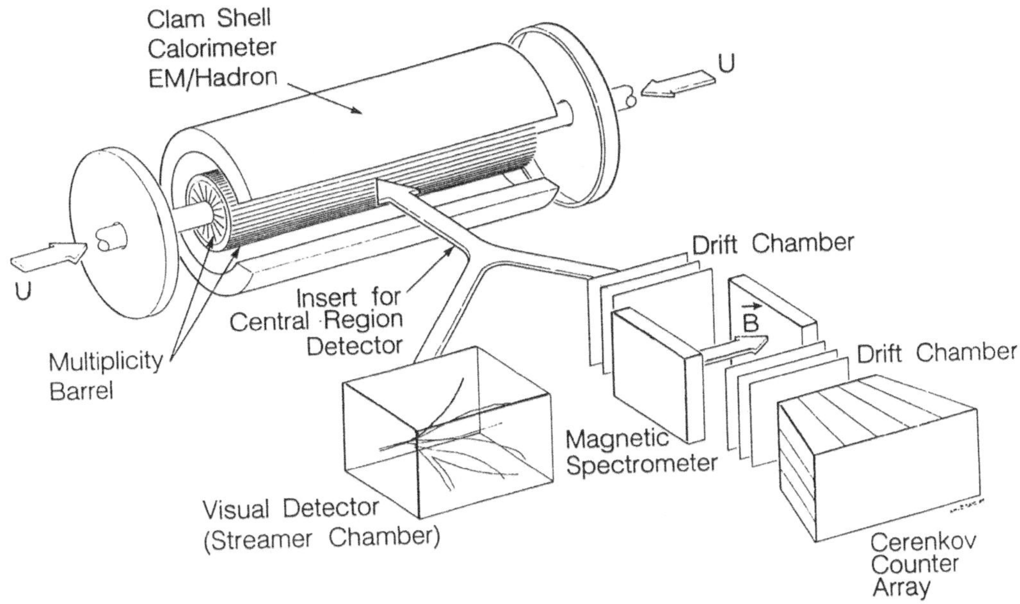

Figure 13: General layout for a collider beam detector with nearly 4π calorimetry and tracking capability at midrapidity.

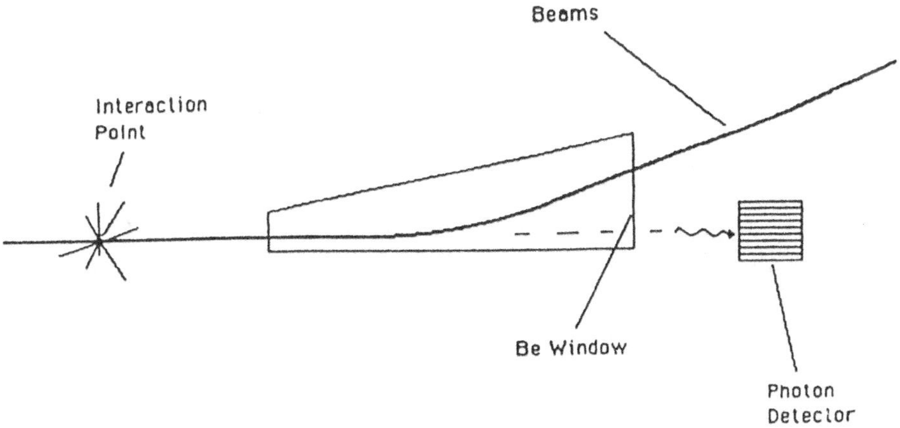

Figure 14: Layout of the zero degree photon detector to measure bremsstrahlung at a collider.

particle identification capabilities. Note that the bending will be in the vertical plane, allowing equal acceptance for positive and negative particles. Such a device, concentrating on the physics of the central region, should have a large dynamic range of operation, i.e., work as well at a few GeV/nucleon as at tens of GeV/nucleon; since the plasma once achieved should be characterized by a single temperature (relatively independent of incident energy).

The group performed a simple Monte Carlo study of the magnetic spectrometer arm's acceptance for HBT experiments. They assumed: i) 10^5 pairs needed to provide a sensitivity of Δ (radius) $\approx$ 0.2 fm, Δ (lifetime) $\approx$ 0.7 fm/c and Δ (intercept-degree of coherence) $\approx$ 0.1, ii) luminosity of 10^{24} $cm^{-2}sec^{-1}$ for uranium-uranium collisions, and iii) rate $\propto \sigma \mathcal{L} (dn/dy)^2$. From the Monte Carlo they obtained 1.5×10^4 like sign π-pairs/day, so that about one week's run would provide detailed HBT information.

Since one cannot place a detector directly in the beam for collider experiments (as is done for fixed target work) other ways of defining centrality are needed. Following the suggestion of Bjorken and McLerran[21] that bremsstrahlung photons emitted as two nuclei pass through each other, could be used as an indicator of the number of charged participants $\left(N_{tot} \approx \frac{Z^2\alpha}{\pi} \ln \frac{\omega_{max}}{\omega_{min}}\right)$, one finds that $N_{tot} \sim 50$ photons can be expected in the energy interval of 100-1500 MeV at a mean angle of ~0.6°, at 100 + 100 GeV/nucleon for a Au + Au collision. The neutrons from projectile break-up are well inside this cone (i.e., at ~1/10 its value). Figure 14 shows such a photon detector, downstream of the interaction region. A lead-glass array, consisting of ~250 cells (to reduce double-hit probability), located $\gtrsim$30 meters from the interaction point, would be suitable.

The group identified the following areas for R&D: i) studies of calorimeter responses to intense fluxes of soft photons and hadrons, ii) issues related to machine design--identifying backgrounds for 0° photons and access to 0° for both charged and neutral particles, and iii) more thinking about what to do with several hundred fully identified tracks, if you have them.

11. Summary

A number of lessons were learned from the Workshop, including:

o Experiments appear do-able, but

i) in high flux environment one will probably have to give up tracking all particles and go to calorimetry

ii) selective tracking will often be required (e.g., to

distinguish between multiple vertices in collider mode).

- o R&D issues identified
  - i) particle identification (thin TPC's, RICH, TRD, ...) and the related question of segmentation of detectors
  - ii) low-mass detectors needed to reduce delta rays, conversions of γ's, secondary interactions, ...
  - iii) use and development of high-technology items (e.g., Si strip detectors)
- o Need for high energy physics input
  - i) interest and participation (the quark-gluon plasma is both an intellectual and instrumental challenge)
  - ii) experience and expertise (e.g., use experience with large detector like UA1 as a guide post)
- o More theoretical work/input on question of signatures of quark-gluon plasma (theorist's R&D).

Much work has occurred since the time of the Bielefeld and Brookhaven Quark Matter meetings on the question of detecting high energy central nucleus-nucleus events. Many new proposals for experiments are being submitted to the new high energy facilities. The message is clear: there is a community of particle and nuclear physicists who want to use nuclear beams to probe the new phenomena expected in the high energy density environment of nucleus-nucleus collisions. The experiments are do-able, but if history teaches us any lessons (and if we are willing to listen), the process of unraveling the signal from the background will not be an easy one. It will tax our ingenuity as experimentalists--but it is a challenge we will be happy to accept.

Acknowledgments

I am grateful to the Quark Matter'84 organizing committee for inviting me to review the Detector Workshop. My many thanks to the conveners of the individual working groups; without their efforts, the Workshop would not have been the success it was.

References

1. Proceedings of the Workshop on Detectors for Relativistic Nuclear Collisions, edited by L. Schroeder--LBL-18225, UC-37, CONF-8403137 (September 1984).
2. See the talk of K. Foley elsewhere in these proceedings.
3. See the talk of R. Stock elsewhere in these proceedings.
4. A Long Range Plan for Nuclear Science: A Report by the DOE/NSF

Nuclear Science Advisory Committee (December 1983).

5. See the talk of T. Ludlam elsewhere in these proceedings.

6. L. S. Schroeder, The Berkeley Mini-Collider (LBL-17990 (1984)), to be published in the Proceedings of the Conference on the Intersections Between Particle and Nuclear Physics, Steamboat Springs, Colorado (May 23-30, 1984).

7. See the talk of H. Pugh elsewhere in these proceedings.

8. G. R. Young, J. A. Martin, and J. B. Ball, Nucl. Phys. A418, 361 (1984).

9. Proceedings of the Bielefeld Workshop, "Quark Matter Formation and Heavy Ion Collisions," May 1982, M. Jacob and H. Satz (Eds.), World Scientific, Singapore, 1982.

10. Proceedings of the Third International Conference on Ultra-Relativistic Nucleus-Nucleus Collisions (Brookhaven, September 1983), T. W. Ludlam and M. E. Wegner (Eds.), published in Nucl. Phys. A418 (1984).

11. Proceedings of the Workshop on SPS Fixed-Target Physics in the Years 1984-1989, Vol. I and II, edited by J. Mannelli, CERN 83-02 (February 10, 1983).

12. M. Gyulassy, Signatures of New Phenomena in Ultrarelativistic Nuclear Collisions, Nucl. Phys. A418, 59 (1984).

13. See the talk on Signatures for New Phenomena by L. McLerran in these proceedings.

14. W. Vernon Jones, Nucl. Phys. A418, 139 (1984) and T. H. Burnett et al., Phys. Rev. Lett. 50, 2062 (1983).

15. I. Otterlund and E. Stenlund, "A Simulation of Ultra-Relativistic Nucleus-Nucleus Collisions," pg. 105 of the Proceedings of the Workshop on Detectors for Relativistic Nuclear Collisions (Ref. 1).

16. C. Gruhn et al., SPS proposal, CERN/SPSC 84-13 (1984).

17. J. Rafelski and M. Danos, GSI-83-6 (1983) and references therein.

18. W. Zajc, Detector Requirements for Intensity Interferometry Measurements, pg. 121 of the Proceedings of the Workshop on Detectors for Relativistic Nuclear Collisions (Ref. 1).

19. T. Ludlam, private communication.

20. T. Ludlam, F. Paige, L. Madansky, "Very High Energy Probes of the Quark-Gluon Plasma, pg. 115 of the Proceedings of the Workshop on Detectors for Relativistic Nuclear Collisions" (Ref. 1).

21. J. D. Bjorken and L. McLerran, "Coherent Photon Radiation from Nuclei as a Probe of Impact Parameters and Nucleon Velocity Distributions in Ultrarelativistic Nuclear Collisions," pg. 95 of the Proceedings of the International Workshop on Gross Properties of Nuclei and Nuclear Excitations XII, Hirschegg, Austria (January 16-21, 1984), edited by H. Feldmeier, Darmstadt 1984 (ISSN 0720-8715).

DIRECT PHOTON AND LEPTON PRODUCTION IN HIGH ENERGY COLLISIONS

Hans J. Specht

Physikalisches Institut der Universität Heidelberg
Heidelberg, Germany

Abstract

A short survey of direct photon and lepton pair production in high energy collisions is given. It addresses both the high-p_T and low-p_T regime, relates the phenomenology in hadron-hadron collisions to the expectations in nucleus-nucleus collisions for the case of a transient quark-gluon plasma, and contains also some discussion of the experimental techniques involved.

1. Introduction

For many years, the study of direct photons and lepton pairs produced in high-energy hadron collisions has been particularly profitable in clarifying the nature of the hadron constituents. In the case of high energy nucleus-nucleus collisions, photons [1-8] and lepton pairs [7-11] may continue to be attractive probes. More specifically, due to the Stefan-Boltzmann behaviour of the production rates ($\sim T^4$) and the absence of final-state interactions, they could provide direct information on the hottest stages of the collision where, for sufficiently high temperatures and particle densities, a transient state of locally unconfined quarks and gluons is expected to occur. In terms of the transverse momenta or transverse masses $(M^2+p_T{}^2)^{1/2}$ involved, the region of 200-500 MeV/c would be of special relevance for thermal production, but other processes both above and below may be of equal importance.

This note is organized as follows. We very shortly survey, in turn, the phenomenology of real photon and lepton pair production in hadron collisions, subdividing each case into two domains: a high-p_T (high mass) part, which is reasonably understood on the basis of single parton processes, and a low-p_T (low mass) part, which appears to not be understood at all. In both cases, we relate this empirical knowledge to the expectations for nuclear collisions and point out some difficulties. We finish, in the last chapter, with the prospects of the forthcoming NA34 experiment for photon and lepton detection in pp, pA and AA collisions.

2. Real Photons

The order-of-magnitude of the inclusive direct photon cross sections in high energy hadron collisions is illustrated in Fig.1. In the range 10 MeV/c $\leq p_T \leq$ 10 GeV/c, a decrease of about 12 orders of magnitude is observed. A detailed review of the experimental results at high transverse momenta is given in ref. [12]. At ISR energies, the level of photon production compared to that of π^0's rises from a few percent at $p_T \leq$ 4 GeV/c to nearly 1 for $p_T \geq$ 10 GeV/c [13]. These high-p_T photons are interpreted as radiation created in a single hard scattering process of a quark on a gluon or antiquark. Of the lowest order ($\alpha\alpha_s$) QCD diagrams, the QCD quark-gluon Compton graph, illustrated in Fig.1, has theoretically been demonstrated to give the dominant contribution (~90%) to the total γ-rate in the case of high energy proton - proton reactions [14]. Studies of the associated event structure [15] support this picture: a jet of hadrons exists on the opposite side, and the high-p_T photon itself is essentially unaccompanied by other particles, leading to an upper limit of 30% (2σ) for the bremsstrahlung contribution. More recent investigations refer to the

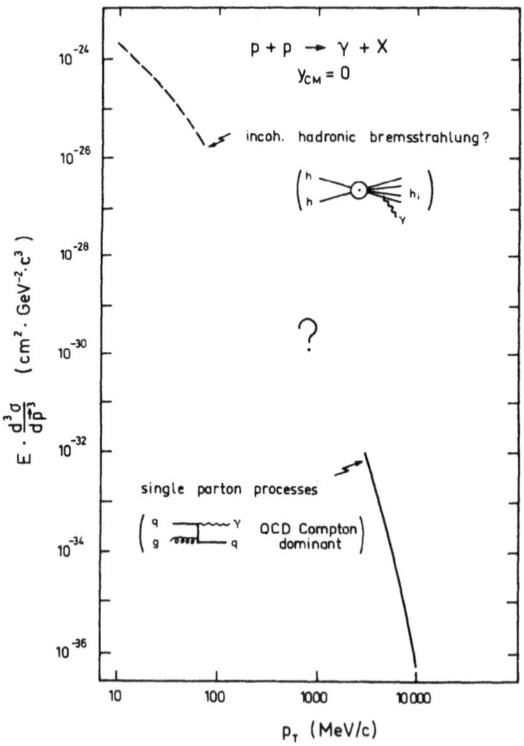

Figure 1: Order-of-magnitude illustration of inclusive direct photon cross sections. The high-p_T part has, e.g., been measured for pp at $\sqrt{s}$ = 63 GeV [13], the low-p_T part for K^+p at $\sqrt{s}$ = 12 GeV [17].

dependences on centre-of-mass energy, rapidity and quark composition of the colliding hadrons. Once the process is quantitatively understood, a unique way is opened to directly extract the structure function of the gluon. Work in this direction is presently proceeding in the R807/808 collaboration at CERN.

The experimental information on very soft photons, in the momentum range $10 \leq p_T \leq 100$ MeV/c, is extremely limited. Only two bubble chamber measurements, one studying $\pi^- p$ at 10 GeV/c [16], the other K^+p at 70 GeV/c [17], report a statistically significant excess of photons at very small values of $x < 0.01$ compared to what could be attributed to the radiative decay of known hadrons. The results from the latter, forming the basis for the dashed line in Fig.1, are reproduced in Figs.2 and 3. Fig.2 has been deliberately selected to demonstrate the procedure employed. In a first step, the contribution of photons emitted in the decay of hadrons other than π^0's (η,η'...) is calculated and subtracted from the data. In a second step, the remaining spectrum is <u>fitted</u> over the region $|x| \geq 0.02$ with a parametrization of π^0-production and decay; alternatively, the measured

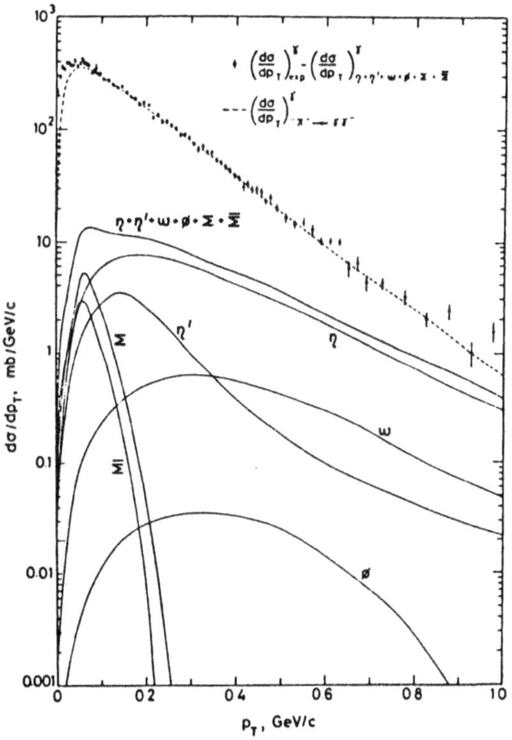

Figure 2: $d\sigma/dp_T$ for inclusive γ production in K^+p at 70 GeV/c [17]. The solid lines are calculated in the Lund model with π^0 production used as normalization. The dashed line represents a fit to the π^0-part on the basis of the measured π^--spectrum.

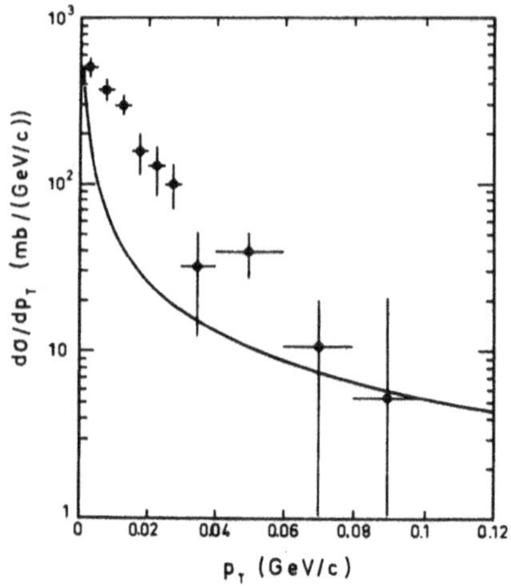

Figure 3: The $d\sigma/dp_T$ spectrum for inclusive γ production in K^+p at 70 GeV/c, remaining after subtraction of all known hadronic decays [17]. The solid curve is the predicted spectrum of inner bremsstrahlung.

π^--spectrum is used in that fit. In a final step, the fit is extrapolated to $|x|<0.02$ and subtracted from the spectrum, resulting in a residual signal both as a function of x [17] and p_T (comp. Fig.3). An equivalent procedure has actually been used in ref. [16]. It is precisely the origin of the large question mark in Fig.1: The production of direct photons relative to π^0's in the region $100 \leq p_T \leq 1000$ MeV/c is presumably suppressed by a factor of order 1/137, and an absolute measurement of π^0-photons on that level of accuracy is extremely difficult, leaving a chance (so far) only below the Jacobian peak (Fig.2).

Theoretically, at very low p_T, the direct photon spectrum should be dominated by the inner bremsstrahlung arising from the acceleration of the final charged hadrons produced. In the infrared limit, with the wavelength of the radiation large compared to the size of the interaction region, the yield should be calculable in a model-independent way, i.e. be insensitive to the details of the hadronic process [18]; only the outer particle lines should matter (Fig.1). Following previous calculations in this field [19-21], specific calculations have also been performed for the present two cases. Surprisingly, while the 10 GeV/c data seem to agree with the expectation [16], the 70 GeV/c data do not (Fig.3), but systematically lie high by a factor of 4-5 compared to the predicted curve. This is intriguing in a double sense. Other, unknown processes may contribute, and there may well exist a physical connection to the intermediate-mass "anomalous" lepton pair continuum to be discussed below. Real and virtual photons in this region can be regarded as probes of the dense plasma of hadron constituents in the non-perturbative regime, a field of rising interest in itself and of obvious relevance for the nuclear case. On the other hand, a transverse momentum of 10 MeV/c already corresponds to a wavelength of 20 fm, and at sufficiently low p_T Low's theorem [18] has to apply. New and more extensive data on soft real photons are therefore urgently needed to clarify the issue.

Turning now to the expectations for nucleus-nucleus collisions, we discuss the case $p_T \leq 100$ MeV/c first. Incoherent bremsstrahlung or whatever else is involved in the central region $x \approx 0$ surely will continue to be of great interest. In addition, bremsstrahlung should also be emitted in the fragmentation regions, as the colliding nuclei get decelerated. If the wavelength of the radiation (in the frame of the respective nucleus) is large compared to the nuclear dimensions, all charged nucleons may contribute to the amplitude, leading to a coherent photon radiation with $N \approx Z^2\alpha/\pi \cdot \ln(p_{T_{max}}/p_{T_{min}})$ photons produced per collision [5,6]. The Z^2-enhancement should greatly ease the detection of this radiation compared to hadron collisions, where nothing is known in the fragmentation region at all. It is hoped that studies of the photon yield and angular distribution in connection with further event information would not only probe the impact parameter of the collision, but would also give some insight into the nuclear stopping mechanism.

In the region $p_T \geq 50$-100 MeV/c, thermal radiation from the deconfined state (if reached) may be the dominant part of direct photon production. The unique feature of this radiation as a "thermometer" has repeatedly been emphasized [1-4,8], although quantitative calculations appear to be quite difficult at this point [8]. On a qualitative level, however, the severe masking by $\pi^0 \rightarrow \gamma\gamma$, which led to the hole in Fig.1, may be eased by two effects. Whereas the π^0's will come from the surface of the interaction region, the photons will be created in the interior and escape that interior without much rescattering, equivalent to a volume/surface gain for the γ/π^0 ratio which counterbalances the coupling constant 1/137. In the limit of large volumes, the γ/π^0 ratio may then reach 10% [2] or even approach one (gammaization [1]). In addition, the spectral shape of the π^0's will reflect the surface temperature at reconfinement, whereas yield and spectral shape of the photons will reflect the hot interior, profiting from the Stefan-Boltzmann behaviour $\sim T^4$ of the production rates. One of the numerous open questions concerns the upper end of the interpretable momentum range: at some point, pre-equilibrium phenomena and ultimately single parton processes will exceed the tails of thermal spectra.

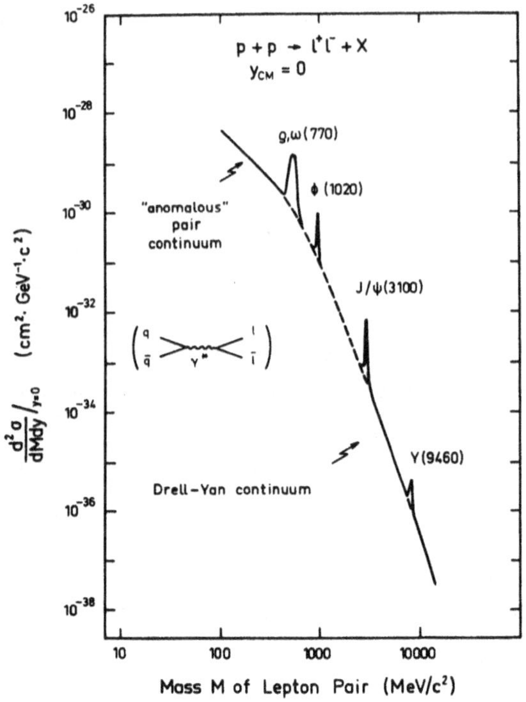

Figure 4: Order-of-magnitude illustration of inclusive lepton pair cross sections. The high-mass part actually applies to $\sqrt{s}$ = 28 GeV (Fig.5), the low-mass part has been estimated from Fig.9 and is rather insensitive to $\sqrt{s}$. The vector mesons are only drawn in a qualitative way.

3. Lepton Pairs

In analogy to photons, Fig.4 first illustrates the order-of-magnitude of the inclusive lepton pair cross sections in high energy hadron collisions over the whole mass range of interest. A structured continuum is observed which sharply falls with the invariant mass of the pairs, spanning again many decades. The superposed structures reflect resonant production of the vector mesons with the same quantum numbers as the photon (1^{--}) and the flavors u,d (ρ,ω), s (ϕ), c (J,ψ) and b (Y). The high-mass continuum is thought to be entirely due to the Drell-Yan mechanism [22] and can be described in terms of the simple parton model or perturbative QCD, while the low-mass part, where most of the cross section lies, is still open to discussion and almost certainly associated with non-perturbative QCD. Detailed reviews of the whole region, though with heavy emphasis on the high-mass part, can be found in refs.[23,24].

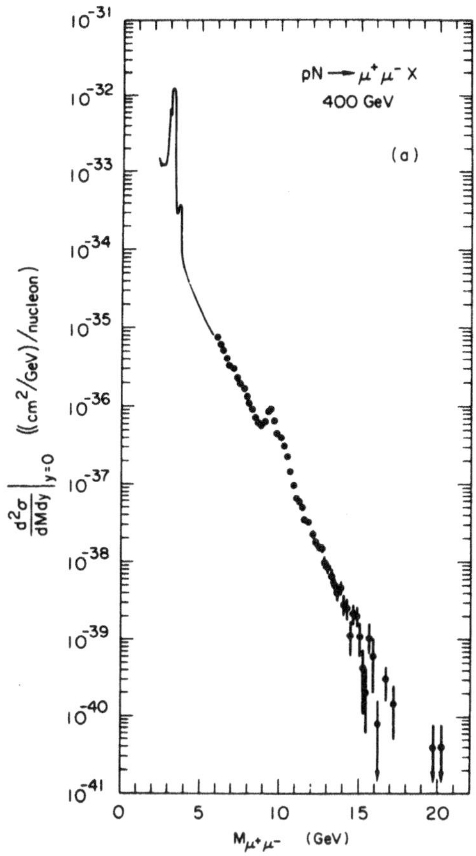

Figure 5: Dimuon spectrum measured by the CFS Collaboration. The solid line describes earlier results (from [24]).

We discuss the well-understood high-mass continuum first. Data from 400 GeV/c pN collisions are shown in Fig.5 as a representative example. In the framework of the Drell-Yan model [22], a valence quark from one of the incoming hadrons annihilates with an antiquark from the sea of the other hadron, creating a virtual photon γ^* which then decays into a pair of leptons (upper part of Fig.6). In the simpliest approach, the cross section is solely determined by the electromagnetic vertices (point-like cross section) and some folding of the probability distributions for the momentum fractions $x_1\vec{p}_A$ and $x_2\vec{p}_B$ (and vice versa) carried by the quark and antiquark, respectively; color conservation is taken care of by an extra factor of 1/3. Ignoring transverse momenta, the only variables necessary to describe the virtual photon are then $M^2 = sx_1x_2$ and $x_F = 2p_L/\sqrt{s} = x_1-x_2$, where s is the total CMS energy squared, and M is the invariant mass, x_F the Feynman-x and p_L the longitudinal momentum of the photon. The cross section then shows scaling with $\sqrt{\tau} = M/\sqrt{s} = \sqrt{x_1 \cdot x_2}$, representing the fraction of the total CMS energy used in the

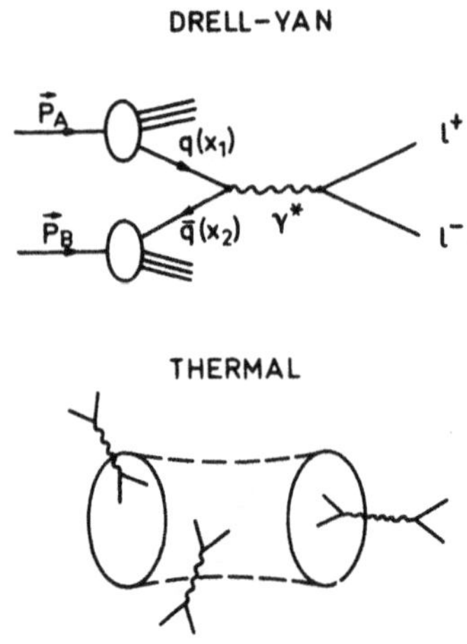

Figure 6: Diagrams for lepton pair production. The Drell-Yan mechanism (upper part) contains the momentum fraction distributions of the quarks inside the colliding hadrons together with the primary two-body kinematics; the thermal mechanism (lower part) does not.

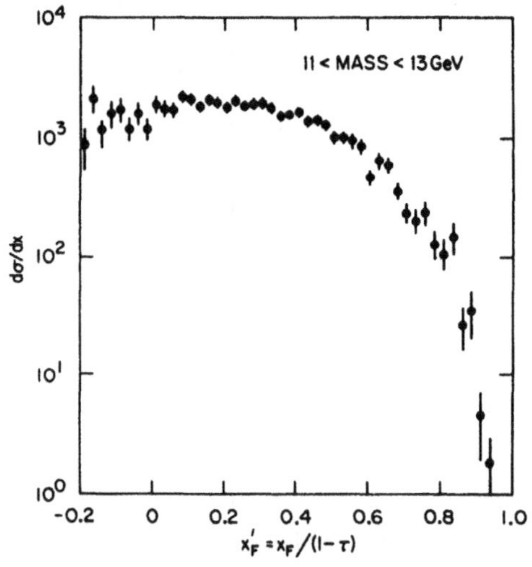

Figure 7: Example for the dependence of the dimuon production cross section on x_F for 400 GeV/c p+W (from [24]).

formation of the photon. The experimentally observed mass spectra are reasonably well reproduced by this approach. Slight scaling violations and discrepancies in the absolute values of the order of 50% (the "K-factor" problem) are attributed to QCD corrections [24]. The experimental x_F-distributions (Fig.7) also follow the model; in particular, a cut-off occurs at $x_F=1$, due to the underlying two-body kinematics, i.e. $x_1, x_2, x_F \leq 1$. Other features noteworthy concern the helicity distribution $(1+C\cos^2\theta^*)$ with $C \approx 1$, consistent with the annihilation of fermions, and complete "transparency" in pA collisions, i.e. $\sigma \sim A^\alpha$ with α close to 1.00 [24]. Apart from the interest in the process itself, studies in the Drell-Yan regime have been extremely useful for extracting structure functions of hadrons which are inaccessible by any other method [24].

Turning immediately to the case of nucleus-nucleus collisions, the question arises, whether and where thermal production of lepton pairs can successfully compete with Drell-Yan production. Assuming independent nucleon-nucleon interactions, the single parton processes â la Drell-Yan can readily be extended to the nuclear case [8]. The rapidity density $\rho(M^2,y)$ is found to be proportional to $A^{4/3}$, scaling with $\sqrt{\tau}$ still occurs and the kinematic constraint of $x_F \leq 1$ remains as well (Fig.8). In contrast,

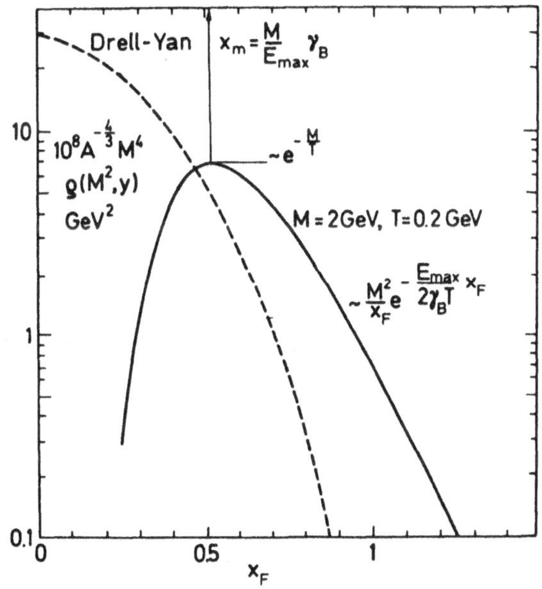

Figure 8: Scaled rapidity density $\rho_{\ell\ell}(M^2,y) = d\sigma_{\ell\ell}/dy \cdot 1/\sigma_{tot}$ (σ_{tot} total inelastic cross section) of lepton pairs of mass M = 2 GeV, produced either thermally at temperature T in the fragmentation region fireball or via the usual Drell-Yan mechanism (dashed line, independent of M or T). E_{max} is the maximum energy/nucleon, $x_F = E_{pair}/E_{max}$, γ_B refers to the target fireball; all quantities in the CMS (from [8]).

thermal production from a plasma of quarks and gluons, if formed, will follow the same basic diagram, but the rate will be governed by the creation of real antiquarks, and the kinematics (lower part of Fig.6) will be that of random emission in the frame of the respective fireball, i.e. the two fragmentation regions or anything in between. For a given fireball position, the distribution in Feynman-x_F (Fig.8) thus yields a (somewhat artificial) maximum, but extends to $|x_F|>1$ in any case, leading to a signal which dominates over the "background" of Drell-Yan pairs in at least some kinematical domain: in a fixed-target experiment either very forward or, as very slow pairs, at even backward angles in the lab [8]. For $|x_F|<1$, pronounced differences still occur in the shape of the rapidity distribution [11]. Scaling with $\sqrt{\tau}$ also does not apply any longer. Thus, critical signitures do seem to exist. As in the case of hard photons, however, questions arise as to the interpretation of far tails (M >> T) in thermal spectra, sensitive to preequilibrium phenomena.

Returning to the phenomenology of hadron collisions, we now address the low-mass regime. Since the first results from the Chicago-Princeton collaboration at Fermilab [25], a number of experiments both on $\mu^+\mu^-$- [25-27] and e^+e^--pairs [28-31] have obtained evidence for a pair continuum in the mass range $200 \leq M \leq 600$ MeV/c^2, i.e. below the ρ,ω resonance region down to threshold (or down to the intense instrumental tail of π^0-Dalitz decays). Irrespective of the severe background problems and the uncertainties in inclusive cross sections and kinematic distributions, the continuum contains an excess over what would be expected from higher mass Dalitz decays or other decays of known hadronic resonances. A compilation of these data, with all known contributions subtracted, transformed to central rapidity densities and normalized to (minimum bias) central pion production, is shown in Fig.9. Several features should be noted:

(i) Most strikingly, a universal behaviour independent of $\sqrt{s}$ is observed for the normalized data, in sharp contrast to the scaling with $m/\sqrt{s}$ for the Drell-Yan mechanism.

(ii) The mass dependence favors $1/M^2$ rather than $1/M^3$ [29] or $1/M$ [31].

(iii) If integrated over the mass range 200-600 MeV/c^2, the normalized rapidity density at $Y_{CM}=0$ is about $6\cdot10^{-5}$, while the total cross section ratio, with integration also over x_F (see iv), is about $3\cdot10^{-5}$, something like 2 orders of magnitude more than to be expected from Drell-Yan.

(iv) The longitudinal and transverse momentum distributions are relatively steep. The distribution in x seems to even steepen with $\sqrt{s}$, pointing in any case to a central phenomenon. The distribution in p_T is possibly consistent with m_T-scaling.

(v) The helicity distribution is largely unknown, with some indications for a flat distribution within large error bars [26].

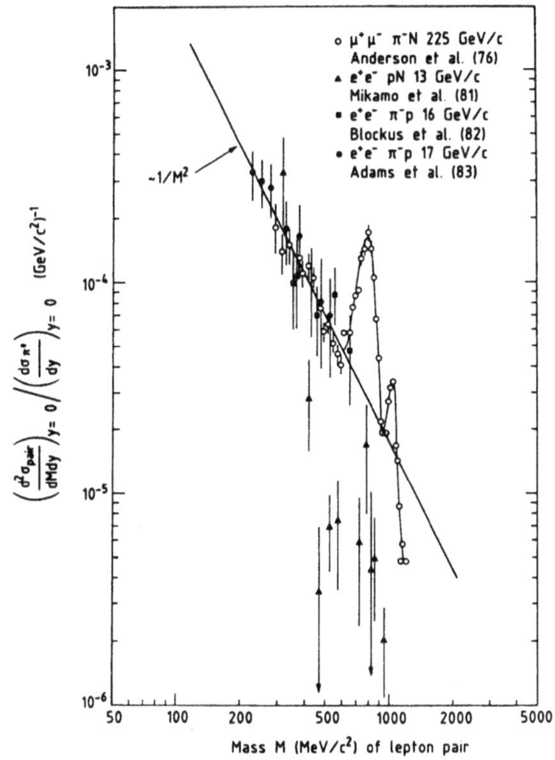

Figure 9: The mass dependence of lepton pair production in pion and proton interactions at different beam energies. The data have been taken from refs.[25, 29-31], extrapolated to $x_F=0$ ($Y_{CM}=0$) and properly normalized to central pion production. The levels of ρ and ϕ production have been plotted assuming the same x-dependence as measured for the continuum.

(vi) The associated production of other particles appears to be unchanged. In particular, the $e^+e^-\gamma$ process suggested in [30] can be ruled out [31].

(vii) In πA or pA collisions, the exponent α in $\sigma \sim A^{\alpha}$ seems to be close to 2/3, although this is not verified at precisely $x_F=0$.

The existence of a central electron pair source with a relative strength of order 10^{-4} brings to mind also the intriguing results on "e/π" at the ISR, i.e. a strong rise in inclusive electron production for $p_T \lesssim 500$ MeV/c [32,33]. The reality of this rise has historically been somewhat controversial, due to the large systematic errors from unrecognized low mass conversion- and π^0-Dalitz pairs and the uncertainties in the contribution from higher mass Dalitz decays (η...). However, a dedicated set-up of the R807/808 collaboration, using the Axial Field Spectrometer in connection with Cerenkov counters, a time-of-flight system and NaI walls, has recently not only been able to confirm the rise, but finds it to continue to a level

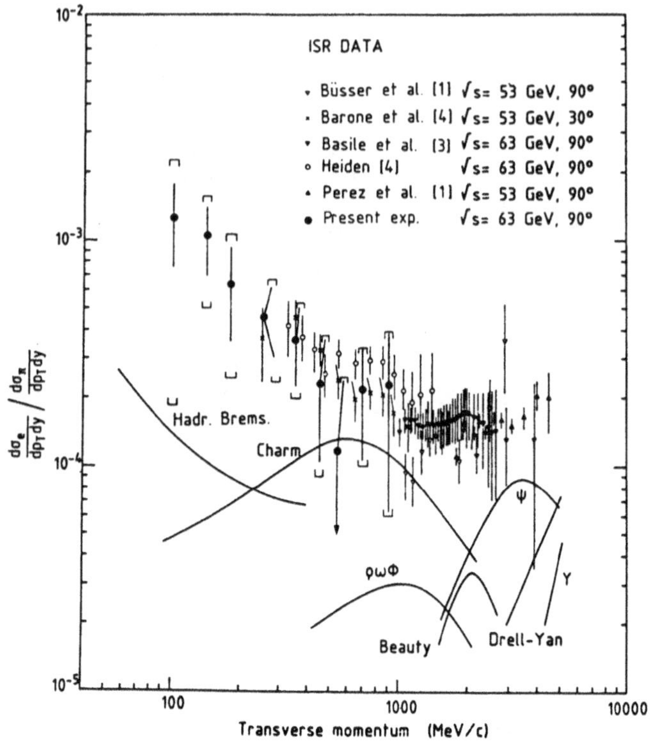

Figure 10: Summary of ISR data on the production of inclusive prompt electrons relative to pions [34]. The original compilation, without the data points "present exp." and the calculation of hadronic bremsstrahlung, is due to ref.[35]. The reference numbers in the figure refer to [34] (the low-p_T data stem from [32,33]).

of $1.2 \cdot 10^{-3}$ at p_T= 100 MeV/c [34]. The summary of the ISR data is shown in Fig.10. Thus, a further lepton signal exists beyond what can be expected from known hadronic decays; in particular, semileptonic charm decays, occasionally suggested [33] to explain the high values of e/π at low p_T, can clearly be ruled out both on the basis of numbers and shape. The low-mass pair source of Fig.9, on the other hand, would have the proper characteristics to naturally account for the rise in the inclusive yield. Whether it can do so quantitatively remains to be verified. An analysis aimed at isolating the pairs themselves is presently proceeding in R807/808.

What then is the origin of the "anomalous" lepton pair continuum? Hadronic bremsstrahlung [20,21] would give a shape both for the mass- and the single lepton p_T-dependence similar to what is being observed, but suspiciously, as in the case of soft real photons, it falls short in absolute magnitude by at least a factor of 5 (comp. Fig.10). Apart from hadronic annihilation [36] and quark bremsstrahlung [37], the Bjorken-Weisberg mechanism [38] has received particular attention. Their picture of "soft annihilation" contains a strong enhancement of the Drell-Yan process by the fact that in hadronic interactions, a large number of quarks and antiquarks may be produced during the collision at rather low relative momenta; a reannihilation of these into leptons before the products of the collision emerge would then be able to draw from a much larger sample. Attempts to quantify this model [39] gave, however, discrepancies with the data, and the suggestion was made that soft-gluon radiation terms are needed in the annihilation to get more than just the 1^{--} channel involved [31,39]. If more complicated diagrams are really needed, one may also go to the other extreme and exploit a simple collective picture. Thermal production of lepton pairs from a quark-gluon plasma created in the hadronic collision has indeed theoretically been investigated by Shuryak [7]; interestingly enough, the temperatures needed to describe the higher-mass end of the continuum were found to be as high as 500 MeV. A further interesting mechanism for soft real and virtual photon production, based on the influence of the QCD vacuum structure on the motion of the hadron constituents during the collision, has recently been suggested by Nachtmann et al. [40]. It is obvious in conclusion, that much more experimental and theoretical work is needed in this difficult field of non-perturbative phenomena to put all of these ideas to a more decisive test.

In view of such uncertainties, the nuclear case becomes even more interesting. It seems clear that the mass region of a few 100 MeV/c^2, i.e. close to the temperatures involved, will be of special relevance for thermal pair production. Out of several calculations in this region [8-10,41,42], a specific example is reproduced in Fig.11 [10]. A straightforward extension to electron pairs demonstrates the true peak in the spectra, free from the influence of the muon rest mass cut-off. A further peak in the region of the ρ reflects a characteristic feature of this model: the quark matter fireball, once created at T_0, is allowed to expand and cool; inside, quark-antiquark pairs annihilate at a rate determined by the local temperature; at the cooler surface, hadronization takes place, and the nascent pions are themselves allowed to annihilate at a rate which is not only governed by their temperature, but also by the pion form factor. The net result is then obtained as a convolution of the rates and the temperature profile function (in space and time), using relativistic hydrodynamics. In absolute terms, assuming a space-time volume of roughly 250 fm^4 as may, e.g., be applicable to a high-energy central ^{16}O + ^{238}U collision, an integration of the mass spectrum of Fig.11 over the interval 200-600 MeV/c^2 readily yields a total pair creation probability per event of $\Delta N \approx 2 \cdot 10^{-4}$ (for T_0= 180 MeV). With the expectation of about 150 π^0's produced under these circumstances [44], this is equivalent to an e^+e^-/π ratio of $1.5 \cdot 10^{-6}$, about a factor of 20 smaller than the value of $3 \cdot 10^{-5}$ quoted above for the anomalous pair continuum in hadron collisions! This somewhat alarming result can hardly be caused by the complications of the temperature profile function, but can directly be traced back to the

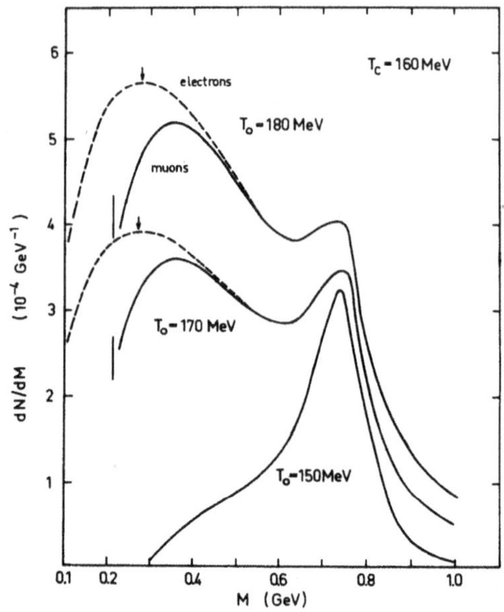

Figure 11: The dilepton mass spectrum for three initial fireball temperatures T_0 and a fixed deconfinement temperature T_C = 160 MeV. The figure is basically due to ref.[10], but has been modified to also include e^+e^- pairs as well as numerical values on the ordinate, assuming a space-time volume of $(\Delta x)^4 \approx 250\ fm^4$.

primary rate equation [7-10] $dN_{\ell\bar{\ell}}/dx^4 = \pi\alpha^2/108 \cdot T^4$ for the annihilation of massless particles. A factor of 20 seems to require either a much larger volume or a higher temperature; the latter is then consistent with Shuryak as indicated before.

Without expanding on it, we finally add for completeness a list of the motivations to also investigate vector meson production in nuclear collisions: the (ρ,ω) as a probe for chiral symmetry restoration [45], the ϕ as a probe of strangeness production [46], the J/ψ as a further probe of thermal production [47]. Any lepton pair experiment, given sufficient resolution, should thus be extremely powerful.

4. The NA34 Capabilities

The forthcoming experiment of the NA34 collaboration consists of two parts, one [43] devoted to pN collisions, the other [44] to AA collisions; the case of pA is, to some extent, incorporated in both. The overall lay-out, shown in Fig.12, consists of five major components: a target-vertex detector system, a 4π-calorimeter with separate sections for electromagnetic and hadronic energy, a compact forward electron spectrometer, a forward muon spectrometer, and an external spectrometer at

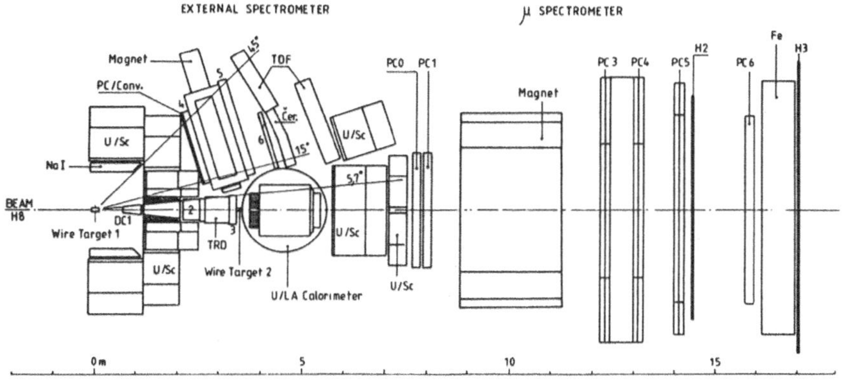

Figure 12: Top view of the experimental configuration of the NA34 collaboration [44].

sideward angles. Smaller components exist, in addition, for the detection of real photons. A sixth major component, an electron spectrometer at larger angles, may be added in the future for the nuclear case.

The first part of the experiment [43] is devoted to a high-accuracy study of the production of electrons, muons and neutrinos. The main physical issues concern the problem of e/μ universality (in particular in charm decay), the contribution of charm decay to the lepton pair yield (in particular in the region between the ϕ and the J/ψ , not discussed above), and the anomalous low-mass pairs and low-p_T leptons. This is indeed the first experiment in the field which combines electron identification, muon identification, missing energy measurement (for neutrinos), and accurate vertex determination (for the direct identification of open charm decays). Within the framework of this note, we restrict the description in the following to the topics of sections 2 and 3.

The muon spectrometer (Fig.12) uses major components of the previous NA3 dimuon experiment, but rearranged in a more compact form. The forward calorimeter also serves the purpose of the necessary hadron filter. Tracks in the muon spectrometer can be matched with upstream tracks inside the electron spectrometer to achieve an excellent mass resolution. The acceptance covers the mid- and forward rapidity part and the whole mass regime of interest.

The forward electron spectrometer (Fig.13) utilizes special drift chambers with high multitrack capability and a calorimetrized weak field (dipole) magnet. Hadron rejection on a level > 10^5 is achieved by combining a set of transition radiation

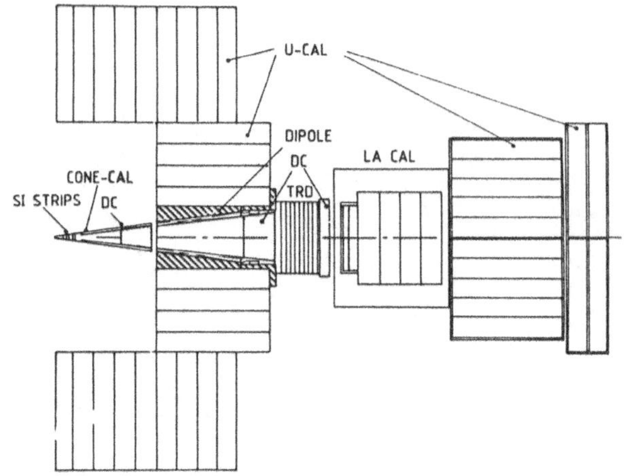

Figure 13: Detail side view of the forward electron spectrometer [43].

detectors with a high-granularity uranium/liquid argon calorimeter. Photon conversions and low-mass Dalitz decays can be vetoed on the trigger level: The low field integral allows a straight-line interpolation between a downstream hit and the target, consisting of a beryllium wire with a diameter of 50 μm and a length of 20 mm (in the beam direction); a Si-pad detector in a distance of 15 cm, matched in accuracy to the overall system, then recognizes close pairs by double dE/dx at the respective intersection point. Several Si-strip detectors, also installed close to the target, serve for accurate vertex determination in connection with charm decays.

Both spectrometers become even more powerful in connection with the 4π-calorimeter. Apart from the forward part, this consists of the uranium/scintillator modules developed and used before in R807/808 at the ISR. For the first time therefore, the dependence of inclusive lepton- and lepton pair production on the hadronic environment, e.g. large transverse energy deposition, can be studied. It does not need to be stressed that pN and pA collisions appear equally attractive in this context.

The nuclear part of the experiment [44] emphasizes the exploratory nature of the first phase. Calorimetry will play a decisive role, both in terms of self-interest (energy flow, rapidity structure) and trigger function (central collision-, high-E_T selection). The external spectrometer at wide angles (Fig.12), containing particle identification with silica aerogel Cerenkov detectors and time-of-flight hodoscopes and viewing the target through a narrow horizontal slit in the calorimeter, allows to study inclusive spectra (strangeness production, $\langle p_T \rangle$ versus E_T etc) as well as identical particle correlations, and to search for long-lived unusual charge or mass particles. The passive single wire target of [43] is replaced by an active multiwire target to minimize secondary interactions and photon conversions, to detect the vertex with high accuracy, and to recognize and eliminate interactions from beam fragments.

The spectra of real photons are investigated with three different approaches. The first uses conversion of the photons in a special converter-wire-chamber arrangement immediately in front of the external spectrometer, thus covering an angular region of 15^{0}-45^{0} (Fig.12). This region is of particular interest because of maximum particle production in asymmetric nuclear collisions. The second method exploits a reuse of two arrays of 600 NaI crystals (each) from the R808 experiment. The third uses a small array of shielded BGO crystals, arranged as to give some directional sensitivity. This array will be movable to scan a wide rapidity region. All set-ups will also be used in connection with pN and pA collisions.

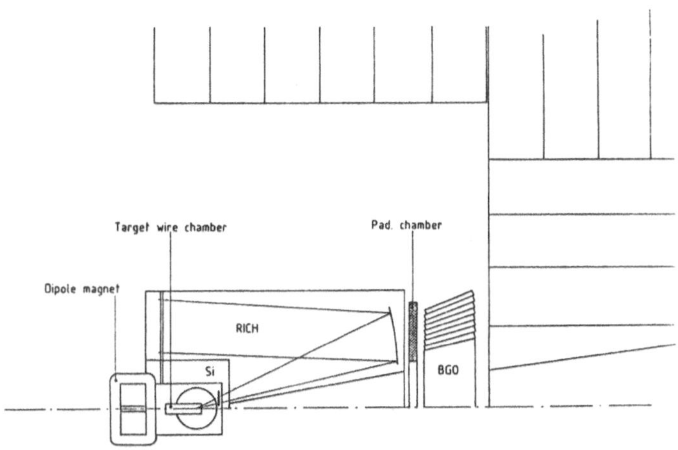

Figure 14: Top view of a possible set-up for an electron spectrometer at larger angles [44].

The muon spectrometer, situated behind the hadron filter and thus not subject to charged particle loads in the hundreds, will be the major tool for lepton studies. It is served by a second active wire target immediately in front of the U/LA calorimeter to minimize the random combinatorial background from $\pi \rightarrow \mu\nu$ decays. Matching to upstream tracks becomes impossible, due to the prohibitive particle load in the (unusable) electron spectrometer, resulting in some loss of mass resolution (limited now by multiple scattering in the calorimeter). The acceptance of the instrument is remarkably good: for the running conditions foreseen with the 200 GeV/c ^{16}O-beam at CERN, > 10^{5} events in the region of anomalous pairs can be accumulated over a running time of 600 h (assuming no new physics to occur).

As an approach complementary to the muon spectrometer, a new electron spectrometer is also under consideration. It is motivated by desires to extend the mass range of dileptons down to 100 MeV/c^2, to extend the rapidity coverage to larger angles, to obtain an improved mass resolution (complete ρ, ϕ separation), and to allow the first target to be used in connection with lepton pairs (more complete event information). The problem one is faced with in this case is twofold: a weak source of rather open electron pairs has to be detected in the presence of hundreds of charged particles; lack of complete detection efficiency for trivial low-mass pairs may create a prohibitive combinatorial background. Yet, extensive Monte Carlo simulations appear to be encouraging. A possible concept is shown in Fig.14. It consists of a ring image Cerenkov counter, a ring of BGO crystals, and a weak-field magnet together with a system of pad detectors. The RICH, operated in a threshold mode, presents the main tool for hadron rejection. Momentum analysis occurs in the electromagnetic calorimeter; the magnetic field is chosen sufficiently low to keep the pairs together, but just high enough ($B\ell \approx 0.01$ Tm) to obtain a detectable deflection for charge determination. Further investigations are being pursued to find the optimal solution to this extremely challenging problem.

Conclusion

The perspectives of photons and leptons look promising indeed.

Acknowledgement

The author has greatly benefitted from a sabbatical leave at CERN in 1983. He owes sincere thanks to the members of the R807/808 collaboration, in particular to C.W. Fabjan and W.J. Willis, for the warm hospitality of the group and countless educative discussion.

References

[1] E.L. Feinberg, Nuovo Cimento 34A, 391 (1976)
[2] F. Halzen and H.C. Liu, Phys. Rev. D25, 1842 (1982)
[3] W.J. Willis, Workshop on Future Relativistic Heavy Ion Experiments, Ed. R. Stock and R. Bock, Darmstadt, GSI 81-6 (1981)
[4] T. Goldman et al., Phys. Rev. D20, 619 (1979)
[5] J. Kapusta, Phys. Rev. C15, 1580 (1977)
[6] J.D. Bjorken and L. McLerran, Fermilab-Pub-84/35-T (1984)
[7] E.V. Shuryak, Phys. Lett. 78B, 150 (1978); Sov. J. Nucl. Phys. 28, 408 (1978)
[8] K. Kajantie and H.I. Miettenen, Z. Phys. C9, 341 (1981) and Z. Phys. C14, 356 (1982)

[9] G. Domokos and J.I. Goldman, Phys. Rev. D23, 203 (1981)
G. Domokos, Phys. Rev. D28, 123 (1983)
[10] S.A. Chin, Phys. Lett. 119B, 51 (1982)
[11] J. Badier et al., CENS preprint, DPhPE 82-11 (1982). Note that eq.A.8 must be multiplied by 2.
[12] T. Ferbel and W.R. Molzon, Rev. Mod. Phys. 56, 181 (1984)
[13] E. Anassontzis et al., Z. Phys. C13, 277 (1982)
[14] E.L. Berger, E. Braaten and R.D. Field, ANL-HEP-PR-83-46 (1983)
[15] T. Åkesson et al., Phys. Lett. 118B, 178 (1982)
[16] A.T. Goshaw et al., Phys. Rev. Lett. 43, 1065 (1979) and Phys. Rev. D24, 2829 (1981)
[17] P.V. Chliapnikov et al., preprint (1984) and Phys. Lett. 141B, 276 (1984)
[18] F.E. Low, Phys. Rev. 110, 974 (1958)
[19] R. Cahn, Phys. Rev. D7, 247 (1973)
[20] R. Rückl, Phys. Lett. 64B, 39 (1976)
[21] N.S. Craigie, Phys. Rep. 47, 1 (1978); N.S. Craigie and H.N. Thompson, Nucl. Phys. B141, 121 (1978)
[22] S.D. Drell and T.M. Yan, Phys. Rev. Lett. 25, 316 (1970) and Ann. Phys. 66, 578 (1971)
[23] L.M. Lederman, Phys. Rep. 26, 149 (1976)
[24] R. Stroynowski, Phys. Rep. 71, 1 (1981)
[25] K.J. Anderson et al., Phys. Rev. Lett. 36, 237 (1976), and
G.G. Henry, Ph.D. Thesis, Univ. of Chicago (1978), quoted in [24]
[26] B. Haber et al., Phys. Rev. D22, 2107 (1980)
[27] K. Bunnell et al., Phys. Rev. Lett. 40, 136 (1978)
[28] J. Ballam et al., Phys. Rev. Lett. 41, 1207 (1978)
[29] S. Mikamo et al., Phys. Lett. 106B, 428 (1981)
[30] D. Blockus et al., Nucl. Phys. B201, 205 (1982)
[31] M.R. Adams et al., Phys. Rev. D27, 1977 (1983)
[32] L. Baum et al., Phys. Lett. 60B, 485 (1976), M. Barone et al., Nucl. Phys. B132, 29 (1978)
[33] M. Heiden, Ph.D. Thesis, Univ. of Heidelberg (1982); CERN EP82-05 (1982)
[34] T. Åkesson et al., subm. to Phys. Lett. (1984)
[35] P. Baillon et al., Proposal CERN/SPSC/82-55
[36] T. Goldman et al., Phys. Rev. D20, 619 (1979)
[37] discarded in [38]
[38] J.D. Bjorken and H. Weisberg, Phys. Rev. D13, 1405 (1976)
[39] V. Černý et al., Phys. Rev. D24, 652 (1981)
[40] O. Nachtmann and A. Reiter, Z. Phys. C24, 283 (1984), and private communication (1984)
[41] L. McLerran, Proc. of this Conference
[42] N. Glendenning, Contrib. to this Conference
[43] H. Gordon et al., Proposal P189 to the SPSC, CERN-SPSC/83-51 (1983); accepted as NA34
[44] H. Gordon et al., Proposal P203 to the SPSC, CERN-SPSC/84-43 (1984); accepted as NA34/1
The present list of names: H. Gordon, T. Ludlam, V. Polychronakos, D.C. Rahm, I. Stumer, C. Woody, T. Åkesson, H. Atherton, H. Breuker, C.W. Fabjan, U. Goerlach, S. Katsanevas, U. Mjornmark, J. Schukraft, W.J. Willis, P. Glässel, A. Pfeiffer, J. Soltani, H.J. Specht, N.J. DiGiacomo, P.L. McGaughey, W.E. Sondheim, J.W. Sunier, S. Almehed, G. Jarlskog, B. Lörstad, L.A. Hamel, C. Leroy, Y. Sirois, G. Beaudoin, J.M. Beaulieu, P. Depommier, H. Jeremie, L. Lessard, A. Lounis, S. Mayburov, A. Shmeleva, V. Cherniatin, B. Dolgoshein, Yu. Golubkov, A. Kalinovsky, V. Kantserov, P. Nevsky, A. Sumarakov, V. Sidorov, W. Cleland, J. Thompson, A. Gaidot, F. Gibrat, G.W. London, J.P. Pansart, D. Bettoni, M. Goldberg, N. Horwitz, G.C. Moneti, L. Olsen, O. Bernary, S. Dagan, D. Lissauer, Y. Oren
(Brookhaven, CERN, Heidelberg, Los Alamos, Lund, Montreal, Moscow, Novosibirsk, Pittsburgh, Saclay, Syracuse, Tel Aviv)
[45] R. Pisarski, Phys. Lett. 110B, 155 (1982)
[46] J. Rafelski, Nucl. Phys. A374, 489 (1982), and Contrib. to this Conference
[47] J. Cleymans and R. Philippe, Z. Phys. C22, 271 (1984)

RHIC AND QUARK MATTER: A PROPOSED HEAVY ION COLLIDER AT BROOKHAVEN NATIONAL LABORATORY

T. Ludlam
Brookhaven National Laboratory

Upton, NY 11973/USA

I. Introduction

In August, 1983 a Working Group of nuclear and particle physicists was convened at Brookhaven to consider the basic parameters for a high energy heavy ion collider facility. The meeting was in response to a unique set of circumstances arising from the termination of the CBA project and the result of the NSAC long range planning study[1] in which an ultra-relativistic heavy ion collider was identified as the important new scientific opportunity for U.S. nuclear physics.

For some time there had been strong interest at Brookhaven in exploiting the existence of the Colliding Beam Accelerator, CBA (nee ISABELLE), for the generation of heavy ion collisions at very high energies.[2] Now the opportunity arose to consider the design of a dedicated heavy ion collider which could take advantage of the work already invested in the CBA project.

In its report[3] the Working Group recommended a top energy of at least 50 + 50 GeV/amu, emphasizing the need to cover a continuous range of energies extending as low 5 + 5 GeV/amu, with ion masses ranging from protons up to $A > 200$, and luminosity $L > 10^{25}\text{cm}^{-2}\text{sec}^{-1}$ initially, ultimately reaching much higher values.

These parameters were the subject of extensive scrutiny at the Quark Matter '83 Conference[4] one month later at which several of the world's laboratories presented possibilities for very high energy nuclear beam facilities. After several days of such discussions a round table panel at the Conference[5] strongly endorsed the recommendations of the August Working Group, with an emphasis on making the top energy as high as possible without sacrificing the capability for exploring the low energy range as well.

In the months since the Quark Matter '83 Conference an intensive design effort has been carried out at Brookhaven for a dedicated Relativistic Heavy Ion Collider (RHIC) using the existing site, tunnel, cryogenic refrigerator and experimental areas of the CBA project. During the course of this study a number of heretofore unappreciated features of accelerator physics came to light regarding the special case of storing intense beams of highly charged ions.[6] Most important of these is the phenomenon of intrabeam scattering[7] which imposes strong limitations on the performance of a heavy ion collider. Nonetheless, the requirements outlined above can be met.

The Brookhaven RHIC design was completed and submitted as a preliminary proposal[8] to the U.S. Department of Energy during June, 1984. It calls for a top beam energy of 100 GeV/amu for A = 200, and the acceleration of ion masses spanning the full periodic table. The proposed accelerator complex utilizes the existing Tandem Van de Graaff and AGS as source and injector, and will provide center-of-mass collision energies from 1.5 + 1.5 GeV/amu to 100 + 100 GeV/amu.

This range is covered, with no gaps, using the AGS beams for fixed target experiments ($\sqrt{s}/A < 5$ GeV/amu); one collider beam striking an internal gas jet or thin wire target (5 GeV/amu $< \sqrt{s}/A <$ 14 GeV/amu), and the colliding beams of the Collider itself (14 GeV/amu $< \sqrt{s}/A <$ 200 GeV/amu). The complete facility will thus provide heavy ion collisions over a spectrum of energies which begins just above the present Bevalac, extends through the range of "maximal stopping"[9] where high energy density is achieved at high baryon number, and on up to the extremely high energies now being glimpsed in rare cosmic ray events.

In the high energy regime achieved with the collider we expect a fully developed central rapidity region: an environment as close as we can achieve in the laboratory to that of the very early universe and the baryon-free conditions assumed in gauge theory calculations of finite temperature QCD and in hydrodynamic theories of the evolution of a quark-gluon plasma. At these energies we have access to a regime in which we know the individual nucleon bags can be penetrated, dislodging the constituent quarks and gluons. Here a potentially powerful probe becomes accessible in the form of jets observed after propagating through the new matter created in the collision.[10]

Fig. 1 shows the range of energies and kinematic terrain covered by the AGS and RHIC complex. In the following sections we discuss the detailed design parameters and performance specifications.

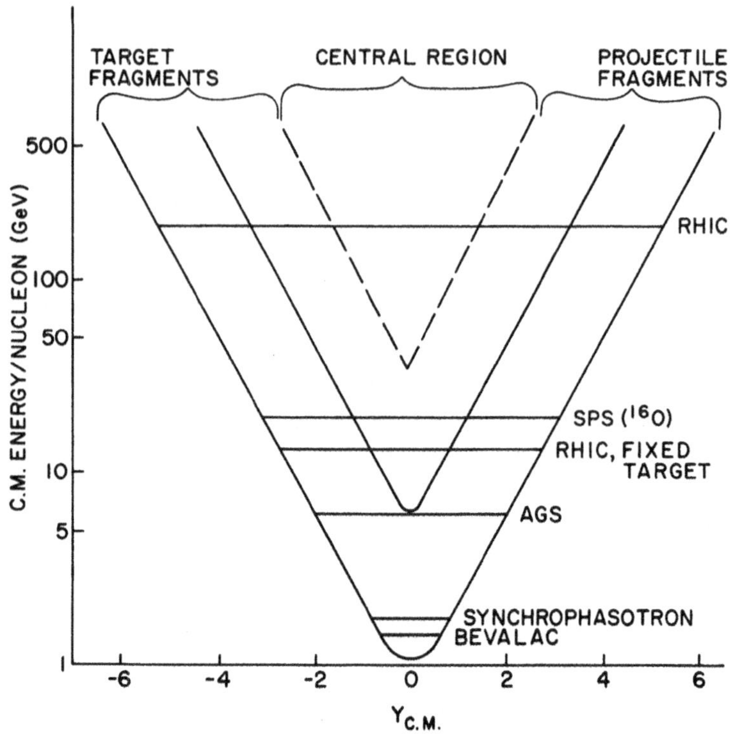

Figure 1.

The kinematic landscape available to the RHIC facility, and some existing machines. The outer "vee" is the phase space limit--the rapidity of the incident nucleons. The inner, solid vee delineates fragmentation regions of width $\Delta y = 2$, as observed in proton-proton collisions. The dashed lines indicate the wider ($\Delta y = 4$) fragmentation regions--and hence narrower central region--expected at a given energy for nucleus-nucleus collisions with $A \approx 200$. The horizontal lines indicate energies reached with the components of the AGS/RHIC complex, as well as in the facilities which presently exist (LBL Bevalac, Dubna Synchrophasotron) and the planned oxygen beams in the CERN SPS.

II. The RHIC Design

II.1 Source and Injection

Figure 2 is a site plan of the Brookhaven accelerator complex, shc ing the major components of the RHIC facility. The existing Tandem Van de Graaff accelerators will serve as source and initial ion accelerator. The ions will then traverse a long (≈2000 ft) transfer line to a small Booster synchrotron, from which fully stripped ions at an energy of ≈320 MeV/amu are injected to the AGS. Each particle bunch transferred from the Booster to the AGS is accelerated to the top AGS energy (28 GeV for protons; 11 GeV/amu for gold) and then transferred to the Collider by a magnet system installed in the existing injection line tunnels.

Construction of the transfer line from the Tandem accelerators to the AGS has now begun, as part of a program to accelerate ion beams in the AGS for fixed target experiments beginning in 1986.[11] These experiments will initally have the Tandem alone as ion source, providing fully stripped ions for injection directly into the AGS up to sulfur (A = 32). Higher masses require the Booster. Brookhaven has requested funds for the Booster as part of a general improvement of the AGS performance for high energy physics experiments. If it is funded construction is expected to begin in 1986, making it operational in 1988.

The Tandem Van de Graaf provides beams of very small transverse emittance (< 1π mm mrad at exit), with an energy stability of one part in 10^4, and so is well suited for injecting heavy ions into a synchrotron. Using a new type of sputter ion source[8] the machine can be operated in a pulsed mode with currents up to 200 μA. The ions pass through two stripping foils before arriving at the booster, where Gold is in charge state +33.

Table 1 shows the characteristics of the beam injected into the Booster for a representative sample of ion species. Here ε_N is the normalized emittance and N_B is the total number of particles in one Booster fill, which will be transferred to the AGS as one bunch. The maximum number of particles which can be injected to the Booster (the space charge limit) is given as N_{SC} in Table 1. It will be seen that the beam intensity is limited by the Tandem currents for the lighter ions, up to Copper, while for the heavier ions the maximum intensity is determined by space charge limits. After extraction from the booster and on their way to the AGS the ions pass through one more stripping target. The ions injected into the AGS are then fully stripped.

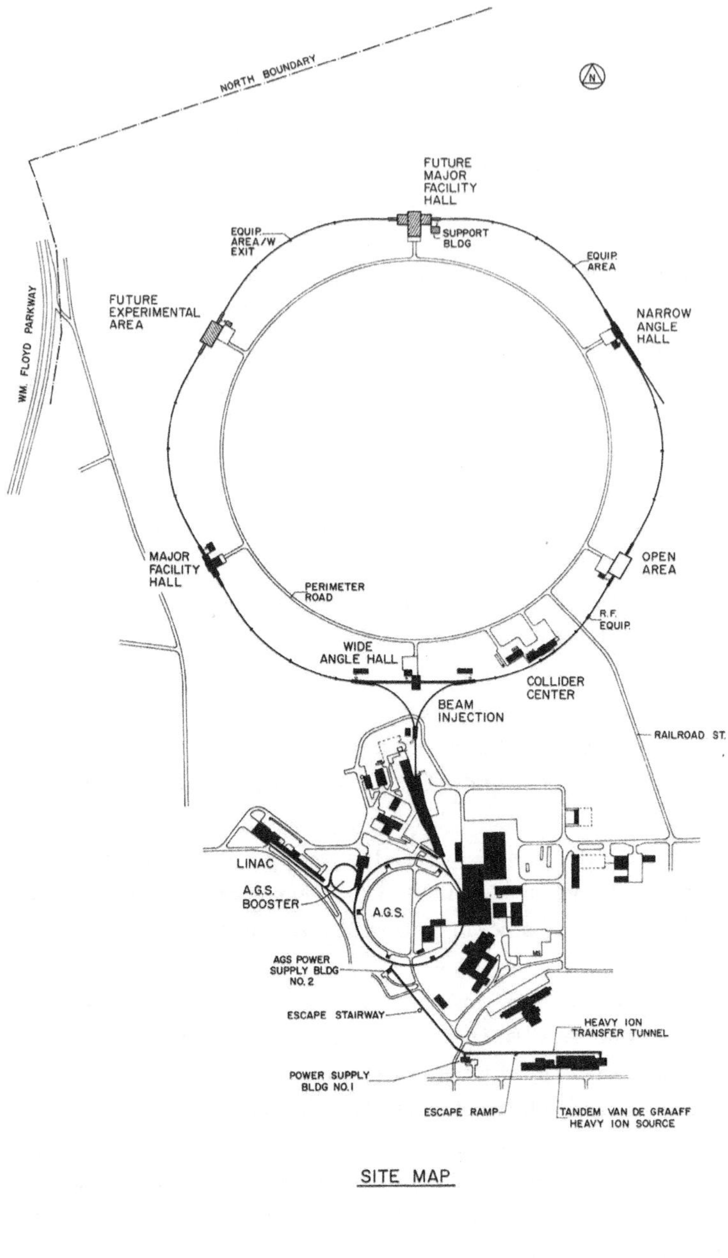

Figure 2.
Site Layout of the RHIC Project.

Table 1

Beam Intensity, Emittance of Injection into the Booster

Element	ε_N $\pi \cdot$mm$\cdot$mrad	T_{rev} μsec	$N_B \times 10^{9*}$	$N_{SC} \times 10^{9**}$
Deuterium	8.8	3.81	100.0	438.0
Carbon	6.3	5.33	22.0	37.0
Sulphur	6.0	6.72	6.7	11.0
Copper	3.9	8.60	4.7	5.5
Iodine	3.0	11.31	3.4	3.2
Gold	2.3	14.53	3.6	2.2

*With 8-turn injection

**At $\Delta\nu = .1$

Since the ions are completely stripped upon injection to the AGS, there is no requirement for improvement of either the vacuum or the rf system of the present accelerator. Ions will be transferred one bunch at a time from the Booster to the AGS, with each bunch then accelerated to the top energy of the AGS and sent to the collider. The cycle time for this process is 0.8 Hz.

II.2 The Collider

The layout of the collider is illustrated in Fig. 3. it is composed of two identical concentric rings in a common horizontal plane. There are six crossing points.

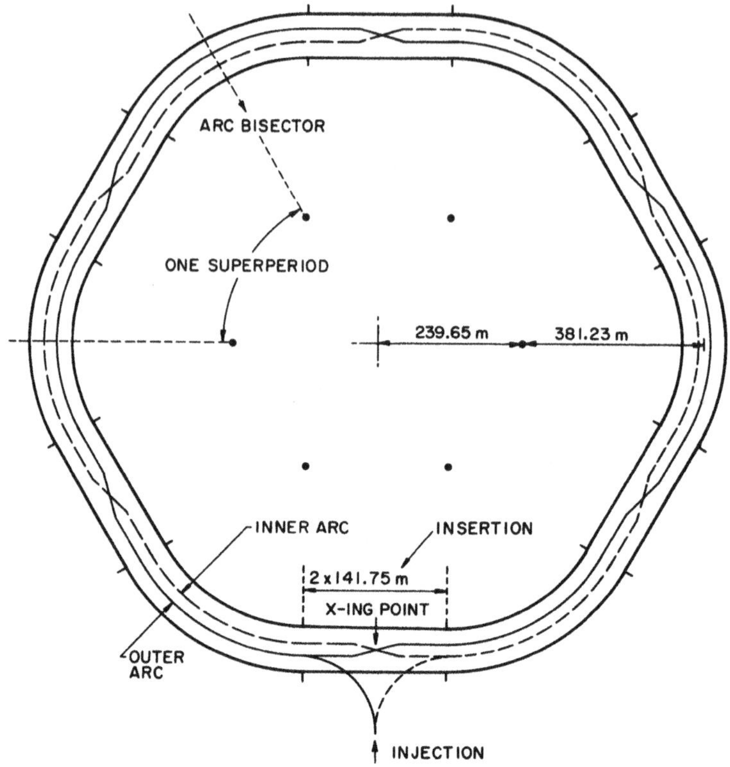

Figure 3.
Geometry of the Collider Beams.

Each ion Bunch accelerated in the AGS is extracted and transferred to one of the two Collider rings. The bunch is then captured by a stationary rf bucket. The two rings are filled in box-car fashion, with a total of 57 bunches per ring. An equivalent number of AGS pulses is required which gives a filling time a little more than one minute per ring. The bunch separation is 67 meters, or about 200 nanoseconds. The major parameters of the collider rings are given in Table 2. Some properties of the beams, at injection, are given in Table 3.

Table 2

General Parameters for the Collider

Circumference	3833.8 m
Revolution Frequency ($\beta = 1$)	78.1972 kHZ
Filling Mode	Box-Car
No. of Bunches/Ring	57
Filling Time/Ring	≈ 1 min
Perodicity	6
Magnetic Rigidity, $B\rho$:	
At injection	9.65 T·m
At top energy	839.50 T·m
Transition Energy, γ_T	26.4
Betatron Tunes, $\nu_{H,V}$	34.4
rf Harmonic Number	342
rf Voltage	1.2 MV
Acceleration Time	1 min

Table 3

General Beam Parameters for the Collider

Element	Proton	Deuterium	Carbon	Sulphur	Copper	Iodine	Gold
Injection: Kinetic Energy, GeV/A	28.0	13.6	13.6	13.6	12.4	11.2	10.7
Bunch Length nsec	±8.6	±8.6	±8.6	±8.6	±8.6	±8.6	±8.6
No. ions/ Bunch, $\times 10^9$	1000.	100.	22.	6.4	4.5	2.6	1.1
Top Energy: Kinetic Energy, GeV/A	250.	124.9	124.9	124.9	114.9	104.1	100.0
Luminosity:* (cm^{-2} sec^{-1})							
α = 0 mrad:	1.2×10^{31}	1.2×10^{31}	6×10^{29}	5×10^{28}	2×10^{28}	7×10^{27}	1.2×10^{27}
α = 2 mrad:	(----------------.23 $\times$ α = 0 Values ------------------)						

*Initial Luminosity at injection.
α = beam crossing angle.

The insertion regions are designed to provide some measure of flexibility in the focussing parameters at the crossing points, so that each insertion may be optimized for a particular experiment or machine function. The half cell insertion layout is shown in Fig. 4. The dipoles BC1 and BC2 are used to define the crossing angle α; BC1 is common magnet for both beams. A free space of 11 meters on either side of the crossing point is available for detector equipment.

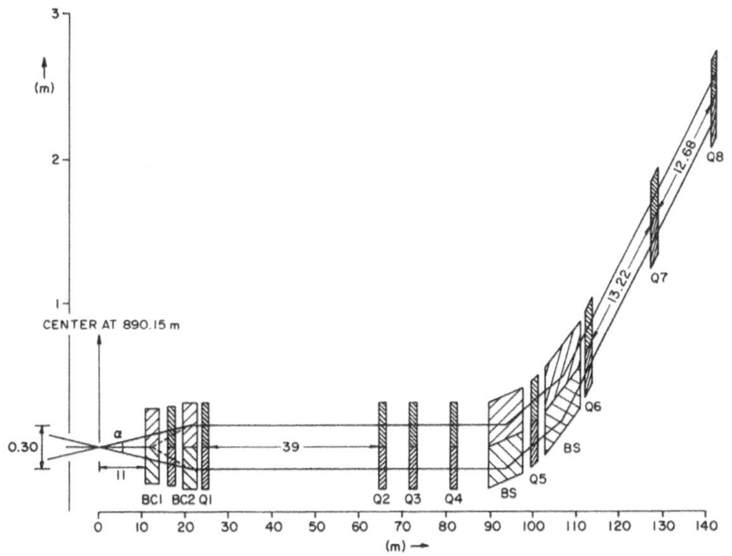

Figure 4.
Optics of a RHIC Half Cell Insertion.

Operation with beams in head-on collisions ($\alpha = 0$) gives the maximum luminosity, and also the maximum length of the luminous diamond. For Au beams at top energy the rms diamond length is ≈20 cm at injection, and grows by nearly a factor of 3, due to intrabeam scattering, after 10 hours of operation. With a crossing angle α = 2 mr the diamond size can be reduced, for those experiments which require it, by a factor of ≈4, with corresponding reduction in luminosity (see Table 3).

The insertion scheme allows for operation with unequal ion species in the two beams. For example, operation with protons in one ring and gold ions in the other can be carried out with luminosity $1.2 \times 10^{29} cm^{-2} sec^{-1}$.

The most severe limitation for maintaining intense beams of heavy ions over long periods of time is the phenomenon of intrabeam scattering.[6-8] This effect, by which particles in the same bunch exchange longitudinal and transverse momenta through Coulomb scattering, depends strongly on the charge-to-mass ratio, being proportional to Z^4/A^2. The phenomenon has been observed in electron and proton storage rings, and is now understood to be the dominant cause of beam diffusion for heavy ion colliders. (By way of comparison, space charge effects such as the beam-beam tune shift and microwave instability go as Z^2/A.)

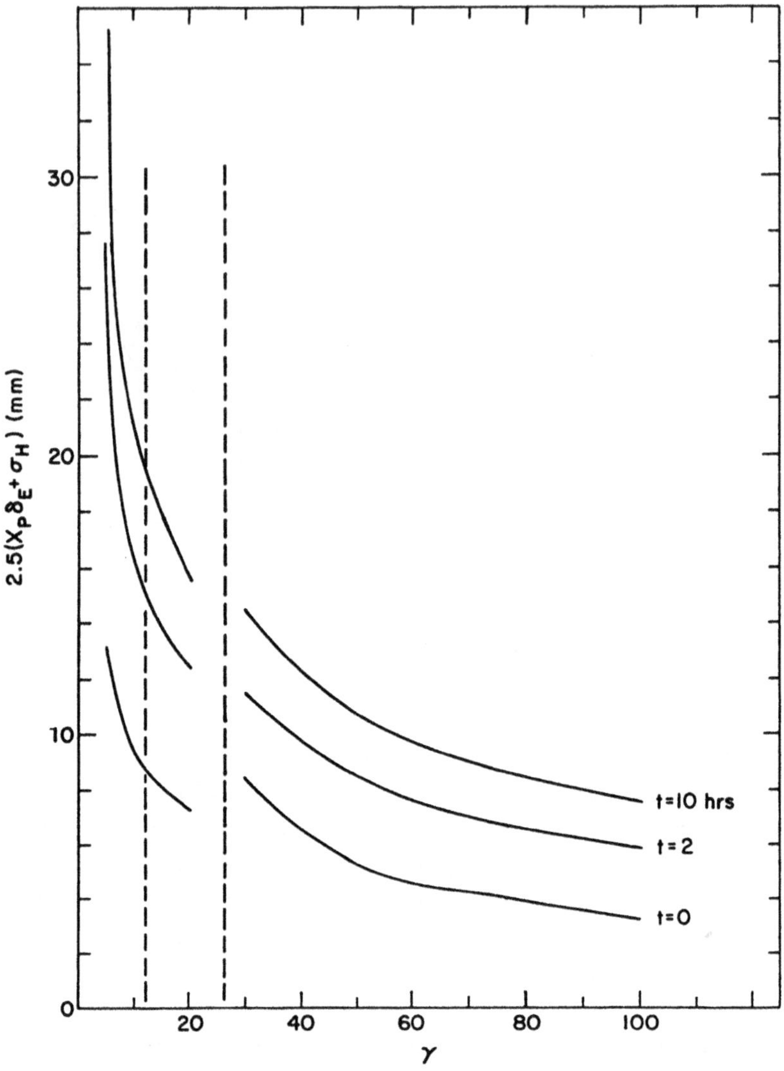

Figure 5.
The half-width for 95% beam containment for gold ions, as a function of the beam energy, showing the growth with time due to intrabeam scattering.

The effect of intrabeam scattering on beams of gold ions in the RHIC lattice is illustrated in Fig. 5. This shows the overall beam dimension as a function of the energy at injection (t = 0), and at the end of two-hour and ten-hour storage periods. It will be seen that for energies below transition (γ_T =26.4) the beams become quickly very large, whereas the beam dimension varies by no more than a factor of two for energies above γ_T. It is the size of the beam envelope, as shown in Fig. 5, which must determine the aperture of the collider magnets. The "good field" region of either quadrupoles or dipoles should be large enough to contain the beam envelope for the required range of energies and storage times. The choice of magnet aperture has a direct influence on the overall cost of the machine as well. For the RHIC design it was decided to set the aperture to adequately contain the beam for energies above γ = 30, after 10 hours of storage. Useful beams are still available down to $\gamma \simeq 7$, but with shorter beam lifetimes. At $\gamma = 7$ the total collision energy is 7 + 7 GeV/amu, equal to the cm energy of a single $\gamma = 100$ beam striking a fixed target. Thus the energy range below that which is accessible in the collider mode can be covered by operating a single ring with one or more internal fixed targets. Suitable targets, with thickness $\approx 10^{-9}$gm cm^{-2}, can be achieved with gas jets or thin wires. In this way the full range, from the lowest AGS energy to the top collider energy can be studied, with no inaccessible gaps. The luminosity which is achieved for the various modes of operation over this energy interval is shown in Fig. 6.

Each collider ring has approximately 200 dipole and 250 quadrupole magnets. These will be superconducting magnets of a design optimized for the heavy ion application, and so will be quite different from the magnets designed for CBA. Because of the lower beam intensity and energy, and a machine design which assumes a short cell, "strong" focussing lattice, the RHIC magnets will be longer ($\approx$10 m), of smaller aperture, and have significantly lower fields (3.5 T at top energy) than was the case for the CBA design. All of these differences contribute to a lower cost for the RHIC magnet system. Candidate solutions now undergoing detailed study call for adjacent magnets in the two rings to share the same vacuum vessel, but without magnetic coupling. These "dual" magnets utilize a single layer dipole design based on experience with the CBA magnets, as well as dipoles under consideration by the Brown-Boveri Corporation for the HERA project at DESY, and dipoles under study for the SSC.

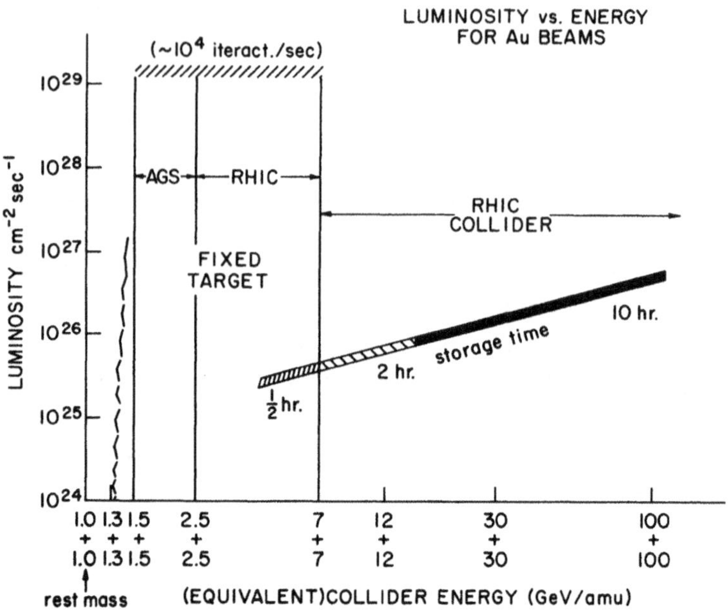

Figure 6.
The average luminosity for ^{197}Au beams over the full range of energy accessible with AGS and RHIC.

III. Experiments and Detectors

An experimental program to explore all of the facets of this new physics implies a measurement capability similar to that which we see in spectrometers for high energy elementary particle experiments. The design of instrumentation for experiments with ultra-relativistic nuclear beams has recently been the subject of extensive workshop activity,[2,3,12] and these discussions have included large, general purpose facilities as well as smaller, more specialized detector systems. Figure 7a shows very roughly the geometry and kinematics of a large solid angle detector, designed for use with colliding beams and having tracking and calorimetery in the central rapidity region and at forward angles. A real detector system might have more limited solid angle coverage. Note that the number of particles in each of the indicated rapidity intervals is expected to be many hundreds. The RHIC insertion design and the experimental halls (three of which are already completed) allow for detector systems of this scale.

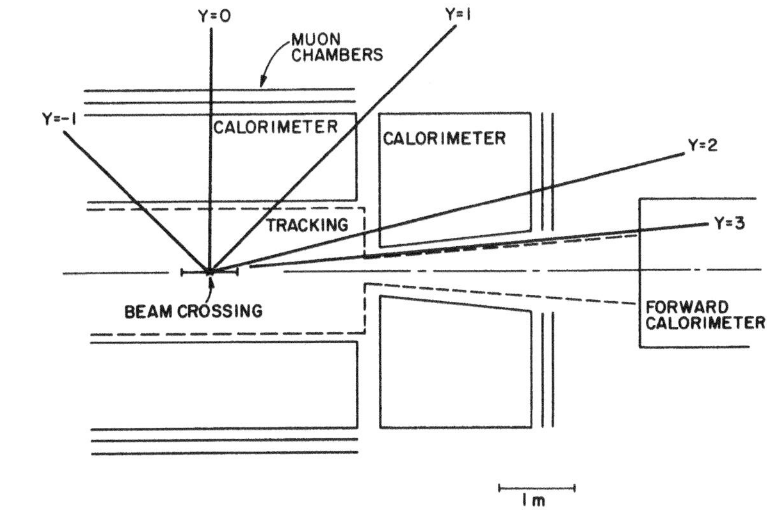

Figure 7.

(a) Schematic layout showing the geometry and kinematics of a detecting system for colling beam experiments. The angular intervals corresponding to various rapidity (y) values are indicated. A full solid-angle detector would have the same coverage on either side of y = 0; this illustration shows coverage of the central region and one of the fragmentation regions.

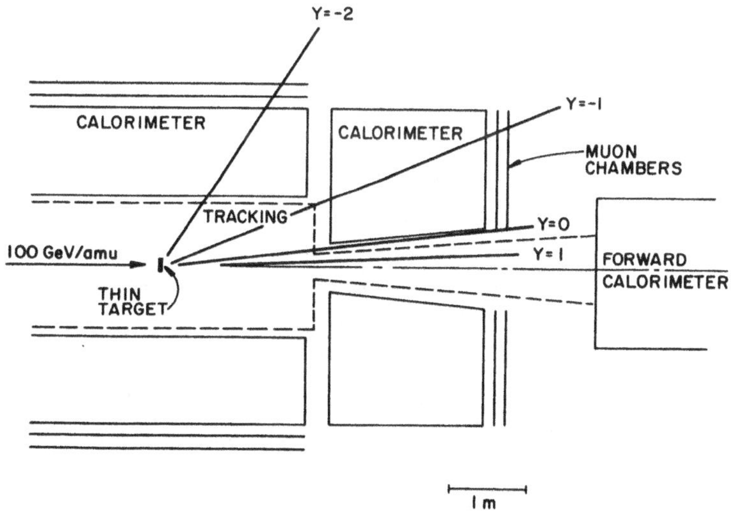

(b) The same detector system, with kinematics of fixed target operation. Here, a single beam at 100 GeV/amu strikes a thin ($\approx 10^{-9}$ gm cm^{-2}) internal target.

Fig. 7b shows the same detector configuation being used in a fixed target mode, with only one of the collider beams accelerated and striking a thin target, such as a gas jet. Operation in the fixed target mode--utilizing, if need be, the same detectors as in the collider mode--allows experiments with good luminosity in the lower energy range where the Collider luminosities would be unacceptably small. This mode of fixed target operation would not, of course, require the expense of extracted beams, and provides in a natural way (as does collider operation) the extremely thin targets which are crucial for nuclear beams experiments and difficult to realize in conventional fixed target operation with external beams.

A clear difference between high energy experiments with elementary particle beams and those with ultra-relativistic nuclear beams is that in the latter case most of the essential measurements involve soft particles, with transverse momenta and pair masses characteristic of the kinetic energies in a thermalized plasma. These range from near-zero (e.g. coherent photons of a few MeV produced in conjunction with large-scale charge fluctuations) through direct radiation from the plasma in the range 0.5-3 GeV. This is in contrast with the elementary particle case where the focus is largely on very rare processes produced in the tails of momentum distributions (P_T > 10 GeV) due to hard scattering events, and thus backgrounds from soft particles characterizing the average properties of events are easily eliminated. For nuclear beam experiments, the signals of interest must generally be extracted from the high multiplicity component of soft particles. This also implies that reasonable samples of useful events can be accumulated without the requirement of extraordinarily high interaction rates.

For the most massive ions ($A \gtrsim 200$) the nuclear interaction cross section is ≈10 barns. This means that a machine with luminosity $L = 10^{23}cm^{-2}sec^{-1}$ would provide an interaction rate of 1 event/sec. This is already a vast improvement over the present rate of accumulation for high energy cosmic ray events (≈12 events/year for the JACEE experiment),[13] but would not allow much flexibility of trigger selection. At $L = 10^{24}cm^{-2}sec^{-1}$ we have available one head-on collision per second. With $L \gtrsim 10^{25}cm^{-2}sec^{-1}$ quite sensitive experiments become possible. An upper limit of usable luminosity (again, for the heaviest ions) is probably in the range $L = 10^{29}cm^{-2}sec^{-1}$, where the detectors become exposed to a flux of particles $>10^7sec^{-1}ster^{-1}$. The design luminosity for the proposed

RHIC facility is $\gtrsim 10^{26}cm^{-2}sec^{-1}$ for Au beams at 100 GeV, as indicated in Fig. 6.

At the very highest energies, and at luminosities $\gtrsim 10^{26}cm^{-2}sec^{-1}$, a new and potentially very powerful probe of nuclear matter becomes accessible in the form of large transverse momentum jets. These may allow us to study the propagation of elementary quarks and gluons through the high temperature plasma in which chiral symmetry is restored, as well as through ordinary nuclear matter. To do these experiments, we must detect jets among the high multiplicity background of soft particles. This background corresponds to ≈320 GeV of transverse energy per unit rapidity at y = 0. The amount of this energy in a cone the size of a jet determines the minimum P_T of an observable jet, which is estimated to be 40 GeV/c for collisions of A = 200 beams at 100 + 100 GeV/amu. The estimated rate for observing jets in heavy ion collisions increases very rapidly with increasing beam energy, as illustrated in Fig. 8.

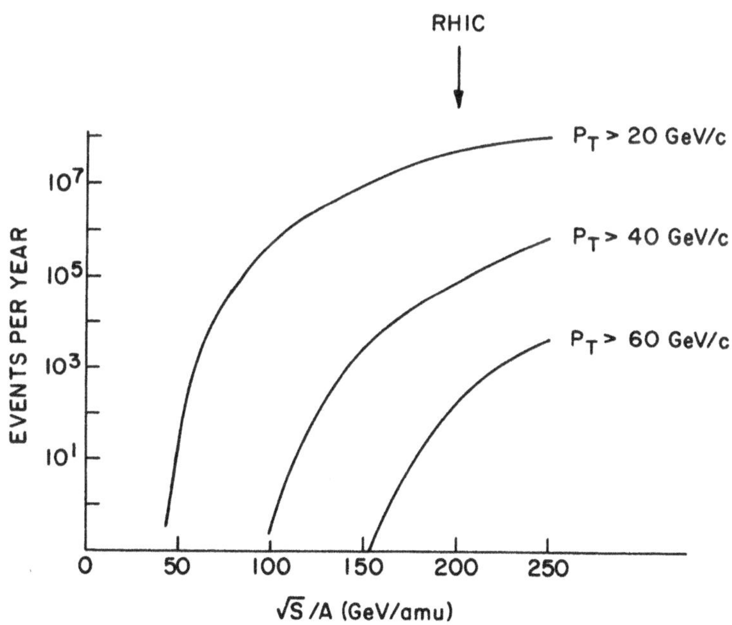

Figure 8

The estimated rates for observing jets at large transverse momentum, as a function of the collision energy. The calculation (Ref. 10) is for gold ions at a luminosity $L = 10^{27}cm^{-2}sec^{-1}$. The curves for three different threshold values of P_T are shown.

Table 4

Experimental Hall Dimensions (Meters)

	Length	Width	Beam Height	Hook Height & Capacity (tons)
2 o'clock				
Central Hall	28	12.0	1.7	6.1/20
Foward Hall	68	7.9	1.7	5.3*
"Stub"	91	2.4	1.0	2.0*
4 o'clock				
Open Area Pad	57	37.0	2.2	---
6 o'clock				
Wide Angle Hall	16	32.0	4.3	10/2 × 20
8 o'clock				
Major Facility Hall	19	15.0	5.2	11/40
2 Forward Halls	16	9.0	3.3	6.6*
Assembly Building	19	19.0	5.2	11/40 + 14/7.5

*No Crane: Ceiling height given.

IV. Conclusion; Costs and Schedules

The availability of completed construction work and other assets from the ISABELLE/CBA project represents an unprecedented opportunity to build, at minimal cost, a dedicated heavy ion facility with beam energies up to 100 GeV/amu. These assets include more than just the ring tunnel. They include also: a complete injection system (the Tandem-Booster-AGS Complex) which will exist independently of the collider; the injection tunnels from the AGS; three completed experimental areas (three more undeveloped experimental areas also exist); a central control building. The cryogenic refrigerator system designed to cool the CBA magnets is complete. In addition, Brookhaven has an experienced superconducting magnet group and very large facilities for constructing, assembling and testing superconducting magnets. The four intersection regions which have already in place the structures necessary for an operating experimental area are indicated in Fig. 2. These areas are at the 2, 4, 6 and 8 o'clock locations. The area at 4 o'clock is an "open area," with a completed concrete pad and auxiliary building. The remaining areas have fully enclosed experimental halls with overhead cranes. The sizes of these experimental areas are given in Table 4. The 10 and 12 o'clock experimental areas were left undeveloped when the CBA project was terminated, and will not be available for first-round RHIC experiments.

The cost estimate obtained for the RHIC proposal assumes the completion of the Tandem-AGS transfer line, which is now under construction, and the Synchrotron Booster, which is planned to start in FY 1986. The total cost estimate for the Collider, including contingency, is $134.4 M in 1984 dollars. This assumes a conservative schedule of 4 years for magnet production (of which the first year goes to R&D and fabrication of tooling) and 5 years for overall project completion. With this schedule, beams could be available for the first experiments during the latter half of 1990.

Acknowledgement

As I have indicated, the development of the RHIC proposal has benefitted from the advise and consultation of a very large number of interested physicists representing laboratories throughout the world. the technical design effort has involved many members of the High Energy Facilities division at BNL, under the direction of

M.Q. Barton and H. Hahn. This group is pleased to acknowledge the significant contributions of A. Ruggiero (Fermilab) and G. Young (ORNL).

This work was supported by the U.S. Department of Energy under contract DE-AC02-76-CH00016.

References

1. "A Long Range Plan for Nuclear Science," A Report by the DOE/NSF Nuclear Science Advisory Committee, December 1983.

2. H.A. Gordon (Ed.), Proceedings of the 1981 ISABELLE Summer Workshop, BNL 51443 (1981); M.Q. Barton, "The Acceleration of Heavy Ions to Very High Energies," Proc. Bielefeld Workshop on Quark Matter Formation and Heavy Ion Collisions (M. Jacob and H. Satz, eds), Word Scientific, Singapore (1982); T.W. Ludlam, "Heavy Ion Physics at CBA," BNL 32716 (1983).

3. "Report of the Task Force for Heavy Ion Physics," Ref. 4, below p. 657.

4. T.W. Ludlam and H.E. Wegner (Eds.), Proceedings of the Third International Conference on Ultra-Relativistic Nucleus-Nucleus Collisions: Quark Matter '83, Nucl. Phys. A418 (1984).

5. "Round Table Discussion of Prospects for Future Experiments." D.A. Bromley, Chairman), Ref. 4, p. 375.

6. G. Young, "Heavy Ion Colliders," Proc. Conference on the Intersections between Particle and Nuclear Physics," May, 1984 Steamboat Springs, Colorado (R. Mischke, Ed.), (To be Published).

7. A. Ruggiero, "Intrabeam Scattering and Heavy Ion Colliders," G. Young, "Heavy Ion Colliders," Proc. Conference on the Intersections between Particle and Nuclear Physics," May, 1984, Steamboat Springs, Colorado (R. Mischke, Ed.) (To be Published).

8. "Proposal for a Relativistic Heavy Ion Collider at Brookhaven National Laboratory," BNL 51801 (June, 1984).

9. W. Busza, These Proceedings; see also W. Busza, Ref. 4, p. 635.

10. T. Ludlam, L. Madansky, F. Paige, "High Energy Probes of the Quark-Gluon Plasma," Proc. of the LBL Workshop on Detectors for Relativistic Nuclear Collisions, March, 1984 (L. Schroeder, Ed.), To be published.

11. K. Koley, These Proceedings.

12. L. Schroeder, Ed., Proceedings of the LBL Workshop on Detectors for Relativistic Nuclear Collisions, March, 1984 (To be Published).

13. O. Miyamura, These Proceedings; See also T. Burnett, et al., Phys. Rev. Lett. 50, 2062 (1983).

A NEW COLLIDER SCHEME AT LBL*

Howel G. Pugh
Nuclear Science Division
Lawrence Berkeley Laboratory
University of California
Berkeley, California 94720

This paper presents current ideas from Berkeley concerning a possible new facility for studying the phase transition from hadronic matter to quark matter. The physics ideas have evolved over a period of more than five years, the VENUS concept for a 25 GeV/nucleon colliding beam facility having been presented in 1979.[1)] The concept for the Minicollider has been, like that of VENUS, the work of Hermann Grunder and Christoph Leemann.

1. Progress Since the Bielefeld Meeting

Our knowledge of heavy ion collisions has increased greatly since my summary talk at Quark Matter '82.[2)]

During that meeting I was able to report that the first uranium beams had been extracted from the Bevalac.[3)] As a result of experiments carried out with the new beams (previous experiments having been limited to projectiles up to the mass-40 region) the importance of using the heaviest nuclei has now been demonstrated. The recent work of Gustafsson et al.[4)] is an excellent illustration. Using the Plastic Ball-Wall combination they studied collisions of equal mass nuclei and measured all the charged particles emitted. They constructed the flow tensor[5)] for each event, and histogrammed the flow angle for various charged particle multiplicity bins. The results for ^{40}Ca + ^{40}Ca and ^{93}Nb + ^{93}Nb are shown in figure 1, together with an intranuclear cascade calculation for comparison. For ^{40}Ca + ^{40}Ca, the flow angle (which represents the most probable direction of energy flow) is peaked in the beam direction, with just a hint of a deflection in the highest multiplicity bin. However, for ^{93}Nb + ^{93}Nb a well-defined deflection is seen. The intranuclear cascade calculation does not reproduce the data. This is clear evidence for relativistic hydrodynamical effects, as has been shown by Graebner et al.[6)] These effects can be discerned in the ^{40}Ca + ^{40}Ca data, taken earlier, but it is only with the increased mass that they emerge

*This work was supported by the Director, Office of Energy Research, Division of Nuclear Physics of the Office of High Energy and Nuclear Physics of the U.S. Department of Energy under Contract DE-AC03-76SF00098.

clearly.

At Quark Matter '83, Busza and Goldhaber[7)] forced a reconsideration of nuclear stopping power. So little has been known about this question that in my review of Bevalac research at Bielefeld[8)] I was unable to provide proof that nuclei stop in each other at 2 GeV/nucleon incident laboratory energy. The fact that they do is now clear[5)] but the results of Busza and Goldhaber on the degradation of high energy protons in nuclei suggest that nuclear stopping may continue to much higher energies. Whereas a few years ago it was believed that nuclear transparency should set in above about 10 AGeV in the laboratory, it now seems quite possible that stopping may occur up to energies of the order of 50 AGeV. This implies that much greater densities may be achieved in the baryon-rich regime than had previously been thought possible.

A major development on the technical side has been a careful study done at Brookhaven on the achievable luminosities in heavy ion colliders.[9)] These have proved to be much lower than previously thought, and are determined primarily by intrabeam scattering. At Berkeley we therefore reexamined the possible use of the Bevalac as an injector. We found to our surprise that the luminosity is not limited by the Bevalac beam quality. We thus already have an operating injector already providing beams of fully stripped uranium at an energy suitable for injection into a collider.

In the two years since Bielefeld there has also been extensive examination and reexamination of possible experiments to look for the quark-gluon plasma. We have a much better appreciation of the conditions under which experiments with present-day techniques are possible, and a better appreciation of the difficulties involved in interpreting the various proposed "signatures" of the quark-gluon plasma.

2. Signatures of the Phase Transition

There has been extensive discussion of signatures of the quark gluon plasma, and none has been found which is free from difficulties of interpretation. The reason is that, because of confinement, there is no first order effect in which we could see the quarks and gluons directly after they have been liberated by the phase transition. We always have to look at re-condensed material (hadrons) in the final state. Experiments involving leptons are partially free from this criticism since they can sense the space- and time-integrated behavior of the collision process, including the period during which the system was in the plasma state. However, their interpretation will require a rather complete understanding of the collision mechanism and will require many subsidiary experiments.

Under the circumstances it seems preferable to start with a number of simpler exploratory studies, and look for signatures of the phase transition. Here we can consider a standard experiment:

a) Choose a variable which is related to the energy density ε. Such a variable is the bombarding energy E, as can easily be checked by the first experiments, by measuring dn/dy (y = 0) as a function of E. Also, ε may be a monotonically increasing function of target and projectile mass A, though perhaps after a certain value of A has been reached only the duration of the quark-gluon phase may increase with A. Or, ε may be a monotonically increasing function of some other variable to be defined by the experimenter.

b) Choose an observable that may be sensitive to the internal properties of the system, such as $\langle p_T \rangle$, or the multiplicity of created particles, or the composition of products in terms of q, $\bar{q}$, s, $\bar{s}$. Many such observables can be imagined.

c) It may then be expected that with the properly chosen variable and observable, the phase transition will be manifested by a curve of the form shown in Fig. 2.

It seems likely that some experiment of such a nature will reveal the phase transition. However, it will be necessary to have a sufficiently wide energy range available for experiments and to be able to identify both regions A and B. An experiment that showed only the lower half of the curve would be unconvincing. Once the phase transition has been observed, the time will be ripe to proceed to study properties of the quark-gluon plasma in region B, perhaps by contrast with similar studies in region A.

3. The Berkeley Minicollider

As a result of the above considerations, we have concluded that a collider for uranium ions in the c.m. region up to about 4 AGeV (50 AGeV laboratory) would provide a wide range of opportunities to study the phase transition to the quark gluon plasma in the baryon-rich regime.

The Berkeley Minicollider would be injected by the Bevalac at 400 AMeV and would accelerate up to 4 AGeV (uranium) or 5.4 AGeV (calcium). Its parameters are shown in Table 1 together with a comparison with the AGS (including the proposed booster).

Table 1
Energies for uranium-uranium collisions

Berkeley Minicollider	AGS with Booster
0.2 - 4.0 AGeV c.m.	up to 1.56 AGeV c.m.
1.0 - 50.0 AGeV lab.	up to 11.4 AGeV lab.

The lower energy limit of the Minicollider is arbitrarily set as that which is presently achieved in fixed target experiments at the Bevalac. It is interesting for orientation purposes to look at estimates of the energy density obtained if the colliding nuclei were to stop exactly in each other,(10) i.e. $\epsilon_g = 2m_p c^2 \rho_o \gamma_{cm}^2$. At Bevalac energies this value is exceeded due to the formation of a shock front. At some energy the energy density will drop below this value. The energy at which the energy density exactly equals ϵ_g might be a suitable definition of the onset of transparency. Table 2 gives values for ϵ_g and the baryon component of it, ϵ_g (baryon) = $2m_p c^2 \rho_o \gamma_{cm}$ at the upper energies of the Minicollider and of the AGS (with Booster).

Table 2
Goldhaber estimates for energy density ϵ_g and the baryon fraction of it at Berkeley Minicollider and AGS (with Booster), for uranium-uranium collisions

	K.E.(lab)	K.E.(cm)	ϵ_g	ϵ_g (baryon)
AGS with Booster	11.4 AGeV	1.56 AGeV	2.03 GeV/fm^3	0.76 GeV/fm^3
Berkeley Minicollider	50.0 AGeV	4.0 AGeV	7.94 GeV/fm^3	1.50 GeV/fm^3

The values for the Minicollider encourage us to believe that it might be possible with that facility not only to reach the phase transition (end of region A in Figure 1) but to exceed it, map out the entire phase transition and reach the region B in Figure 1 where the bulk of the interacting nuclei have been converted into quark-gluon plasma.

The layout of the facility has been planned to have minimal interference with existing projects. The collider can be constructed while the Bevalac is running, and when complete it can operate simultaneously with the ongoing program of therapy, biomedical research, nuclear physics and atomic physics. A possible layout is shown in Figure 3. Here one of the intersection regions is placed conveniently in the high-bay area of Building 64. The opposite intersection area falls in a region of Blackberry Canyon which is already planned to be filled by mid-1985. This layout is in the spirit of the project, which is to make maximum utilization of existing facilities, to use in-hand technology for the magnets, and to use stochastic cooling techniques which LBL has helped develop for Fermilab.

4. Some Details of Performance

The following are approximate numbers from the present stage of development of the planning:

-- There will be single-turn injection into and extraction from the Bevalac, and about 10^6 particles per Bevatron pulse will be accepted into each Minicollider ring;

-- Stochastic cooling will be applied continuously during filling to allow the storage of many injector pulses: approximately 100 pulses, over 10 minutes, will lead to 10^8 ions stored in each ring;

-- The beams will then be accelerated slowly (over a period of about 60 sec) to operating energy;

-- The Minicollider will have high luminosity intersection region design, and stochastic cooling will be applied continuously during experiments to counteract intrabeam scattering and the corresponding fall-off in luminosity and beam quality;

-- The beam lifetime is expected to be of the order of a day, so that the facility hardly takes up any significant fraction of Bevalac time;

-- Luminosity at 4 AGeV (scales as $\beta\gamma A^2/Z^4$)

U - U	$8 \cdot 10^{23}$ cm^{-2} sec^{-1}
Au - Au	10^{24} "
Ca - U	$3 \cdot 10^{24}$ "
Ca - Ca	10^{25} "
p - p	10^{27} "

-- Beam size: <1 mm high, a few mm wide
crossing angle--a few mrad
length of diamond--1 m for stated luminosity

-- The duty factor will be 100%, since a coasting beam will be used; furthermore, the luminosity will remain near its peak value because of the continuous application of stochastic cooling.

5. Experiments with the Minicollider

With the design luminosities, most types of desired experiments can be carried out. For example, the U - U event rate would be ~7/sec. If one takes as a standard the statistics achieved for minimum bias data in recent Plastic Ball experiments[5)] demonstrating collective flow, we find that 50,000 minimum bias events are extremely useful. These could be obtained in 2 hours. For central collisions, applying a selection procedure much more severe than any used to date, i.e. central collision cross section of 40 mb (equal to the p - p cross section), we find one extremely central collision every 25 seconds. Taking as a standard the statistics achieved by recent Streamer Chamber data on central collisions,[5] 1000 such events would be very valuable. These could be obtained in 7 hours.

Such data rates are excellent for central region production, stopping power studies, strange particle production, studies of temperature, entropy and flow, and photon production. They are inadequate for dilepton production as presently foreseen and for high transverse momentum studies.

A consideration of detector needs has been made at the recent Detector Workshop at Berkeley.[11] The salient points are:

-- Extrapolation of Bevalac experience with uranium at 1 AGeV to the Minicollider at 4 AGeV is not a big step. By comparison, extrapolation of experience with 50 GeV protons to uranium at 50 AGeV at an equivalent fixed target facility is a much bigger step;

-- Detectors at the Minicollider will be less sensitive to problems due to track density than the equivalent detectors at a 50 AGeV fixed target facility by a factor of about 25;

-- A variety of basic detectors is feasible, including a time projection chamber, a streamer chamber, and an improved Plastic Ball; however, because of expense a choice will have to be made, and this has not been done yet;

-- The central region detectors at the Minicollider will be similar in many ways to those eventually needed at the Brookhaven 100 AGeV collider.

6. Conclusion

The Minicollider is feasible, economical, and useful. It will be complementary to projects at Brookhaven and CERN.

Acknowledgments

In the preparation of this paper I am indebted to Hermann Grunder and Christoph Leemann for their extraordinary creativity and skill, and to the many physicists who have worked on Berkeley planning for future facilities.

This work was supported by the Director, Office of Energy Research, Division of Nuclear Physics of the Office of High Energy and Nuclear Physics of the U.S. Department of Energy under Contract DE-AC03-76SF00098.

References

1. The VENUS Project, LBL PUB-5025 (1979)

2. H.G. Pugh, (Bielefeld Workshop on) Quark Matter Formation and Heavy Ion Collisions, ed. M. Jacob and H. Satz, World Scientific (1982), p. 319

3. J.R. Alonso et al., Science 217, 1135 (1982)

4. H.A. Gustafsson et al., Phys. Rev. Lett. 52, 1590 (1984)

5. H.G. Pugh, Equilibration, Compression and Flow at the Bevalac, these Proceedings

6. G. Buchwald et al., Phys. Rev. Lett. 52, 1594 (1984)

7. W. Busza, Third International Conference on Ultrarelativistc Nucleus-Nucleus Collisions, Brookhaven (1983), published in Nuclear Physics A418, 635c (1984)

8. H.G. Pugh, (Bielefeld Workshop on) Quark Matter Formation and Heavy Ion Collisions, ed. M. Jacob and H. Satz, World Scientific (1982), p. 185

9. T. Ludlam, Heavy Ion Collider, these Proceedings; K. Foley, Fixed Target AGS Program for Nuclear Beams, these Proceedings

10. A.S. Goldhaber, Nature 275, 114 (1978)

11. L. S. Schroeder, these Proceedings

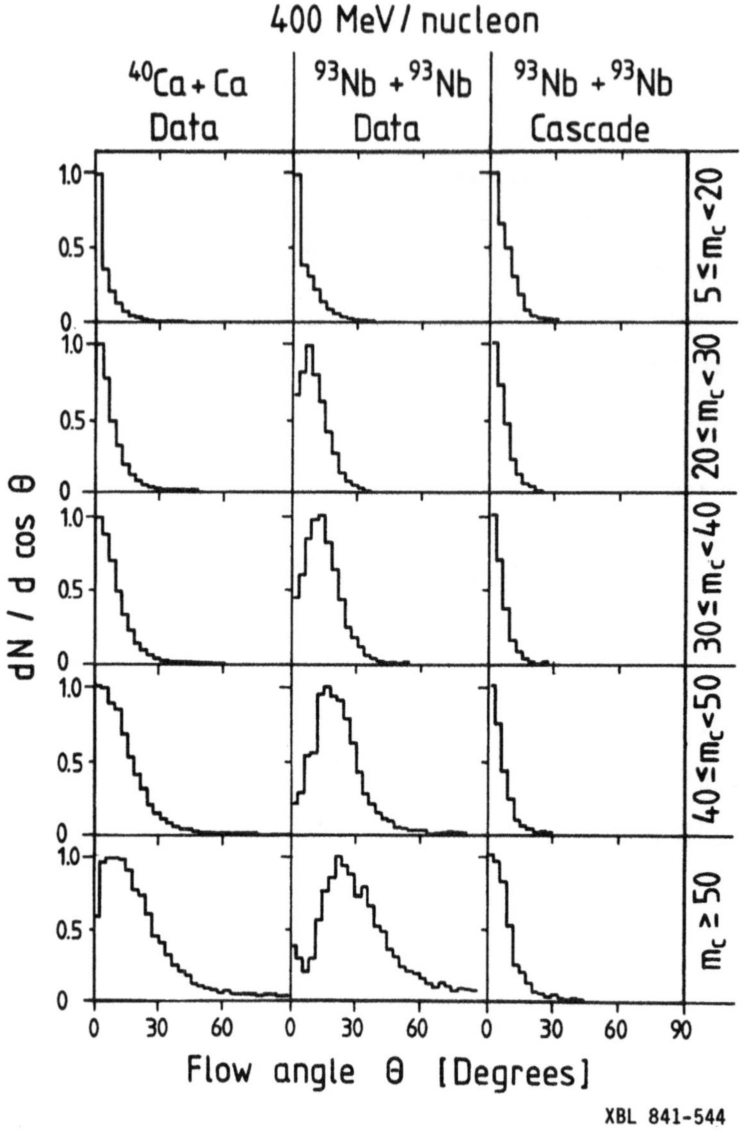

Figure 1

Flow distributions for the reactions Ca + Ca and Nb + Nb, and cascade model predictions for Nb + Nb. The peak in the flow angle (predicted by hydrodynamic calculations) is clearly visible for Nb + Nb but not well-defined for Ca + Ca.

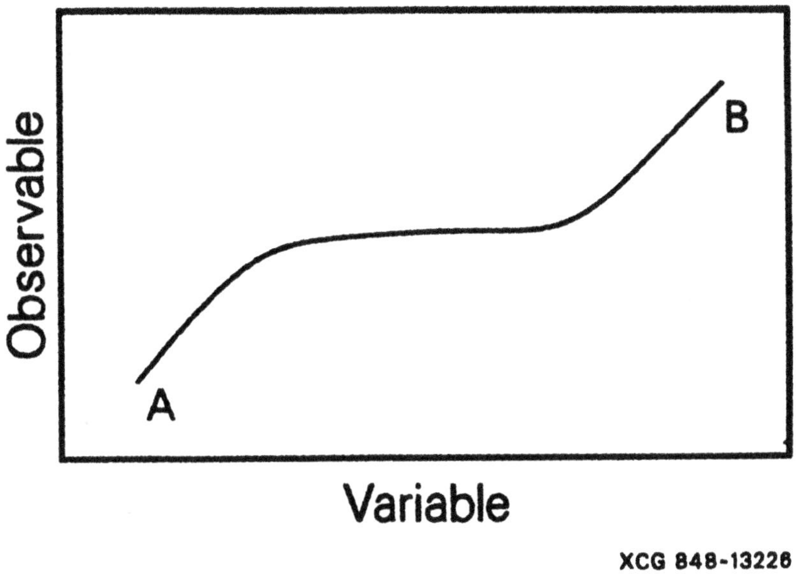

Figure 2

Schematic experiment to observe the phase transition from a hadronic gas to the quark-gluon plasma.

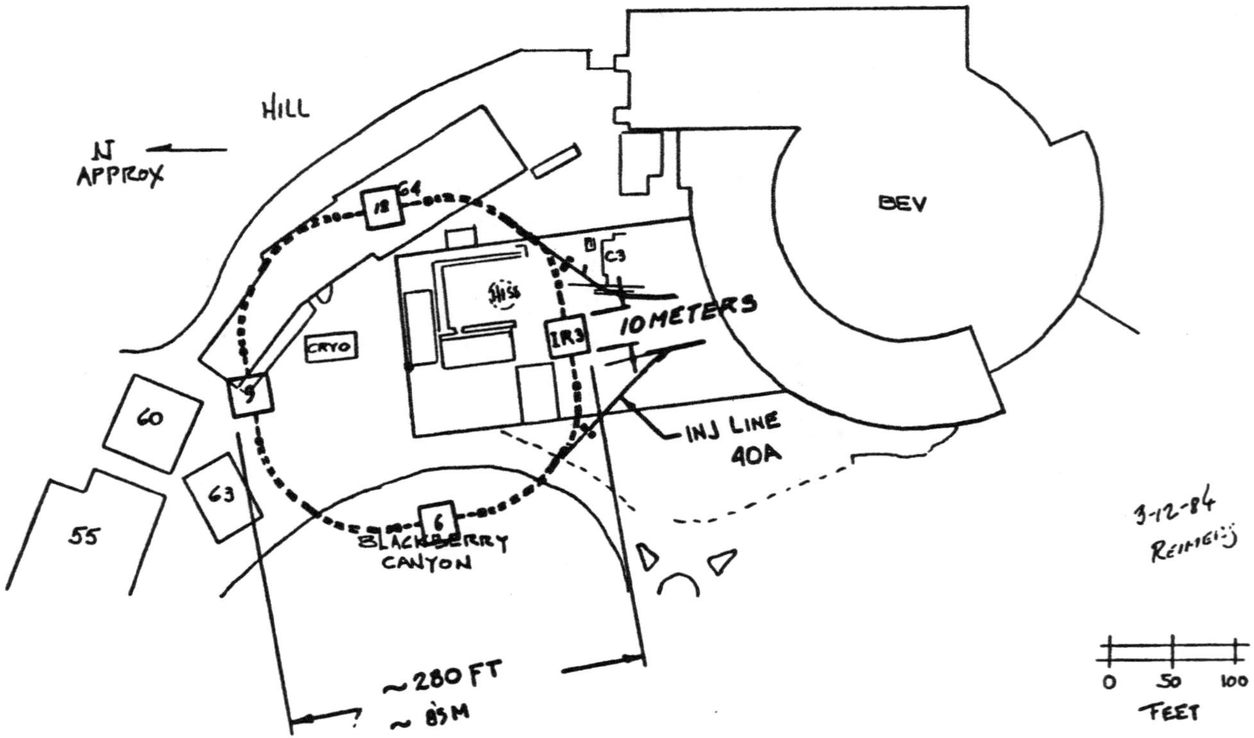

XBL 849-3691

Figure 3

One of the possible layouts for the LBL Minicollider. This makes maximum use of existing facilities and interferes minimally with the ongoing Bevalac program.

THE AGS FIXED TARGET PROGRAM WITH NUCLEAR BEAMS

K.J. Foley
Department of Physics
Brookhaven National Laboratory
Upton, New York U.S.A. 11973

A. Introduction

The recent approval of the beam transfer line from the Tandem Van de Graaf to the AGS signals the advent of a new era of Nuclear and Particle Physics at BNL. High Energy nuclear beams are expected to be available for experiments in 1986.

A photograph of the BNL site is shown in Fig. 1.

Figure 1: A photograph of the BNL site. The Tandem accelerator and the AGS are in the foreground.

I will discuss only the direct link between the Tandems and the AGS. Two other proposed projects, the Relativistic Heavy Ion Collider and the Synchrotron Booster, are discussed in another presentation to this conference.[1]

B. The Accelerators[2]

The plan is to transport full stripped ions from the Tandem machines to the AGS at an energy of 8.A MeV. They will then be accelerated in the AGS to an energy of up to 15.A GeV before extraction for use in up to four simultaneous experiments.

Pulsed beams with the required characteristics have been produced at the Tandems for nuclei up to sulphur; for larger z, higher energy is required to produce sufficiently intense beams of fully stripped ions; a future booster for the AGS would produce such capability. The transporting of the beam from the Tandem to the AGS requires an 1800-foot-long tunnel, with eight quadrupole doublets to keep the beam focussed. The tunnel is of modest size ($\approx$ 10 foot wide) with sufficient space for servicing the various magnet and vacuum systems. The AGS magnetic field required at injection is substantially lower than that currently used for proton beams, but is only 20% lower than that used with the original 50 MeV proton linac; no serious problem is anticipated in achieving the required field. Also, because of the lower injection energy, the present RF system does not have a large enough frequency swing, but the original AGS RF system is available and is being modified to accelerate from Tandem energies to the point where the usual RF can be used. The use of two RF systems is considered a safer route than an alternate scheme involving harmonic switching.

The maximum energy available in the AGS is 15.A GeV and fluxes will range from 4×10^{10} particles per pulse for oxygen to 6×10^{8} particles per pulse for sulphur. The duty factor will be $\approx$ 0.4 (1 sec flat top every 2-1/2 seconds).

The existing external beam system delivers simultaneous beams in the A, B, C, and D lines. Each beam line has the capability of delivering primary beam particles to at least one experimental area and can therefore be used for experiments with nuclear beams. However, since the program with nuclear beams will supplement the existing High Energy Physics research, care will be required to mesh together the two programs.

C. Experiments

Extensive efforts are currently under way in the study of possible experiments with nucleus-nucleus collisions. It is expected that during this summer several experiments will be proposed to the AGS Program Committee ("APC", formerly known as HEAC) requesting running time for first-round experiments in 1986. The people involved in these studies include scientists from Nuclear Physics, Particle Physics and Chemistry at BNL as well as many from outside institutions. I will not attempt a comprehensive review, but will mention a few of the potential experiments.

1. Single Arm Spectrometer[3]

The goal is to measure particle production cross sections in the central rapidity region, corresponding to angles in the lab of 5° - 20°. A possible setup is shown schematically in Fig. 2. The

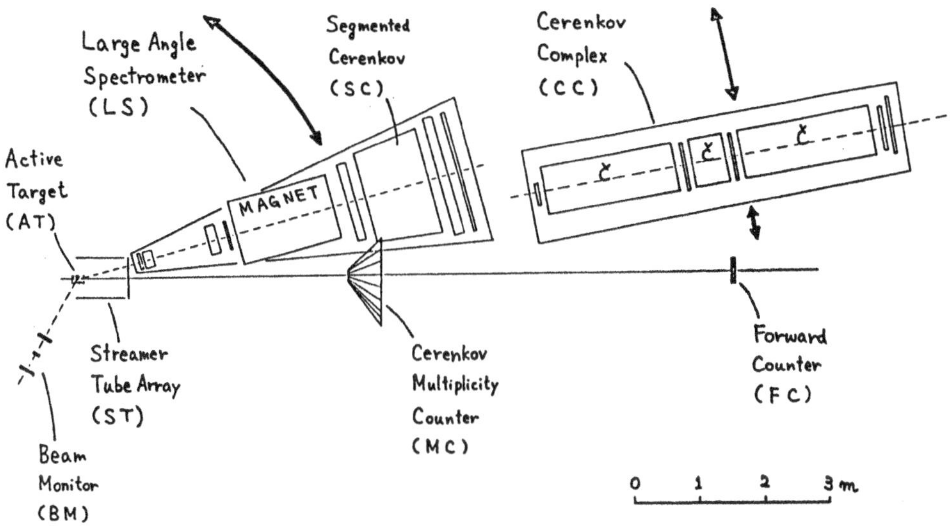

Figure 2: A possible arrangement of the single-arm spectrometer.

solid angle will be about 10 millisteradians. A set of Cerenkov counters and time-of-flight counters will be used to identify p, $\bar{p}$, $\pi^{\pm}$, $K^{\pm}$. Because of the high multiplicities expected in head-on collisions, up to 15 charged particles are expected at small angles even for a 10 ms solid angle, so sophisticated tracking detectors will be required. Additional detectors (probably streamer tubes) will be used to measure outgoing charged particles over a large solid angle (approaching 4π). This will permit the measurement of overall multiplicity and the distribution in pseudo-rapidity associated with particles detected in the spectrometer. The use of a 0° calorimeter is also envisioned.

2. <u>Multiparticle Spectrometer</u>

The Multiparticle Spectrometer (MPS), a user facility in the BNL High Energy Physics program, is a large solid-angle detector based on a C-magnet with a pole area 6 ft. x 15 ft. with a 4 ft. high gap. Typically it has been used for detailed study of events produced by pions and kaons in the momentum range 5 - 25 GeV/c. The detectors currently in use are narrow cell drift chambers[4] designed for high beam rates (few x 10^6/sec) and multiplicities of $\lesssim 15$; these criteria are well-satisfied by drift chambers using projective geometry rather than true "3D" detectors. Several large multi-cell Cerenkov counter hodoscopes are available and a large time-of-flight scintillation counter hodoscope is under construction.[5] Various sophisticated triggering systems are also available.[6]

In order to assess the usefulness of the present MPS in the Heavy-ion Program, events produced by the Hijet[7] Program have been superimposed on the MPS. A typical minimum-bias centrally-produced event, O^{16} on Au at 15.A GeV, is shown in Fig. 3 (a plan view, looking along the magnetic field). The charged particle multiplicity is ~ 130.

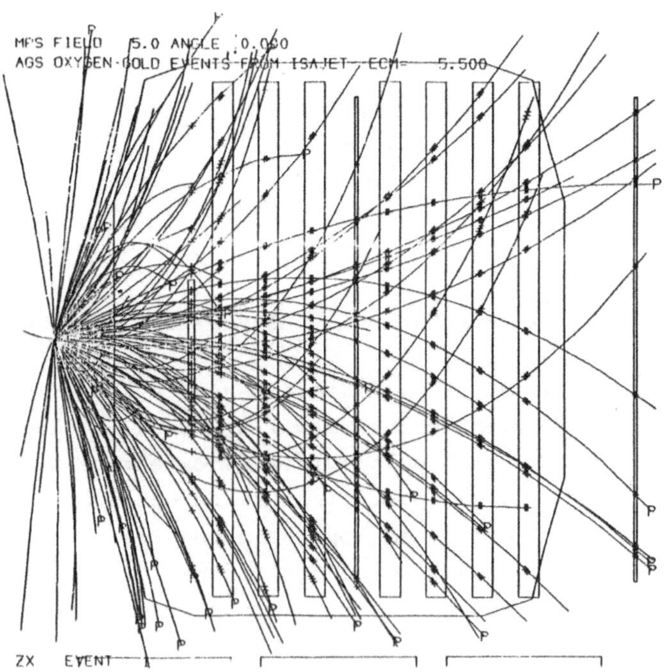

Figure 3: A simulated event superimposed on the MPS.

We conclude that the existing system is not the correct approach for full reconstruction of events of such high multiplicity (although, as will be mentioned later, we can configure the present MPS for use as a very large aperture spectrometer for the negative particles in such events). Instead we propose to build a new detector system based on 3D point measurements to handle the high multiplicities, with a goal of angle measurements for virtually all charged particles, momentum measurement for charged particles emitted forward in the lab frame, and particle identification for charged particles exiting from the MPS magnet.

The setup is shown schematically in Fig. 4. The angles (and hence pseudorapidity, η) of outgoing charged particles will be

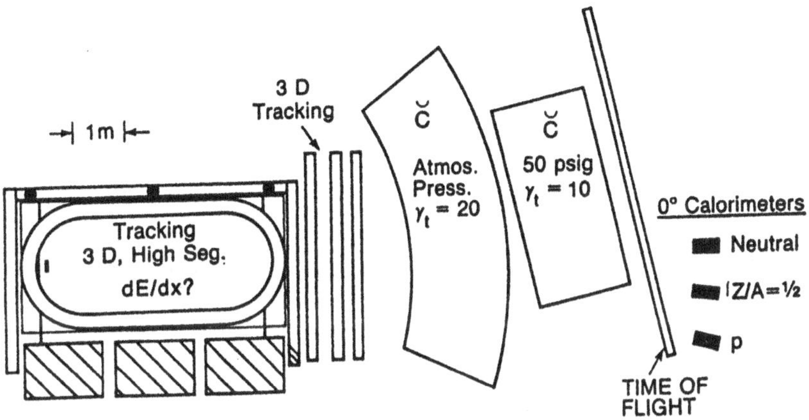

Figure 4: A possible arrangement for experiments with nuclear beams at the MPS.

measured in an array of charge-coupled-devices (CCD's) surrounding the target. These detectors are admirably suited to the purpose. Recent measurements at CERN[8] have demonstrated that CCD detectors can measure minimum ionizing particles with high efficiency (98% ± 2%), extremely good resolution (5μm) and good two-particle separation (40μm). With a typical active size of 1 - 2 cm, one has an effective segmentation of > 50,000 elements per plane with true (x,y) point readout. A potential problem of errors due to particles traversing the detector during readout will be obviated by kicking the beam on to a beam-stop when a trigger is selected -- this is a relatively simple task at the beam rates (~ 10^4/sec) envisioned for this spectrometer.

The strongest candidate for the new detectors in the magnet are Time-Projection Chambers (TPC's). Such a detector would drift electrons along the magnetic field, measuring x by knowledge of the detector elements involved and y by the drift time. In order to make use of the full magnetic field length while keeping both the budget and construction problems manageable, we expect to build a set of relatively thin modules each with perhaps four measurements along the tracks. A possible geometry is shown in Fig. 5. Electrons produced

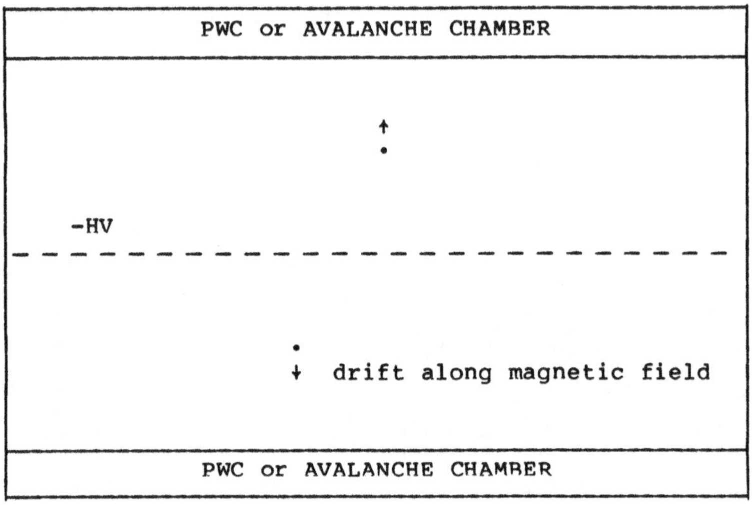

Figure 5: A schematic layout of a TPC for Heavy Ion experiments.

by charged particles in the gas drift up and down away from a negative high voltage electrode near the median plane and are measured by PWC (or multistep avalanche chambers) with short wires parallel to the beam direction at the top and the bottom of the TPC. These detectors are very similar in principle to those in use in the "RICH" counter at the Omega Spectrometer at CERN.[9] One should note that a potential problem due to positive ion buildup from the very high dE/dx of non-interacting beam particles is eased by the proximity of the beam to the negative HV terminal. If we assume a two-particle separation of 1 cm, a somewhat pessimistic value, an effective segmentation of 20,000 elements is achieved in a 2 m x 1m chamber.

Particle identification can be achieved in the downstream region with two Cerenkov counters and a time-of-flight hodoscope. With the Cerenkov counters set at $\gamma_t = 10$ and 20 and a flight path of 10 m πKp separation can be achieved from 1.4 to 10 GeV in one setup. Of course, at low momentum large corrections will have to be made for kaon decays.

As a first phase, to be run in 1986, we propose to use the existing MPS, augmented by a few TPC modules, to study the production of negative particles over a very large solid angle. We will bring in the incident beam alongside the MPS chambers, using the magnetic field to sweep positive particles away from the detectors. In addition to supplying information on the momentum spectra of π^-, K^-, $\bar{p}$ and the correlations therein, we can search for unusual energy flow patterns etc. We also plan to complete a CCD vertex detector in time for this run to give information on multiplicities and pseudorapidity. The knowledge and experience gained will be extremely valuable as we proceed with the full spectrometer.

3. π^0 Production

As is described in a paper[10] presented to this conference, the production of π^0's in α - α collisions at the ISR continues to higher E_T than is predicted by models based on independent nucleon collisions. An experiment presently being considered[11] will use a lead-glass hodoscope to extend these measurements to heavy nuclei at the AGS in a very sensitive probe of collective effects in nuclear collisions. Studies are under way to see if this equipment can be set up in combination with the single arm spectrometer discussed above.

4. ϕ Meson Production

The use of a two-arm spectrometer to study ϕ meson production is being considered.[12] Since the ϕ is almost pure $s\bar{s}$ in quark content it should prove a sensitive probe of enhanced strange quark production - also, it's narrow width increases the probability of its detection in the K^+K^- decay mode. Possible experimental arrangements are being discussed.

5. Lepton Production

There is much discussion of the measurement of lepton pairs as a volume-sensitive probe in nuclear collisions. Because of the beam energy available at the AGS, muons from low mass produced near rest in the center-of-mass are in a region not easily accessible to detection; their low energy (and resultant short range) forces the use of relatively thin pion absorbers with a consequent worsening in the identification of muons. Studies are also under way on measurement of electron-positron pairs; here one has to face a charged particle multiplicity of greater than 100 as well as spurious pairs produced by 50 or so π^0's. Clearly, some good new ideas are needed!

D. Users Organization

A "Heavy Ion Users Group" has been established which is expected to function in a way very similar to the general AGS Users Group, HEDG, by being the contact surface between heavy ion users and the Brookhaven Laboratory management. An Executive Committee has been appointed by the Chairman of the BNL Physics Department, consisting of:

Peter Braun-Munzinger, (Chairman) SUNY-Stony Brook
Eric R. Cosman, Massachusetts Institute of Technology
Ole Hansen, Brookhaven National Laboratory
Tom Ludlam, Brookhaven National Laboratory
Leon Madansky, Johns Hopkins University
Franz Plasil, Oak Ridge National Laboratory
Julia A. Thompson, University of Pittsburgh (member of HEDG Executive Committee)
James Symons, Lawrence Berkeley Laboratory

Among other things, the charge of this committee is to set up an infrastructure for the users organization and to arrange for the election by the users of the next Executive Committee.

REFERENCES

1. T. Ludlam, these proceedings.
2. I wish to thank the project head, H. Foelsche, the Deputy Head, P. Theiberger, and H. Wegner for helpful discussions.
3. BNL/Tokyo/MIT/LBL/Hiroshima Collaboration. I wish to thank O. Hanson for useful discussions.
4. S. Eiseman, A. Etkin, K.J. Foley, R.S. Longacre, W.A. Love, T.W. Morris, S. Ozaki, E.D. Platner, V.A. Polychronakos, A.C. Saulys, C.D. Wheeler, S.J. Lindenbaum, M.A. Kramer, Y. Teramoto. The MPS II Drift Chamber System. (Proc. of the Wire Chamber Conference, Vienna, Austria, February 15-18, 1983). Nucl. Instr. & Methods 217, 140-148 (1983).
5. S.U. Chung and Z. Bar-Yam, private communication.

This work supported by the U.S. Department of Energy under contract de-ac02-76ch00016.

6. E.D. Platner, A. Etkin, K.J. Foley, J.H. Goldman, W.A. Love, T.W. Morris, S. Ozaki, A.C. Saulys, C.D. Wheeler, E.H. Willen, S.J. Lindenbaum, J.R. Bensinger, M.A. Kramer. Programmable Combinational Logic Trigger System for High Energy Particle Physics Experiments. Nucl. Instr. & Methods 140, 549-552 (1977).

7. I wish to thank T. Ludlam and W.A. Love for supplying the simulated events.

8. R. Bailey et al. Nucl. Instr. & Methods 213, 201 (1983).

9. M. Davenport, R.S. Deol, P.S. Flower, D.I. Giddings. G.D. Hallewell, A. Ikiades, J.A.G. Morris, J.V. Morris, P.H. Sharp, C.N. Uden, J. Eades, D. Newton, M. Ibbotson, R.J. Thompson, J. Waterhouse, J.J. Bunn, S. Danaher, R.H. McClatchey. A Ring Image Cerenkov Detector for the CERN OMEGA Spectrometer. IEEE Transactions on Nuclear Science, NS-30, 35 (1983).

10. M. Tannenbaum, these proceedings.

11. M. Tannenbaum, private communication.

12. A. Shor, to be published.

QUARK MATTER 1984: A SUMMARY

H. Satz
Fakultät für Physik
Universität Bielefeld
F.R. Germany

We survey the present state of statistical quantum chromodynamics and its application to high energy nuclear collisions.

I. Introduction

We are at present witnessing a rather striking growth of interest in the study of nuclear collisions at very high energies. Should you want to measure this, then you only need to count the number of people engaged in proposals for high energy heavy ion experiments, or the number of meetings dealing with the quark-gluon plasma, and compare your results to the count of four years ago, when the first meeting of our series took place.

One reason for this growth is the realization that energetic nuclear collisions are our only tool to study in the laboratory the condensed state of matter in strong interaction physics. It may not be an ideal tool, perhaps not even a sufficient one - but it does appear to be the only one we have.

Another reason is the prediction, based on QCD as the theory of the strong interaction, that strongly interacting matter can exist in different phases - as nuclear matter at low density, as a chromoplasma of quarks and gluons at high density. So it is our hope to observe these states, and the transition between them, in energetic nuclear collisions.

Let us, however, approach the subject slowly and begin by asking whether the experimental data available today provide us with any evidence that there exist collective features in the interaction of nuclei. After looking at some possible cases of such behaviour we shall then return to the most fascinating instance, the quark-gluon chromoplasma.

II. Collective Phenomena in Strong Interactions

Is there any indication that when a leptonic, hadronic or nuclear projectile hits a heavy nucleus, the whole is more than the sum of the parts - that we get more than just a superposition of single nucleon interactions? Some experiments suggest that this may indeed be the case.

Probably the first instance of "joint action" in nuclear matter is the cumulative effect[1)]. The collision of hadrons with a variety of nuclear targets leads to the production of secondaries, e.g. pions, which recoil from the target with a higher kinetic energy than would be kinematically allowed in a nucleon-nucleon collision (fig. 1). The effect is stronger the heavier the nucleus is. To provide it, some of the nucleons in the target nucleus must have joined forces, simultaneously or successively, to give the secondary such a high backward momentum.

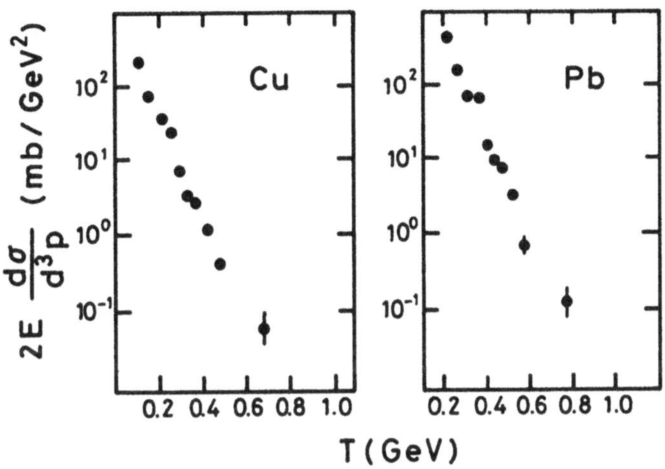

Figure 1 : Data for the reaction $pA \rightarrow \pi^{+}(180^{0}) + X$ at 8.4 GeV/c , for copper and lead; from ref. 1).

Another example is the anomalous enhancement of large p_T production in hadron-nucleus collisions[2)] (Cronin effect). Here the production rate of secondaries from p-A collisions, normalized to that of A p-p collisions, becomes greater than unity for sufficiently high p_T (fig. 2); at fixed large p_T , it increases with A . Data from α-α collisions support the enhancement at large p_T [3)].

A very striking demonstration of collective behaviour in nuclei was discovered two years ago at CERN - the EMC effect[4)]. The momentum distribution of quarks in nuclei, as measured by deep inelastic lepton scattering, is shifted towards noticeably lower values in comparison to that in nucleons (fig. 3). In other words, quarks in nuclei

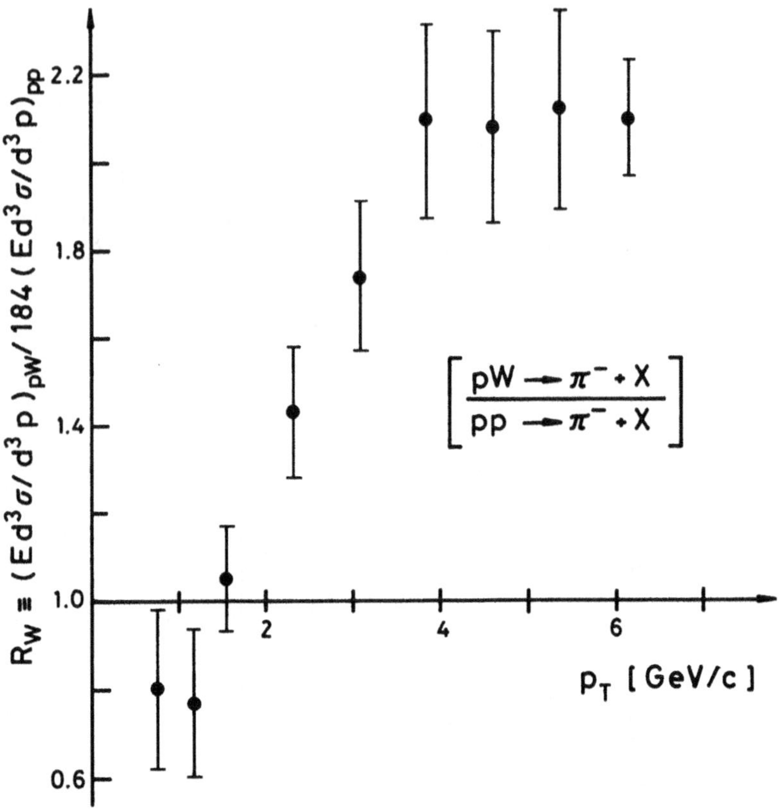

Figure 2 : Data for π^- production on tungsten and on hydrogen; from ref. 2).

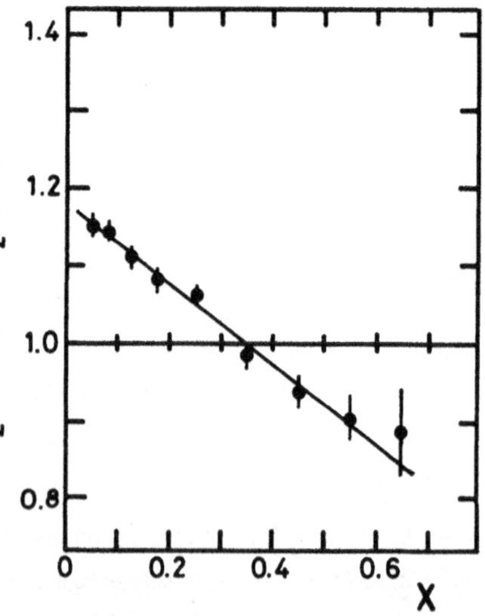

Figure 3 : Structure functions on iron and on deuterium; from ref. 4).

are "softer" than in nucleons. The relative suppression of hard quarks in nuclei moreover seems to increase with A [5].

Very recently, nuclear collision data from LBL have given us evidence for collective behaviour of nuclear matter[6]; the experimental results were presented in detail at this meeting[7]. Nuclei colliding at comparatively low energies are seen to fragment predominantly sideways for large A - in contrast to p-p or small A collisions, where the fragmentation occurs along the beam axis (fig. 4). In addition, one also finds collective "bounce-off" effects.

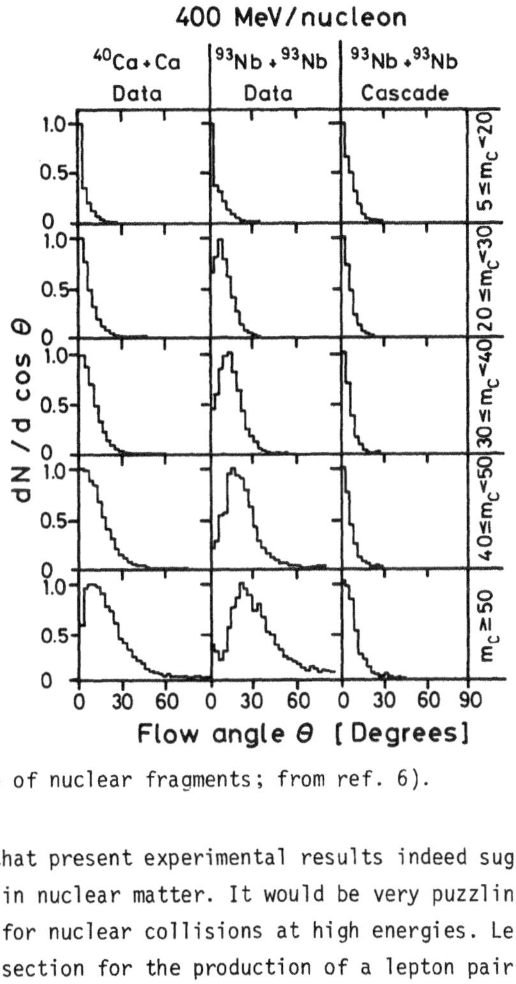

Figure 4 : Flow angle of nuclear fragments; from ref. 6).

We thus conclude that present experimental results indeed suggest some type of collective behaviour in nuclear matter. It would be very puzzling if such behaviour would not also occur for nuclear collisions at high energies. Let me give one simple example[8]. The cross section for the production of a lepton pair of mass M in the collision of two nucleons (or nuclei) is given by the Drell-Yan form

$$M^2(d\sigma/dM^2) \sim \int_{M^2/s}^{1} dx\, x^{-1}\, q(x)\, \bar{q}\left(\frac{1}{x}\frac{M^2}{s}\right) \quad , \tag{1}$$

provided M is large enough ($M \gtrsim 2$-3 GeV) . Here $q(x)$ ($\bar{q}(x)$) denotes the probability that a quark (antiquark) carries a fraction x of the incident momentum of target or projectile. The EMC experiment just mentioned measures the ratio of $q(x)$ in a nucleus to that in a nucleon. The observed behaviour - more low x , fewer intermediate x quarks in nuclei, as seen fig. 3 - clearly will have consequences for the production of lepton pairs in nuclear collisions, compared to that in nucleon collisions: there will be a relative enhancement of pairs at low M^2/s , a suppression of those at intermediate M^2/s . The prediction of a simple thermal model[8)] for the ratio

$$R_A(y) \equiv (d\sigma/dM^2)_{AA} / A(d\sigma/dM^2)_{NN} \quad , \tag{2}$$

with $y \equiv M/\sqrt{s}$, is shown in fig. 5, illustrating such behaviour. It should be emphasized, however, that any other model which provides the quark softening in nuclei must give qualitatively similar results.

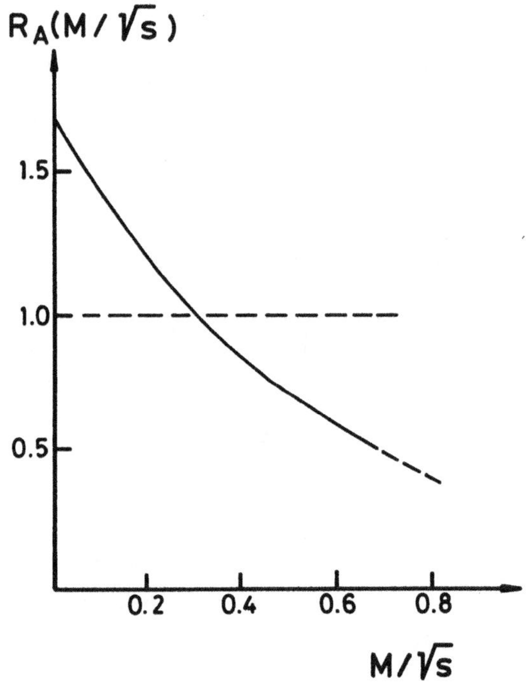

Figure 5 : Predicted lepton pair spectrum ratio from AA vs. pp collisions; from ref. 8).

So we certainly expect interesting collective phenomena in high energy nuclear collisions. Will they also suffice to produce the most fascinating such possibility, the transition to the chromoplasma of quarks and gluons? Before we can address that question, we must see what predictions QCD makes about the states of strongly interacting matter.

III. Statistical Quantum Chromodynamics

The problem we want to consider here is the classical topic of statistical physics: given microscopic dynamics, derive the equilibrium thermodynamics which describes a macroscopic system. In our case the microscopic theory is QCD, and we want to obtain the thermodynamics of strongly interacting matter. Let us note already at this point that two basic features of what we're about to study - the equilibrium condition and the arbitrarily large volume of the systems in question - may restrict the applicability of the results to nuclear collisions.

QCD describes the interaction between coloured quarks and gluons[9] through the Lagrangian density

$$\mathcal{L}(\psi,\bar{\psi},A) = -\frac{1}{4}(\partial_\mu A^a_\nu - \partial_\nu A^a_\mu - gf^a_{bc}\, A^b_\mu A^c_\nu)^2 + \bar{\psi}^a_f(i\not\partial - g\not{A}_a\, \lambda^a_{\alpha\beta})\,\psi^\beta_f \quad . \qquad (3)$$

Here ψ and A denote quark and gluon fields, with $\alpha,\beta,\ldots$ and $a,b,\ldots$ for the corresponding colour indices; f denotes the quark flavour $(u,d,s,\ldots)$ and g is the dimensionless coupling parameter of the theory. The first term on the r.h.s of eq. (3) describes the interaction of gluons; this interaction is possible - in contrast to QED, where photons don't interact directly - because the gluons also carry a colour charge. The second term in eq. (3), for the interaction of quarks and gluons, in fact has basically the same structure as its electromagnetic counterpart. The specific feature of QCD is thus the interaction between gauge fields, and one may study pure gauge field thermodynamics as a meaningful model of statistical QCD.

From the Lagrangian density (3) we now construct - at least formally - the Hamiltonian H needed to calculate in canonical fashion the partition function

$$Z(T,V) \equiv \mathrm{Tr}\,\{\exp - H(\mathcal{L})\,/\,T\} \quad ; \qquad (4)$$

the trace is to be carried out over all physical states of the system in a volume V. From the partition function we then obtain all thermodynamic quantities, such as the energy density

$$\varepsilon(T) = \lim_{V\to\infty} \{-\frac{1}{V}\,(\partial \ln Z\,/\,\partial(1/T))_V\} \quad , \qquad (5)$$

the pressure

$$P(T) = \lim_{V\to\infty} \{(\partial \ln Z\,/\,\partial V)_T\} \quad , \qquad (6)$$

etc. Thus, in principle we know what do.

In practice we immediately encounter two major obstacles. Perturbative calculations lead to the usual divergences of quantum field theory; we thus have to renormalize to obtain finite results. What is worse, however, is that we want to cover the entire possible range of behaviour of the system, from confinement to asymptotic freedom, i.e., from nuclear or hadronic matter to ideal chromoplasma. That is not possible perturbatively, and we must therefore find a new way to approach the problem, one more generally applicable than the perturbative scheme of QED.

The only such approach available up to now is the lattice regularization proposed by Wilson[10]. By evaluating the partition function on a large but finite lattice, whose points are separated by multiples of some lattice spacing a, we have $1/a$ as largest and $1/(Na)$ as smallest possible momentum; here Na is the linear size of the lattice. Hence neither ultraviolet nor infrared divergences can occur for finite N and non-zero a. We are left with two questions, however: how do we actually carry out the evaluation, and how can we be sure that physical quantities calculated are independent of the lattice? As the functional form of the partition function on the lattice turns out to be quite similar to what one encounters in the evaluation of spin systems, we can use Monte Carlo simulation techniques[11] from statistical physics to calculate the thermodynamic quantities of interest. Moreover, the renormalisation group theory of statistical physics tells us how the lattice spacing a and the coupling g of the Lagrangian (3) must be related, if the results are to be independent of the choice of lattice. So we do indeed have today what seems to be a viable procedure to evaluate the interesting observables of statistical QCD - and this is certainly a major reason for the great interest which lattice gauge theory has attracted in recent years.

Before we look at the more recent results of finite temperature lattice QCD, let us see what we should expect to find. In table I, we compare the basic features of QED and QCD. In particular, we note that the transition from hadronic or nuclear matter

QED	QCD
electric charge	colour charge
atom, positronium (electrically neutral)	hadron (colour neutral)
insulator (no electric conductivity)	hadronic matter (no colour conductivity)
metal (electric conductor)	chromoplasma (colour conductor)

Table I : QED vs. QCD

to a chromoplasma corresponds to an insulator-metal transition in solid state physics. Deconfinement thus appears as the Mott transition[12] of QCD; it occurs because of the Debye screening of a given colour charge due to the presence of many other such charges[13].

We anticipate another effect also familiar from solid state physics. The conduction electrons in a metal have an "effective" mass different from the electron mass in vacuum, as a collective effect due to the overall field present in the system. Similarly, we expect quarks in the plasma (current quarks with $m_q \simeq 0$) to have a different mass than those inside hadrons or hadronic matter (constituent quarks with a mass $m_q^{eff} \simeq 200 - 300$ MeV). The vanishing of the effective quark mass is related to the restoration of the chiral symmetry, which the Lagrangian (3) has but which is broken for the actual state of the system at low temperatures. We thus want to study deconfinement and chiral symmetry restoration, as well as the relation of the two phenomena.

The past year has brought quite a number of interesting results in the thermodynamics of lattice QCD with dynamical fermions, using several different types of lattice formulations[14-19]. All find a very abrupt change in the energy density ε , in the average thermal Wilson loop $<\bar{L}>$ as some measure of confinement, and in $<\bar{\psi}\psi>$ as a measure of chiral symmetry. The sudden change of these quantities takes place at the same value of the temperature, suggesting that deconfinement and chiral symmetry restoration occur at the same point. We show some representative results of this behaviour in fig. 6 to 8. The main other conclusion to be drawn at present is that there is no indication for any disappearence of the deconfinement transition with decreasing quark mass, as had been suggested previously[20].

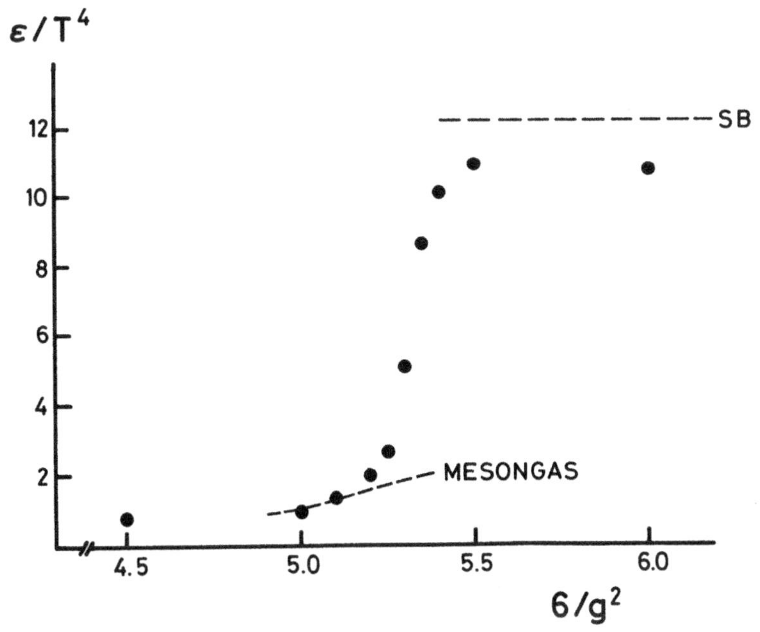

Figure 6 : Energy density of the quark-gluon system, with Wilson fermions, on an $8^3 \times 3$ lattice; from ref. 15).

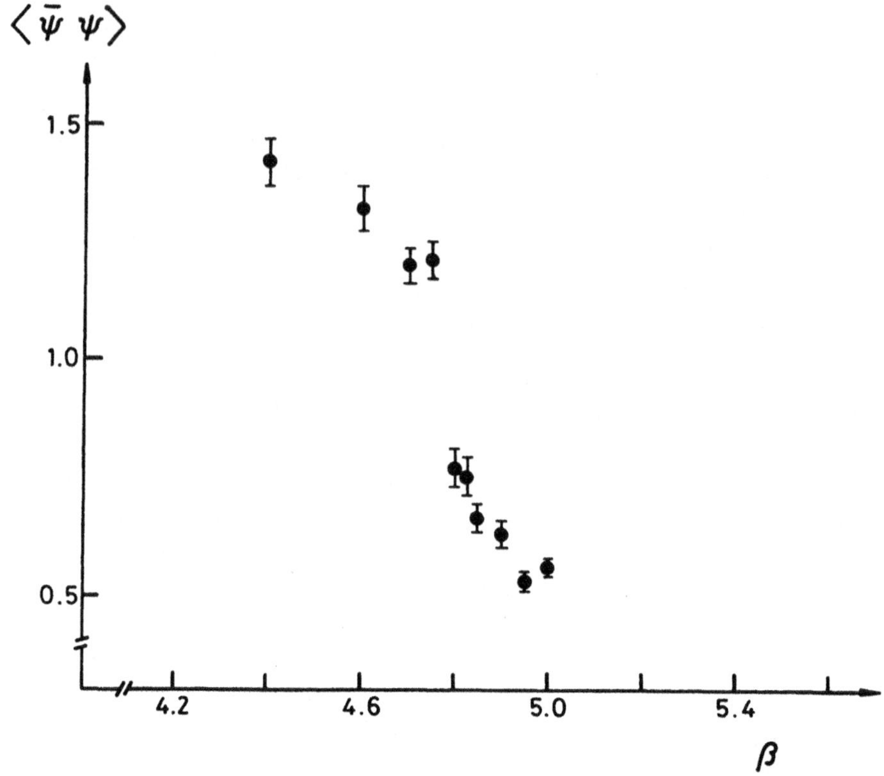

Figure 7 : Chiral order parameter of the quark-gluon system, with Kogut-Susskind fermions (canonical), on a $6^3 \times 2$ lattice; from ref. 18).

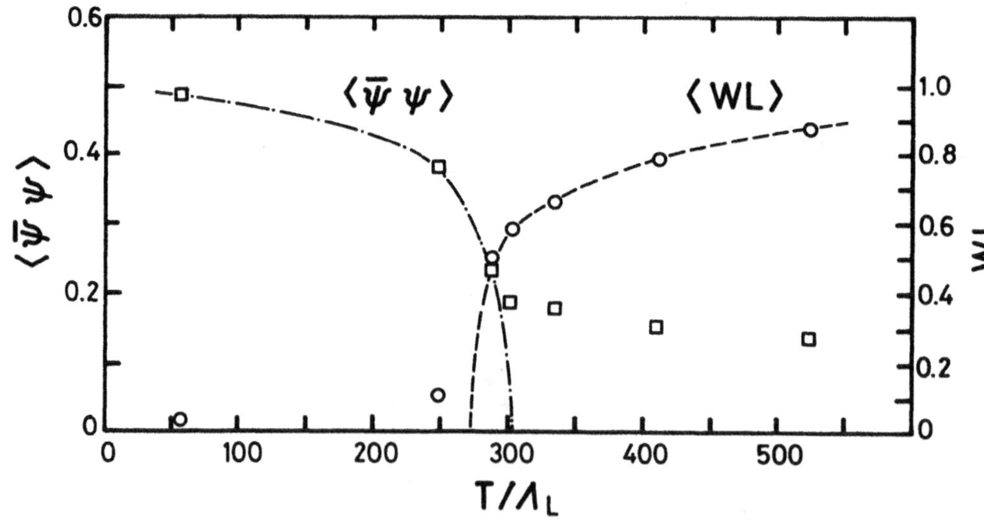

Figure 8 : Average thermal Wilson loop (< WL >, o) and chiral order parameter (< $\bar{\psi}\psi$ >, □) of the quark-gluon system, with Kogut-Susskind fermions (microcanonical); from ref. 19).

We are still facing several important open questions, however. What is the order of the transition? For the pure gluon system $(m_q \simeq \infty)$, deconfinement is a first order transition[21], but with decreasing quark mass, the transition seems to become second order[15]. There are arguments that chiral symmetry restoration is also of first order for massless quarks[22]. This would imply in both cases a persistence of first order behaviour for some range of quark masses. What is the relation between deconfinement and chiral symmetry restoration? Does one somehow drive the other? Can chiral symmetry breaking stabilize the first order character of deconfinement[23]? The present state of affairs is summarized in fig. 9.

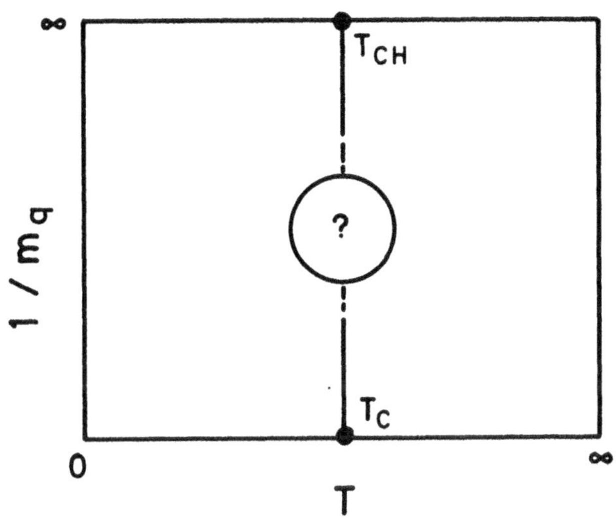

Figure 9 : Schematic view of the relation between deconfinement temperature T_C and chiral symmetry restoration temperature T_{CH}, as function of the quark mass m_q.

For all experimental applications, we certainly want to obtain the values of the critical temperatures for deconfinement and chiral symmetry restoration in physical units. QCD, however, only predicts the ratios of physical observables, since it contains no dimensional parameter. We must therefore calculate the critical temperature T_c and some other, measured quantity, e.g. m_ρ, both in lattice units, and then take the ratio to obtain a numerical prediction of T_c/m_ρ. Work on the calculation of hadron masses, with dynamical fermions included, is in progress; so far, there is no evidence that the deconfinement temperature differs significantly from the value $T_c \simeq 200$ MeV of the pure gluon system. For the existence of a quark-gluon chromoplasma, this would imply an energy density $\varepsilon_{PLASMA} \gtrsim 2.5\ GeV/fm^3$.

Before leaving statistical QCD, let us ask whether the presently used lattices are really sufficient to give valid continuum results. Up to now, this question has been studied in detail only for the pure gluon system. We show in fig. 10 a compilation of lattice results for the deconfinement temperature T_c and the string tension σ , both measured in terms of the (arbitrary) lattice scale Λ_L . They are plotted as a function of the inverse coupling; the smaller the lattice spacing, the larger $6/g^2$. We note that for $g^2 \gtrsim 1$, "asymptotic scaling" - the lattice size independence of T_c and σ - seems to be a reasonable assumption. The ratio $T_c/\sqrt{\sigma}$ appears constant, with

$$T_c/\sqrt{\sigma} \simeq 0.5 \quad , \qquad (7)$$

already at lower values, in accord also with recent block-spin calculations[24]. With $\sqrt{\sigma}$ = 400 MeV , this yields the value of T_c quoted above.

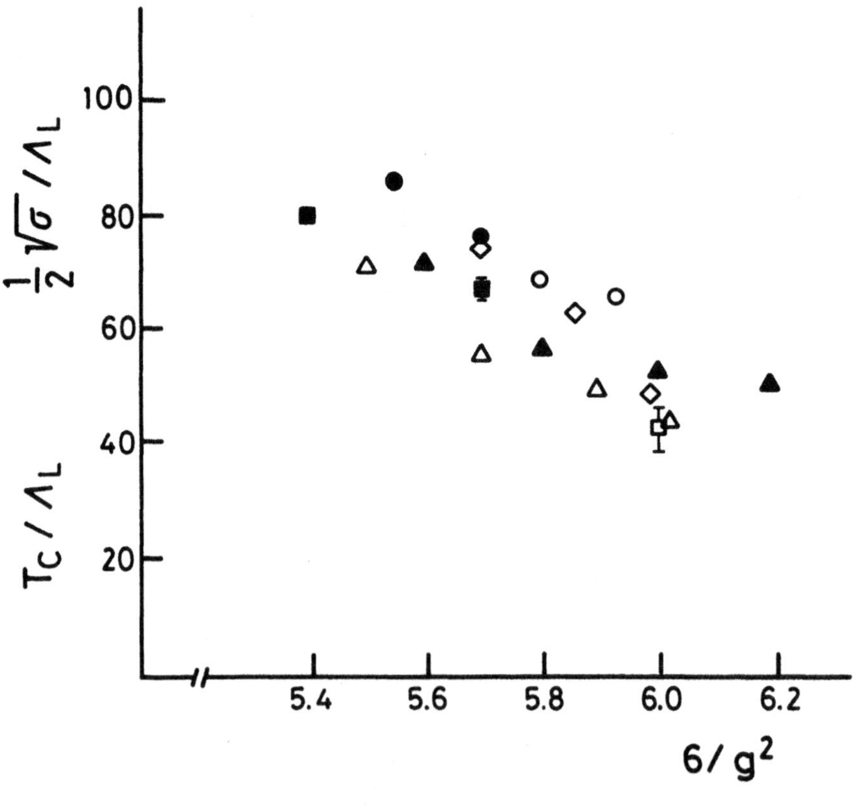

Figure 10 : Deconfinement temperature (● from ref. 25a, ○ from ref. 25b, ◇ from ref. 25c) and string tension (■ from ref. 25d, ▲ from ref. 25f, □ from ref. 25e, Δ from ref. 25g) for the pure SU(3) gauge system.

IV. Excitation, Thermalisation, Expansion

In the last section, we considered systems of arbitrarily large spatial extension and in equilibrium. Can such a theory be applied to nuclear collisions? The "history" of a nuclear interaction at high energy is schematically shown in fig. 11[26)]. First we find an excitation regime, during which energy is deposited in a given spatial region. How large is the energy density in comparison to that needed for plasma formation? The chromodynamic flux present will after some time lead to the spontaneous formation of quark-antiquark pairs (we concentrate here on the central region, where the baryon number density is essentially zero). Is the mean free path of these constituents in the state of the system at that time short enough to have thermal equilibrium with subsequent hydrodynamic expansion? Such an expansion results in cooling and will thus bring us to the transition regime studied in the last section. At sufficiently low temperature, finally, all strong interactions cease and we have free hadrons.

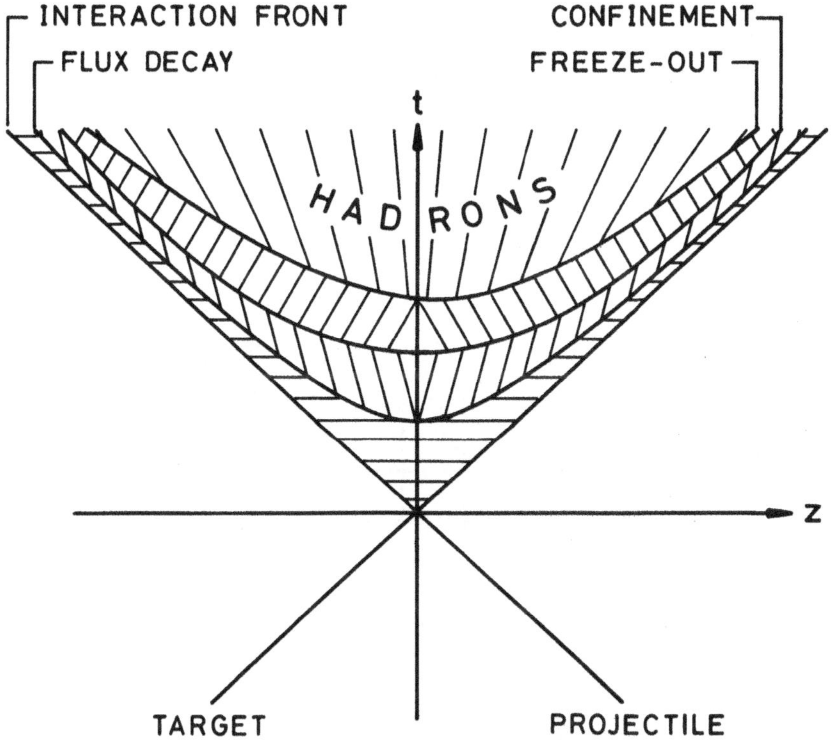

Figure 11 : Schematic view of the space-time development of a nucleus-nucleus collision.

What can we say about the energy deposited in the collision? In the scaling central region of an energetic nucleon-nucleon collision, one finds on the average per unit rapidity two secondaries of about 1/2 GeV CMS energy each. For a cylindrical interaction region, of radius and longitudinal height of one fermi (see fig. 12a), this means an energy density

$$\varepsilon_0 \simeq 1/3\ \mathrm{GeV/fm^3} \quad . \tag{8}$$

In a central nucleus-nucleus collision, we have within such a cylinder $A^{1/3}$ superimposed nucleon-nucleon collisions, due to the nucleons "sitting behind each other" in the Lorentz-contracted colliding nuclei[27] (see fig. 12b). This leads to an energy deposit

$$\varepsilon_A \simeq A^{1/3}\,/\,3\ \mathrm{GeV/fm^3} \tag{9}$$

and hence implies e.g. about 2 GeV/fm³ for the collision of two ^{238}U nuclei[26], bringing us at best to the threshold for plasma formation.

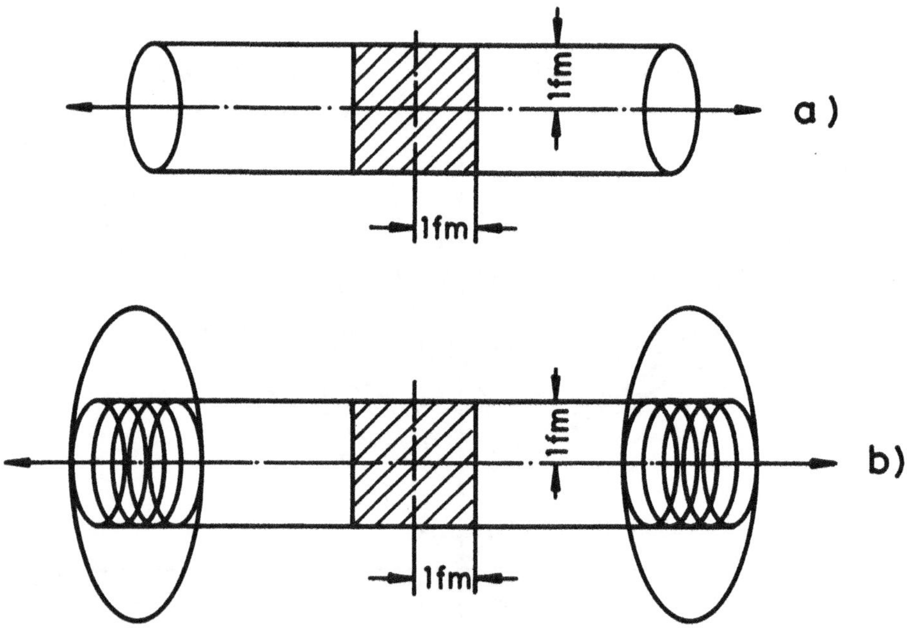

Figure 12 : Schematic view of central production in pp (a) and AA (b) collisions.

Present estimates are, however, considerably more optimistic than this lower bound. Both the increase in multiplicity[28] and the enhanced nuclear stopping power[29] suggest a higher energy deposit. Let us write the energy density ε_A as[30]

$$\varepsilon_A = \bar{\omega}_A (dN/dy)_A \,/\, (\pi\, R_0^2\, A^{2/3}\, t_A) \quad , \tag{10}$$

where $(dN/dy)_A$ denotes the average multiplicity per unit rapidity, $\bar{\omega}_A$ the average energy per secondary, and t_A the time (or length) of formation for hadronic secondaries, all in A-A collisions. As before, $R_0 \simeq 1$ fm . We recover eq. (9) with

$$(dN/dy)_A = A(dN/dy)_0 \simeq 2A \tag{11}$$

$$\bar{\omega}_A = \bar{\omega}_0 \simeq 1/2 \text{ GeV} \tag{12}$$

$$t_A = t_0 \simeq 1 \text{ fm} \tag{13}$$

where the subscripts 0 refer to the corresponding result in a nucleon-nucleon collision. A decrease in formation time leads, by the uncertainty relation, to an increase in $\bar{\omega}$. We take $\bar{\omega}_A = \bar{\omega}_0 \; (t_0/t_A)$[30] and retain eq. (11), assuming the multiplicity enhancement to be a reflection of the decreased formation time. Certainly the interrelation of the A dependence in $\bar{\omega}$, t and (dN/dy) needs further clarification, however. With the indicated form of $\bar{\omega}_A$ and $t_A \simeq 1/3$ fm [30], we obtain

$$\varepsilon_A \simeq 3A^{1/3} \text{ GeV/fm}^3 \tag{14}$$

for the energy density - an increase by a factor $t_A^{-2} = 9$ over the lower bound (9). For a collision of ^{238}U ions this gives $\varepsilon_{238} \simeq 18$ GeV/fm^3 , which would be well within the plasma range.

The estimates of ε_A obtained so far were essentially predicted from results of nucleon-nucleon collisions. Do they agree with what one would expect from a reconstruction using A-A data? If the observed secondaries there are the result of an isentropic hydrodynamic expansion[26,31], part of the incident energy goes into collective hydrodynamic flow[32]. For an ideal gas, we can combine the relation between entropy and particle density, $s = 4n$, with that between entropy and energy density, $\varepsilon \simeq 0.3\, s^{4/3}$, to obtain

$$\varepsilon_A \simeq 1.9 \; [(dN/dy)_A / (\pi R_0^2 A^{2/3} t_A)]^{4/3} \tag{15}$$

in place of eq. (10). If

$$\bar{\omega}_A = 1.9 \; [(dN/dy)_A / (\pi R_0^2 A^{2/3} t_A)]^{1/3} \tag{16}$$

then the stronger multiplicity dependence of ε_A in eq. (15), due to hydrodynamic flow, would be obtained in eq. (10) by a corresponding increase of $\bar{\omega}_A$.

Up to now we have only talked abount average head-on nuclear collisions. Triggering on a higher than average multiplicity would of course increase ε_A . As an illustration: we had $\varepsilon_0 \simeq 1/3$ GeV/fm^3 for an average nucleon-nucleon collision. For the highest central multiplicities observed at the CERN $p\bar{p}$ collider - about twenty charged secondaries per unit rapidity - this value is increased by more than a factor ten.

It thus seems reasonable to expect that nuclear collisions will provide energy densities sufficient for plasma formation. What about equilibrium and subsequent hydrodynamic cooling?

Let us assume an incoherent production of quark-antiquark pairs at the spontaneous breakdown of the chromodynamic flux[33]. Is the mean free path of quarks and gluons in the resulting medium sufficiently small in comparison to the size of the system, so that we can expect hydrodynamic expansion and cooling? The mean free path is

$$\lambda = 1/(n\sigma) \quad , \tag{17}$$

where n is the density of scatterers and σ the corresponding cross section. For an ideal gas of quarks and gluons, $n \sim T^3$; for dimensional reasons, $\sigma \sim T^{-2}$. Hence the mean free path behaves as

$$\lambda \sim 1/T \sim 1/\varepsilon_A^{1/4} \tag{18}$$

for the asymptotically free system, so that it will be small in comparison to the size of the system provided ε_A is large enough. It is not so clear, however, what happens at lower ε_A . Non-perturbative corrections to asymptotic freedom and/or chromomagnetic effects are then expected to play a significant role[34]. At the confinement point, the transverse dimension of the system will be about $(\varepsilon_A/2.5)^{1/3}\ A^{1/3}$ fm , with 2.5 GeV/fm³ as the energy density at the onset of confinement. For ε_A in the range discussed above, this is considerably larger than the mean free path of about a fermi, which we would expect in hadronic matter.

Nevertheless, the question of sustained thermal equilibrium in nuclear collisions appears at this time one of the more open interesting problems in the study of plasma formation.

V. Signatures

Before we survey the possible observable consequences of chromoplasma formation in nuclear collisions, let us once more return to the purely exploratory level. Since we are looking for collective features in high energy nuclear interactions, two obvious things to do are therefore the following:

(i) measure observables X for the same range of parameters in both AA and pp collisions. How does the ratio $R_A \equiv X_{AA}/X_{pp}$ behave, as function of the kinematic variables of the problem, and as function of A ? The EMC effect and its extension to nuclear collisions illustrate what such studies could provide.

(ii) measure observables X in AA collisions for parameter values for which they should vanish in pp collisions. Non-vanishing X_{AA} then provide evidence for some type of collective behaviour inside the nucleus. The cumulative effect is an example of this type of phenomenon.

The outcome of such studies will in general not say anything about plasma formation, but it may tell us about interesting collective effects. Since the first experiments at really high energy will probably be performed with rather light projectiles, such effects may prove valuable not only for their own sake, but also to justify an extension to heavier ions. In addition it should also be kept in mind that none of the plasma signatures discussed so far would provide unambiguous evidence for plasma formation.

In table II we then list the main tests for chromoplasma production that have been considered; in each case, the primary aim of the test is also indicated. All tests must be checked against contamination of "faking" from non-plasma mechanisms, by both theoretical and experimental checks. The former require the calculation of non-plasma competition, such as direct Drell-Yan production of lepton pairs; the latter requires a comparison of AA with pp data, as just mentioned above. In addition, there is an interrelation between tests; the specific type of hydrodynamic expansion involved will certainly have consequences on lepton pair distributions, to quote just one example.

TEST	AIM
Lepton pair production, direct photons, heavy exotics	state of the primordial environment at $T > T_c$
Strange particle production	"chemical" equilibrium in the chromoplasma
p_T spectra	transition signal
Meson, baryon momentum distributions, fluctuations	hydrodynamic expansion
Interferometry	plasma "ball" size
Jets	stopping power of plasma
Resonance "fade-out"	chiral symmetry restoration

Table II : Some signatures of plasma formation

In closing this brief survey, let us consider one specific case in a little more detail. The p_T distributions of secondaries have long been taken as a measure of some type of hadronic temperature, most systematically in the work of Hagedorn[35]. The multiplicity per unit rapidity, on the other hand, provides us with a measure of the initial energy deposit. Thus a plot of $\bar{p}_T$ vs. (dN/dy) in the central region might be expected to look like that shown in fig. 13[36]: at low energy deposit, we obtain the canonical 300 MeV ; an increase in energy density leads to an increase in p_T , because of transverse hydrodynamic expansion. In the transition region, increasing the energy deposit does not (first order transition), or very little (second order transition), change the temperature and hence does not increase the transverse expansion; thus $\bar{p}_T$ remains constant. Above the transition region, both the hydrodynamic flow and hence $\bar{p}_T$ increase again. - In fig. 14 we show cosmic ray data presented at this meeting by Miyamura[37]. Is this a first experimental indication for the chromoplasma?

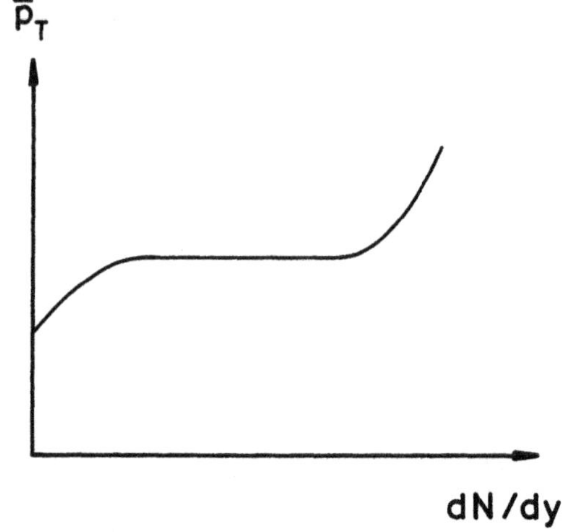

Figure 13 : Schematic view of the average transverse momentum in hadronic collision as function of the average number of secondaries in the central region; from ref. 36).

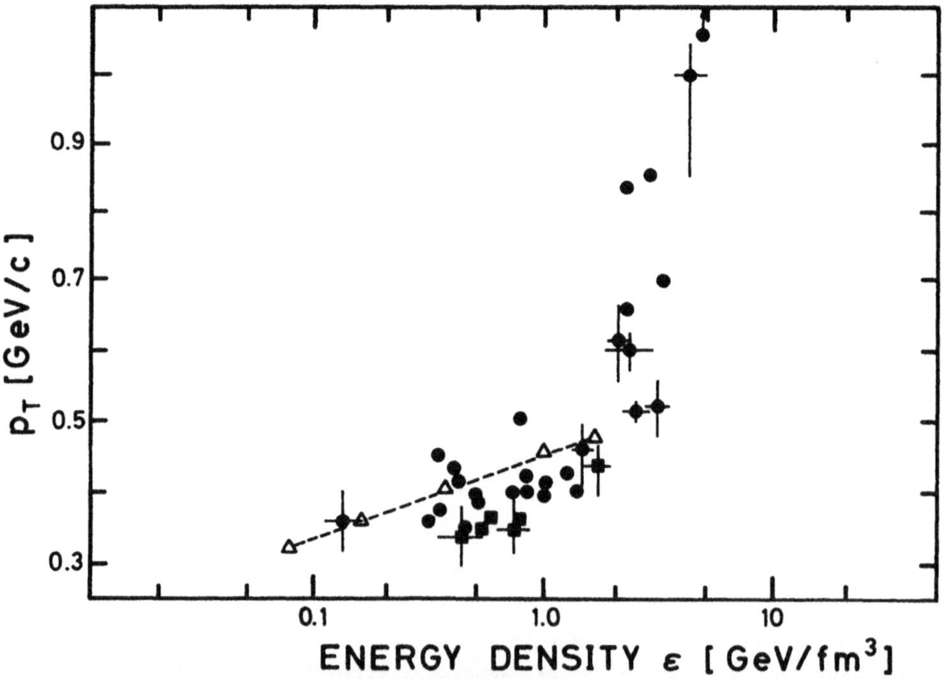

Figure 14 : Average transverse momenta in cosmic ray data (solid symbols) and from CERN collider data (open triangles); from ref. 37).

VI. The Experimental Situation

Two years ago in Bielefeld, when it was first decided to hold the Quark Matter Meetings as regularly recurring events, the basis for experimental study looked not so promising. The ISR at CERN had just been "removed" as one possible heavy ion machine; there was one propbsal for the CERN-PS, by GSI-LBL; and it was often asked if there existed a community really interested in carrying out experiments of this type.

Today there are definite plans for accelerating ^{16}O up to 50 GeV/nucleon at the CERN-SPS, and ^{32}S up to 15 GeV/nucleon in the AGS at Brookhaven, with the needed transfer line at BNL now funded. At CERN, one two-part experiment (GSI-LBL) is approved, five further proposals are submitted at this time, and at least three more are expected before the end of the summer. The proposals so far submitted were surveyed here by Specht; they come from more than thirty different institutions all over the world.

For the more distant future, we have in the USA considerable discussion and planning of a high energy colliding beam facility ($\sim$50-100 GeV/nucleon per beam), as a result of last year's NSAC recommendation. We heard at this meeting from Ludlam how such a machine could be built in the existent tunnel at Brookhaven. For the LBL in Berkeley, Pugh reported considerations of a mini-collider (4 GeV/nucleon) dedicated to the study of baryon-rich environments. In Europe, there are speculations about injecting heavier ions (up to ^{56}Fe) into the CERN-SPS, and a recent workshop on possible hadron physics in the CERN-LEP tunnel suggests a facility for which experiments with "heavy ions are also possible"[38].

VII. Concluding Remark

Since we are entering a new field of research, we should do so with the right mixture of optimism and humility. The best illustration I could find for the necessary frame of mind is shown in a little sketch which appeared in the New York Herald Tribune some months ago. I present it as my closing remark.

Acknowledgement

It is a pleasure to thank many colleagues for helpful discussions before, during and after Quark Matter 1984. Particular thanks go to T. Çelik, J. Engels, R.V. Gavai, M. Gyulassy, T. Matsui, L. McLerran, J. Polônyi and H. Specht. The written version of this talk was completed at LBL, Berkeley; the hospitality of J. Randrup and his colleagues there was much appreciated.

Last but most certainly not least I want to thank K. Kajantie and his colleagues in Finland for organizing such a stimulating, informative and enjoyable meeting.

References

1) A.M. Baldin, in Proc. of the XVIth International Conference on High Energy Physics, FNAL, Batavia/Illinois 1972
A.M. Baldin et al., Yad. Fiz. 20 (1974) 1201

2) L. Kluberg et al., Phys. Rev. Lett. 38 (1977) 670
D. Antreasyan et al., Phys. Rev. D19 (1979) 764

3) See the survey by M. Faessler, in Quark Matter Formation and Heavy Ion Collisions, M. Jacob and H. Satz (Eds.), World Scientific, Singapore, 1982

4) J.J. Aubert et al., Phys. Lett. 123B (1983) 275

5) R.G. Arnold et al., Phys. Rev. Lett. 52 (1984) 727

6) H.A. Gustafsson et al., Phys. Rev. Lett. 52 (1984) 1590

7) See the report by H. Pugh, in these proceedings

8) H. Satz, Bielefeld Preprint BI-TP 84/13, June 1984

9) For a summary, see e.g. W. Marciano and H. Pagels, Phys. Reports 36 (1978) 137

10) K. Wilson, Phys. Rev. D10 (1974) 2445

11) Introduced in this context by M. Creutz, Phys. Rev. D21 (1980) 2308

12) See N.F. Mott, Rev. Mod. Phys. 40 (1968) 677

13) See H. Satz, Nucl. Phys. A418 (1984) 447c

14) P. Hasenfratz, F. Karsch and I.O. Stamatescu, Phys. Lett. 133B (1983) 221

15) T. Çelik, J. Engels and H. Satz, Phys. Lett. 133B (1983) 427 and Bielefeld Preprint BI-TP 84/09, 1984

16) F. Fucito and S. Solomon, Cal. Tech. Preprint CALT-68-1084, January 1984

17) F. Fucito, C. Rebbi and S. Solomon, Cal. Tech. Preprints CALT-68-1124 and 68-1127, May 1984

18) R.V. Gavai, M. Lev and B. Petersson, Bielefeld Preprint BI-TP 84/10, May 1984

19) J. Polônyi, H.W. Wyld, J.B. Kogut, J. Shigemitsu and D.K. Sinclair, Phys. Rev. Lett. 53 (1984) 644
See the report of J. Polônyi in these proceedings.

20) T. Banks and A. Ukawa, Nucl. Phys. B225 [FS9] (1983) 145

21) T. Çelik, J. Engels and H. Satz, Phys. Lett. 125B (1983) 411
J. Kogut, M. Stone, H.W. Wyld, W.R. Gibbs, J. Shigemitsu, S.H. Shenker and D.K. Sinclair, Phys. Rev. Lett. 50 (1983) 353

22) R. Pisarski and F. Wilczek, Santa Barbara Preprint NSF-ITP-83-152, December 1983

23) V. Alessandrini, Orsay Preprint LPTHE 84/14, March 1984

24) A. Hasenfratz, P. Hasenfratz, U. Heller and F. Karsch, Phys. Lett. 143B (1984) 193

25) a) ● T. Çelik, J. Engels and H. Satz, Phys. Lett. 129B (1983) 323
b) ○ F. Karsch and R. Petronzio, Phys. Lett. 139B (1983) 403
c) ◇ A. Kennedy, J. Kuti, S. Meyer and B.J. Pendleton, Santa Barbara Preprint NSF-ITP-84-61 (1984)
d) ■ F. Gutbrod, P. Hasenfratz, Z. Kunszt and I. Montvay, Phys. Lett. 128B (1983) 415
e) □ G. Parisi, R. Petronzio and F. Rapuano, Phys. Lett. 128B (1983) 418
f) ▲ D. Barkai, K.J.M. Moriarty and C. Rebbi, Brookhaven Preprint BNL 34462 (1983)
g) △ M. Teper, private communication

26) J.D. Bjorken, Phys. Rev. D27 (1983) 140
For further references, see the reports of G. Baym and L. McLerran in these proceedings.

27) H. Satz, Nucl. Phys. A400 (1983) 541c

28) See e.g. S. Frederikson, G. Eilam, G. Berlach and L. Bergström, Royal Institute of Technology/Stockholm Preprint TRITA-TFY-84-06, May 1984 (to appear in Phys. Reports)

29) W. Busza and A.S. Goldhaber, Stony Brook Preprint ITP-SB-83-58, December 1983

30) See the report of L. McLerran at this meeting.

31) K. Kajantie and L. McLerran, Phys. Lett. 119B (1982) 203
K. Kajantie, R. Raitio and P.V. Ruuskanen, Nucl. Phys. B222 (1983) 152

32) M. Gyulassy and T. Matsui, Phys. Rev. D29 (1984) 419
P.V. Ruuskanen, LBL Preprint (1984)

33) F.E. Low, Phys. Rev. D12 (1975) 163
S. Nussinov, Phys. Rev. Lett. 34 (1975) 1286
H. Ehtamo, J. Lindfors and L. McLerran, Z. Phys. C 18 (1983) 341

34) A.D. Linde, Phys. Lett. 96B (1980) 289
D. Gross, R. Pisarski and L. Yaffe, Rev. Mod. Phys. 53 (1981) 43
M. Gyulassy, private communication
A. Hosoya and K. Kajantie, Helsinki Preprint HU-TFT-83-62 (1983)
See the report of O. Kalashnikov in these proceedings.

35) R. Hagedorn, Nuovo Cim. Suppl. 3 (1965) 147
See the report of R. Hagedorn in these proceedings.

36) L. Van Hove, Phys. Lett. 118B (1982) 138

37) See the report of O. Miyamura in these proceedings.

38) Summary Report: Workshop on the Feasibility of Hadron Colliders in the LEP Tunnel, G. Brianti et al., CERN Yellow Report, to appear

PARALLEL SESSIONS

Tuesday 19 June, Experimental parallel sessions

Chairman: I. Otterlund

W. Willis:

Study of high energy densities over extended nuclear volumes via nucleus-nucleus collisions.

M. Heiden:

The production of strange baryons and antibaryons with relativistic light ions at the CERN SPS.

R. Zitoun:

Study of high energy nucleus-nucleus interactions using the Ω' spectrometer equipped with a multiparticle high p_T detector.

H. Sann:

Particle identification in nucleus-nucleus collisions at the CERN SPS.

Search for anomalous fragments in relativistic nucleus-nucleus collisions.

S. Garpman, G. Baroni, and I. Lund:

Search for anomalous fragments in nucleus-nucleus collisions.

T.A. Parnell:

JACEE nucleus-nucleus interaction data above 20 GeV.

Y. Takahashi:

NASA/Marshall Space Flight Center Search for ϕ-mesons in heavy ion collisions.

R. Weiner:

Friction between spectator and participant systems as a signal for QGP.

Tuesday 19 June, Theoretical parallel sessions

Chairmen: T. DeGrand, V. Ruuskanen and H. Satz

P. Damgaard:

Chiral symmetry restoration at finite temperature.

M. Gross:

Phase structure of gauge and spin systems.

E. Hilf:

Thermal Wilson loops with fixed boundary conditions.

J. Zimanyi:

Relativistic mean field theory with real pions.

M. Biyajima:

Stochastic theories of KNO scaling and $\pi\pi$-interferometry.

M. Kutschera:

Dense matter in chiral soliton model.

N. Glendenning:
Signals from the evolution of the QCD-plasma with confinement and phase transition.
B. Müller:
Quark-gluon plasma cooling through pion emission in a chiral bag model.
S. Pratt:
Pion interferometry of the QCD-plasma.
J. Parikh:
Pion charge correlations and hadronization of the QCD-plasma.
R. Weiner:
Correlations between transverse momentum and multiplicity; a signal for quark-gluon plasma formation.
J. Polonyi:
Deconfinement and chiral symmetry restoration in the microcanonical approach.
R. Gavai:
Finite temperature QCD with staggered fermions.
J. Engels:
Finite temperature QCD with Wilson fermions.
O. Kalashnikov:
Quark-gluon plasma through perturbative and non-perturbative calculations.

Wednesday 20 June, Experimental parallel sessions

Chairman: T. Ludlam
T. Ludlam, S. Garpman:
Event simulation.
M. Šumbera:
Hints for collective effects in collisions of relativistic nuclei.
S. Katsanevas:
Soft photon production at ISR energies.
S. Frankel:
High p_T nuclear cross sections.
M. Heiden:
Composition of particles emitted at large p_T and medium angles in pp, dd and $\alpha\alpha$ collisions.
A.T. Goshaw:
Search for deconfined quark-gluon phase of strongly interacting matter in $p\bar{p}$ interactions at $\sqrt{s}$ = 2 TeV.
A. Hahn:
Search for fractional charge production in heavy ion collisions.
J.B. Carrol:
Recent Bevalac results on production of subthreshold K^-, $\bar{p}$ and direct electrons.

Wednesday 20 June, Theoretical parallel sessions

Chairmen: P. Ruuskanen and R. Raitio

C. DeTar:

The modes of excitation of the quark-gluon plasma.

J. Kapusta:

Birth of the QCD-plasma in a superheated hadron vapour.

D. Strottman:

Ultrarelativistic collisions of heavy ions in a two fluid hydrodynamic model.

H. Sumiyoshi:

Baryon and energy stopping power of nuclei.

C.Y. Wong:

Baryon distribution in relativistic heavy ion collisions.

R. Hwa:

Degradation of the quark momentum and the thermalization problem

R. Hagedorn:

Statistical thermodynamics in relativistic particle and ion physics: canonical or grandcanonical?

J. Rafelski:

Strange hadrons from quark-gluon plasma.

K. Kajantie:

Quark-hadron phase transition in the early universe.

I. Lovas:

Gluon condensation.

List of participants

Dr. L. Abbott
Dept. of Physics
Brandeis University
Waltham, MA 02254
USA

Prof. Giustina A. Baroni
University di Roma
La Sapienza
Piazzale A. Moro
00100 Roma
ITALY

Dr. J. Bartke
Inst. of Nuclear Physics
ul. Kawiory 26a
30-055 Krakow
POLAND

Prof. G. Baym
Dept. of Physics
Univ. of Illinois
Urbana, IL 61801
USA

Dr. M. Biyajima
Faculty of Liberal Arts
Shinshu University
390 Matsumoto
JAPAN

Dr. J.-P. Blaizot
SPLT
CEN-Saclay
F-91191 GIF-sur-YVETTE
CEDEX
FRANCE

Dr. P. Bond
K.V.I.
Universiteitscomplex
Paddeloel
9747 AA Groningen
NETHERLANDS

Prof. W. Busza
Phys. Dept. and Lab.
for Nuclear Science
M.I.T.
Cambridge, MA 02139
USA

Dr. D. Callaway
CERN/TH-Division
CH-1211 Geneva 23
SWITZERLAND

Dr. B. Callen
Dept. of Physics
Univ. of Pennsylvania
Philadelphia 19104
USA

Prof. D.D. Carmony
Purdue University
Purdue Building
West Lafayette
Indiana 47907
USA

Dr. J. Carrol
Lawrence Berkeley Lab.
Univ. of California
Berkeley, CA 94720
USA

Dr. P. Carruthers
Los Alamos National Lab.
P.O. Box 1663, T-8 MS B285
Los Alamos
New Mexico 87545
USA

Dr. T. Celik
Dept. of Physics (Fizik-83)
Hacettepe University
Ankara
TURKEY

Dr. J. Cleymans
Fakultät für Physik
Univ. Bielefeld
D-4800 Bielefeld
W.-GERMANY

Dr. I Cugnon
Université de Liège
Inst. de Physique
Bat. B5
B-4000 Liège 1
BELGIQUE

Dr. P.H. Damgaard
NORDITA
Blegdamsvej 17
DK-2100 København
DANMARK

Prof. T. DeGrand
Dept. of Physics
Univ. of Colorado
Boulder, CO 80305
USA

Prof. C. DeTar
Dept. of Physics
201 North Physics Bldg.
Univ. of Utah
Salt Lake City 84112
USA

Dr. J. Engels
Fakultät für Physik
D-4800 Bielefeld 1
W.-GERMANY

Dr. K.J. Foley
Brookhaven Nation. Lab.
Upton, Long Island
New York 11973
USA

Dr. S. Frankel
Physics Dept.
Univ. of Pennsylvania
Philadelphia, PA 19104
USA

Dr. B. Friman
Dept. of Physics
Åbo Akademi
Porthansgatan 3
20500 Åbo
FINLAND

Dr. S.A.I. Garpman
Dept. of Physics
Sölvegatan 14
S-223 62 Lund
SWEDEN

Dr. R.V. Gavai
Fakultät für Physik
D-4800 Bielefeld 1
W.-GERMANY

Dr. J.-F. Gilot
Lousiana State Univ.
Baton Rouge, LA 70803
USA

Dr. N.K. Glendenning
Nuclear Theory Group
LBL
Berkeley, CA 94720
USA

Dr. P. Gorodetzky
Centre de Rech. Nucleaires
BP 20, 67037 Strasbourg
Cedex
FRANCE

Prof. A. Goshaw
Dept. of Physics
Duke Univ.
Durham, NC 27706
USA

Dr. M. Gross
Dept. of Theor. Physics
1 Keble Road
Oxford OX1 3NP
ENGLAND

Dr. C.R. Gruhn
CERN/EP
CH-1211 Geneve 23
SWITZERLAND

Dr. M. Gyulassy
Nuclear Science Div.
LBL
Berkeley, CA 947280
USA

Prof. R. Hagedorn
CERN
CH-1211 Geneve 23
SWITZERLAND

Prof. A. Hahn
Univ. of California
Irvine, CA 92717
USA

Dr. M. Heiden
LBL
Berkeley, CA 947210
USA

Dr. E. Hilf
GSI Darmstadt
Planckstrasse 1
6100 Darmstadt 11
W.-GERMANY

Dr. R. Hwa
Inst. of Theoretical
Science
Univ. of Oregon
Eugene, OR 97403
USA

Dr. S. Kahana
Brookhaven Nat. Lab.
Upton,
Long Island, NY 11973
USA

Prof. K. Kajantie
Dept. of Theor. Physics
Siltavuorenpenger 20 C
00170 Helsinki
FINLAND

Dr. O. Kalashnikov
Lebedev Phys. Institute
Leninsky Prospect 53
Moscow 117924
USSR

Prof. J. Kapusta
Tate Lab. of Physics
116 Church Street S.E.
Minneapolis, MN 55455
USA

Mr. M. Kataja
Dept. of Physics
Univ. of Jyväskylä
Seminaarinkatu 15
40100 Jyväskylä
FINLAND

Dr. S. Katsanevas
CERN/EP-div.
CH-1211 Geneve 23
SWITZERLAND

Mr. H. Kurki-Suonio
Dept. Physics
RLM 5.218
Univ. of Texas at Austin
Austin, TX 78712
USA

M. Kutschera
Institut für Theor. Physik
Philosophenweg 19
D-6900 Heidelberg 1
GERMANY

Prof. P. Landshoff
Dept. of App. Mathem.
and Theor. Physics
Silver Street
Cambridge CB3 9EW
ENGLAND

Prof. K. Lassila
Dept. of Physics
Ames, IA 50011
USA

Dr. L. Lovas
Central Research Inst.
for Physics
P.O. Box 49
H-1525 Budapest
HUNGARY

Dr. T. Ludlam
Physics Dept.
Brookhaven Nat. Lab.
Upton, NY 11973
USA

Dr. I. Lund
Dept. of Physics
Sölvegatan 14
S-223 62 Lund
SWEDEN

Dr. J. Maalampi
Dept. of High Energy Pys.
Siltavuorenpenger 20 C
00170 Helsinki
FINLAND

Mr. J. Markkanen
Dept. of Physics
Jyväskylä University
Seminaarinkatu 15
40100 JYVÄSKYLÄ
FINLAND

Dr. H. Matis
Lawrence Berkeley Lab.
Bl. 70 A/3307
1 Cyclotron Road
Berkeley, CA 94720
USA

Prof. L. McLerran
Fermilab
P.O. Box 500
Batavia, IL 60510
USA

Dr. H.I. Miettinen
Research Inst. for Theor.
Physics
Siltavuorenpenger 20 C
00170 HELSINKI
FINLAND

Dr. O. Miyamura
Faculty of Engineering
Science
Osaka Univ.
1-1, Machikaneyama
Toyonaka, Osaka 560
JAPAN

Dr. C. Montonen
Dept. of Theor. Physics
Siltavuorenpenger 20 C
00170 HELSINKI
FINLAND

Dr. B. Müller
Inst. for Theor. Physics
Univ. of Frankfurt
D-6000 Frankfurt Main 1
W.-GERMANY

Prof. W.D. Myers
LBL-Nuclear Science
Berkeley, CA 94720
USA

Prof. S. Nagamiya
Univ. of Tokyo
Dept. of Physics
7-3-1 Hongo Bunkyo-Ku
Tokyo 113
JAPAN

Dr. W. Nörenberg
GSI Darmstadt
Planckstrasse 1
6100 Darmstadt 11
W.-GERMANY

Prof. I. Otterlund
Dept. of Physics
Univ. of Lund
Sölvegatan 14
S-22362 Lund
SWEDEN

Dr. I.C. Parikh
Texas A & M Univ.
College Station
Texas 77843-3366
USA

Dr. T. Parnell
ES-52, Space Science
Lab./HEP
NASA/Marshall Space Flight Cen.
Huntsville, AL 35812
USA

Dr. J. Polonyi
Dept. of Physics
Univ. of Illinois
Urbana, IL 61801
USA

Dr. S. Pratt
Tate Lab. of Physics
116 Church Street S.E.
Minneapolis, MN 55455
USA

Prof. H. Pugh
Lawrence Berkeley Lab.
Berkeley, CA 94720
USA

Dr. E. Quercigh
CERN/EP
CH-1211 Geneva 23
SWITZERLAND

Prof. J. Rafelski
Physics Department
Univ. of Cape Town
Rodebosch 7700, Cape
Rep. of South Africa

Dr. R. Raitio
Dept. of Theor. Physics
Siltavuorenpenger 20 C
00170 Helsinki
FINLAND

Dr. G. Ravenhall
Dept. of Physics
Univ. of Illinois
Urbana, IL 61801
USA

Prof. V. Ruuskanen
Dept. of Physics
Jyväskylä University
Seminaarinkatu 15
40100 Jyväskylä
FINLAND

Prof. N. Samios
BNL
Upton, NY 11973
USA

Dr. H. Sann
Max-Planck-Institute
P.O. Box 103980
D-6900 Heidelberg
W-GERMANY

Prof. H. Satz
Dept. of Physics
Univ. of Bielefeld
D-48 Bielefeld
W.-GERMANY

Dr. L. Schroeder
Lawrence Berkeley Lab.
Berkeley, CA 94720
USA

Dr. M. Soyeur
SPhT
CEN-Saclay
F-91191 GIF-sur-YVETTE
Cedex
FRANCE

Prof. H. Specht
Inst. für Theor. Physik
Philosophenweg 16
D-69 Heidelberg
W.-GERMANY

Prof. R. Stock
GSI
Postfach 110541
D-6100 Darmstadt 11
W.-GERMANY

Dr. H. Stroebele
GSI Darmstadt
Planckstrasse 1
6100 Darmstadt 11
W.-GERMANY

Dr. D. Strottman
T-9, MS B279
Los Alamos Nat. Lab.
Los Alamos, NM 87545
USA

Prof. E. Suhonen
Dept. of Theor. Physics
University of Oulu
Linnanmaa
90570 Oulu
FINLAND

Dr. M. Sullivan
Univ. of Pittsburgh
Pittsburgh, PA 15260
USA

Dr. M. Sumbera
Nuclear Physics Inst.
CS 250 68 REZ
Near Prague
CZECHOSLOVAKIA

Dr. H. Sumiyoshi
Nucl. Science Div. LBL
Univ. of California
Berkeley, CA 94305
USA

Dr. R. Szwed
IFD
Universytet Warszawski
00-681 Warszawa
ul. Hoza 69
POLAND

Dr. Y. Takahashi
Marshall Space Flight Cen.
Alabama 35812
USA

Prof. M.J. Tannenbaum
BNL
Upton, Long Island
New York 11973
USA

FL T. Toimela
Dept. of Theor. Physics
Siltavuorenpenger 20 C
00170 Helsinki
FINLAND

FK E. Turunen
Dept. of Theor. Physics
University of Oulu
Linnanmaa
90570 Oulu
FINLAND

Dr. A. Valadakis
N.R.C. "Democritos"
Aghia Paraskevi
Athens
GREECE

Dr. H. Wegner
BNL
Upton, NY 11973
USA

Prof. R. Weiner
Fachbereich Physik
Mainzer Gasse 33
D-3550 Marburg
W.-GERMANY

Dr. W. Willis
CERN/EP-Div.
CH-1211 Geneve 23
SWITZERLAND

Dr. M. Winter
Centre de Recherches
Nucléaires
B.P. 20 67037
Strasbourg-Cedex
FRANCE

Dr. C.-Y. Wong
Oak Ridge Nat. Lab.
Physics Div.
P.O.B.
Oak Ridge, TN 37830
USA

Dr. H. Wyld
Phys. Dept.
Univ. of Illinois
Urbana, IL 61801
USA

Prof. W.A. Zajc
Dept. of Physics
Univ. Pennsylvania
Philadelphia, PA 191104
USA

Dr. J. Zimányi
Central Res. Inst.
for Physics
H-1525 Budabest 114
P.O.B. 49
HUNGARY

Dr. R. Zitoun
LPNHE
4 Place Jussieu
75230 Paris Cedex 05
FRANCE

Vol. 195: Trends and Applications of Pure Mathematics to Mechanics. Proceedings, 1983. Edited by P. G. Ciarlet and M. Roseau. V, 422 pages. 1984.

Vol. 196: WOPPLOT 83. Parallel Processing: Logic, Organization and Technology. Proceedings, 1983. Edited by J. Becker and I. Eisele. V, 189 pages. 1984.

Vol. 197: Quarks and Nuclear Structure. Proceedings, 1983. Edited by K. Bleuler. VIII, 414 pages. 1984.

Vol. 198: Recent Progress in Many-Body Theories. Proceedings, 1983. Edited by H. Kümmel and M. L. Ristig. IX, 422 pages. 1984.

Vol. 199: Recent Developments in Nonequilibrium Thermodynamics. Proceedings, 1983. Edited by J. Casas-Vázquez, D. Jou and G. Lebon. XIII, 485 pages. 1984.

Vol. 200: H.D. Zeh, Die Physik der Zeitrichtung. V, 86 Seiten. 1984.

Vol. 201: Group Theoretical Methods in Physics. Proceedings, 1983. Edited by G. Denardo, G. Ghirardi and T. Weber. XXXVII, 518 pages. 1984.

Vol. 202: Asymptotic Behavior of Mass and Spacetime Geometry. Proceedings, 1983. Edited by F. J. Flaherty. VI, 213 pages. 1984.

Vol. 203: C. Marchioro, M. Pulvirenti, Vortex Methods in Two-Dimensional Fluid Dynamics. III, 137 pages. 1984.

Vol. 204: Y. Waseda, Novel Application of Anomalous (Resonance) X-Ray Scattering for Structural Characterization of Disordered Materials. VI, 183 pages. 1984.

Vol. 205: Solutions of Einstein's Equations: Techniques and Results. Proceedings, 1983. Edited by C. Hoenselaers and W. Dietz. VI, 439 pages. 1984.

Vol. 206: Static Critical Phenomena in Inhomogeneous Systems. Edited by A. Pękalski and J. Sznajd. Proceedings, 1984. VIII, 358 pages. 1984.

Vol. 207: S. W. Koch, Dynamics of First-Order Phase Transitions in Equilibrium and Nonequilibrium Systems. III, 148 pages. 1984.

Vol. 208: Supersymmetry and Supergravity/Nonperturbative QCD. Proceedings, 1984. Edited by P. Roy and V. Singh. V, 389 pages. 1984.

Vol. 209: Mathematical and Computational Methods in Nuclear Physics. Proceedings, 1983. Edited by J. S. Dehesa, J. M. G. Gomez and A. Polls. V, 276 pages. 1984.

Vol. 210: Cellular Structures in Instabilities. Proceedings, 1983. Edited by J. E. Wesfreid and S. Zaleski. VI, 389 pages. 1984.

Vol. 211: Resonances – Models and Phenomena. Proceedings, 1984. Edited by S. Albeverio, L. S. Ferreira and L. Streit. VI, 369 pages. 1984.

Vol. 212: Gravitation, Geometry and Relativistic Physics. Proceedings, 1984. Edited by Laboratoire "Gravitation et Cosmologie Relativistes", Université Pierre et Marie Curie et C.N.R.S., Institut Henri Poincaré, Paris. VI, 336 pages. 1984.

Vol. 213: Forward Electron Ejection in Ion Collisions. Proceedings, 1984. Edited by K.O. Groeneveld, W. Meckbach and I.A. Sellin. VII, 165 pages. 1984.

Vol. 214: H. Moraal, Classical, Discrete Spin Models. VII, 251 pages. 1984.

Vol. 215: Computing in Accelerator Design and Operation. Proceedings, 1983. Edited by W. Busse and R. Zelazny. XII, 574 pages. 1984.

Vol. 216: Applications of Field Theory to Statistical Mechanics. Proceedings, 1984. Edited by L. Garrido. VIII, 352 pages. 1985.

Vol. 217: Charge Density Waves in Solids. Proceedings, 1984. Edited by Gy. Hutiray and J. Sólyom. XIV, 541 pages. 1985.

Vol. 218: Ninth International Conference on Numerical Methods in Fluid Dynamics. Edited by Soubbaramayer and J. P. Boujot. X, 612 pages. 1985.

Vol. 219: Fusion Reactions Below the Coulomb Barrier. Proceedings, 1984. Edited by S.G. Steadman. VII, 351 pages. 1985.

Vol. 220: W. Dittrich, M. Reuter, Effective Lagrangians in Quantum Electrodynamics. V, 244 pages. 1985.

Vol. 221: Quark Matter '84. Proceedings, 1984. Edited by K. Kajantie. VI, 305 pages. 1985.

Selected Issues from

Lecture Notes in Mathematics

GPSR Compliance

The European Union's (EU) General Product Safety Regulation (GPSR) is a set of rules that requires consumer products to be safe and our obligations to ensure this.

If you have any concerns about our products, you can contact us on ProductSafety@springernature.com

In case Publisher is established outside the EU, the EU authorized representative is:

Springer Nature Customer Service Center GmbH
Europaplatz 3
69115 Heidelberg, Germany

Batch number: 10704763

Printed by Printforce, the Netherlands